BIOLOGICAL PERFORMANCE OF MATERIALS

FUNDAMENTALS OF BIOCOMPATIBILITY

Second
Edition

JONATHAN BLACK
Department of Bioengineering
Clemson University
Clemson, South Carolina

Marcel Dekker, Inc.   **New York • Basel • Hong Kong**

Library of Congress Cataloging-in-Publication Data

Black, Jonathan
    Biological performance of materials : fundamentals of
biocompatibility / Jonathan Black. — 2nd ed.
        p.   cm.
    Includes bibliographical references and index.
    ISBN 0-8247-8439-1 (alk. paper)
    1. Biomedical materials. 2. Biomedical materials—Testing.
3. Biocompatibility. I. Title.
R857.M3B59 1992
617.9'5'0028—dc20   92-6363
                                                     CIP

This book is printed on acid-free paper.

Marcel Dekker, Inc.
270 Madison Avenue, New York, New York 10016

Current printing (last digit):
10 9 8 7 6 5 4 3 2 1

PRINTED IN THE UNITED STATES OF AMERICA

Preface to the Second Edition

The preparation of a new edition of a book is always a time to evaluate the intentions and achievements of the original work, before attempting to make improvements. In this case, it appears that *Biological Performance of Materials* has achieved acceptance by students and professionals alike and that revisions should be undertaken primarily to acknowledge progress made in the past decade in the field of biomaterials, especially in understanding aspects of biocompatibility.

The work was originally intended for use as an undergraduate text for a one-term junior–senior level course in biocompatibility. It has been used by myself and others in this role. However, it has also proven useful, with the assignment of selected articles, as the central text in undergraduate survey courses on biomaterials and on artificial organs. It has also been used, again with additional reading material, both from the scientific and clinical literature and from materials science texts, as the focus of a first-year graduate course in biomaterials for students with engineering (but not biological or medical) backgrounds. Finally, engineers working in medical device development and evaluation have found it a useful ready reference book. The sparsity of reference to actual materials and specific applications has apparently made this diversity of use possible; the revised edition attempts to maintain the versatility of the work.

All chapters have been revised and include new material. Tables have been added after Parts II and III, respectively, to provide generic materials properties and observations related to diagnosis of host response in animals and patients. Finally, a new chapter (Chapter 20) has been added dealing with the design and selection of biomaterials for specific applications.

I wish to thank my students and colleagues whose challenges to the ideas in the original work have contributed to many of the revisions incorporated here. The work of my tireless editorial research assistant, Lynda Overcamp, made the undertaking of this project possible and contributed in many ways, large and small, to its hoped-for success.

Inappropriate host response to implants and premature device failure secondary to materials degradation continue to impose unwanted limits on the engineering solutions to biological and medical problems. It is hoped that ideas and information contained in this revised work will contribute to the further improvement of biomaterials in their application to the alleviation of human disability and disease.

Jonathan Black

Preface to the First Edition

Biocompatibility of materials increasingly occupies the consciousness of engineers dealing with medical and biological problems. The engineer has long been accustomed to dealing with materials limits on design. These limits, such as yield stress, endurance limit, and rupture life, are reflected in design margins tailored to the criticality of the specific application. In situations involving biological interactions as a portion of the design problem, the additional materials limit of biocompatibility must be considered.

Failure of compatibility—that is, incompatibility—is proving to be the ultimate limit to the engineering solution of many biomedical problems. As a result, it is necessary to incorporate a thorough grounding in the aspects of biocompatibility into the training of bioengineers.

This book is designed as the central text for a junior–senior level, one-term undergraduate engineering school course in biocompatibility. Primary training in materials science and biology is assumed. This book is designed to be used in conjunction with undergraduate texts in materials science and biology so as to accommodate variations in individual degrees of preparation.

We begin with an examination of the concept of biocompatibility. Two major sections are devoted to the effect of materials on biological systems. Selected additional readings are provided at the end of many chapters.

This volume does not deal with the effects of transplantation of viable or processed natural tissues. This is a field in itself and is somewhat removed from the normal consideration of engineers. However, acceptance and performance of processed materials of biological origin as well as synthetic organic biomaterials are discussed.

The reader will note an emphasis on methods for determination of biocompatibility. This reflects the relatively early stage of disciplined study of biocompatibility as well as the continued need to select new and modified materials for specific applications. The general problem of selection, qualification, and specification of materials is dealt with in the final section of this work.

The practicing engineer will find this a useful source of references, test methods, and approaches to the problem of establishing biological performance of materials. The chapters dealing with qualification, standardization, and regulation of implant materials will be of special assistance to the professional.

The author wishes to thank his many undergraduate and graduate students whose ideas, questions, and discussions have contributed significantly to the scope and content of this work. Special thanks is due to G. K. Smith and J. L. Woodman for their contributions to Chapters 13 and 14.

It can be hoped that greater sensitivity to biological aspects of materials performance will lead to improved solutions to biological and medical engineering problems.

Jonathan Black

Contents

I
GENERAL CONSIDERATIONS

1
Biocompatibility: Definitions and Issues

1.1 INTRODUCTION

The issue of biocompatibility rises from a recognition of the profound differences between living tissues and nonliving materials. In both a historical and a practical perspective, we are aware of a wide range of interactive behavior between tissues and materials. In any of these interactions we also observe both beneficial and detrimental effects. Thus, materials we consider as foods and beverages can be nutritious or nonnutritious. From another viewpoint, they can be considered either toxic or nontoxic. Such judgments are relative to use or abuse rather than to an absolute scale. Alcohol, although a central nervous system depressor, has a positive virtue as a disinhibiting stimulant and social drug in small doses. In large doses it is toxic and, in still larger doses, lethal.

It is desirable to extend this sort of relativitism to examination of the interactions between biomaterials and living systems. *Biomaterials* are materials of natural or manmade origin that are used to direct, supple-

3

ment, or replace the functions of living tissues. When these materials evoke a minimal biological response, they have come to be termed *biocompatible*.

The term "biocompatible" as used here is inappropriate and defective of content. Compatibility is strictly the quality of harmonious interaction. Thus, the label "biocompatible" suggests that the material described displays universally "good" or harmonious behavior in contact with tissue and body fluids. It is an absolute term without any referent.

Furthermore, the traditional ideas of biocompatibility refer essentially to the effect of the material on the biological system. Effects of biological processes on materials are rarely included in the meaning, unless the results of material changes elicit a change in biological response. The effects of the biological system on the material are usually lumped in the term *biodegradation*; this term implies "bad" behavior, again without a referent.

One can protest that this is a semantic discussion without content. On the contrary, I think that the terminology used and the assumptions inherent in that terminology tend to condition the approach taken in both experiment and analysis. Thus, at present, the most common approach to establishing the biocompatibility of a material is to establish the *absence* of deleterious effects due to its use in biological applications. Once such tests are completed, the material is regarded as *qualified*. It can be argued that the absolute nature of the language employed has led to the use of absolute criteria.

However, the real issues in the use of biomaterials in medical and surgical devices are not absolute, any more than is the choice of a material for any other engineering application. The choice of materials for construction of a device or machine is made early in the design process. The properties of the candidate materials, particularly those properties that bear on the intended function of the complete mechanical assembly, then interact strongly with the design. The ultimate test of the appropriateness of the choice of materials is the performance of the completed design. In this performance we can see the interaction between design choices (shape, size, linkage, etc.) and materials properties (strength, density, conductivity, etc.). (Chapter 20 deals with some of these points more fully.)

The real issue of biocompatibility is not whether there are adverse reactions to a biomaterial, but whether that material performs satisfactorily (that is, in the intended fashion) in the application under consideration. This should lead directly to the traditional engineering design process of considering the advantages and disadvantages inherent in the selection of a particular material for a design in a specific application.

Among the factors considered must be the interaction of the material with the biological processes in its intended site of operation.

1.2 BIOLOGICAL PERFORMANCE

We shall adopt the term *biological performance* as a descriptor of materials in order to replace the present idea of biocompatibility. Biological performance and two closely related terms are defined as follows:

Biological performance: The interaction between materials and living systems. The two aspects of this performance are:
Host response: The local and systemic response, other than the intended therapeutic response, of living systems to the material.
Material response: The response of the material to living systems.

The generality of these terms is obvious. There are no value judgments in their definition nor do they suggest absolute qualities.

However, these terms are not sufficient for a full discussion. I have stressed the need for consideration of interactions of materials and living systems on a relative, rather than an absolute, basis. This suggests the need for a system of grading based upon the results of tests. Additional terms are needed to implement such a concept. The first three definitions are supplemented with several others.

Reference (or control) material: A material that, by standard test, has been determined to elicit a reproducible, quantifiable host or material response.

Note that this definition carries no implication of "good" or "bad" behavior on the part of the material. A reference material might be a material with minimal host response (a negative reference) or an extreme host response (a positive reference). Reference materials may or may not be selected from those conventionally used for implant fabrication.

Level of host (or material) response: The nature of the host (or material) response in a standard test with respect to the response obtained with a reference material.

A *standard test*, as referred to in these two definitions, is simply any well-defined, repeatable test. The requirements for such tests are given in Chapters 16 and 17.

Finally, it is suggested that we retain the use of the term "biocompatibility" for historical reasons, but with a narrow and careful redefinition:

Biocompatible (-ity): Biological performance in a specific application that is judged suitable to that situation.

So, at the end of the consideration, when host and material response are known and the particular device application is examined, a final value judgment can then be made that leads to the acceptance or rejection of the material. Such a selection and a resulting record of adequate performance does not "qualify" a material. Rather, it increases the confidence in the use of the material and points to *possible* successful use in similar applications.

1.3 CONSENSUS DEFINITIONS

A major attempt has been made to reach consensus concerning definitions related to biocompatibility. A working consensus conference, sponsored by the European Society for Biomaterials, was held in 1986 to discuss these matters in an international setting (Williams, 1987). Thirteen terms gained consensus definitions; those that are relevant to this discussion are:

Biomaterial: A nonviable material used in a medical device, intended to interact with biological systems.
Host response: The reaction of a living system to the presence of a material.
Biocompatibility: The ability of a material to perform with an appropriate host response in a specific situation.

Although terse, these are not bad definitions. They preserve the idea of interaction, of relative rather than absolute attributes. They are limited in that they specifically exclude living tissues from the spectrum of biomaterials. Since tissues, in addition to exhibiting active physiological processes, are materials with definable physical structure and properties, this exclusion seems unwarranted.

It is unfortunate that the conferees chose to deprecate the terms *biological performance* and *material response*. However, words become part of language by the repetition of their use or are abandoned, in the same way that paths broken through the wild may or may not become superhighways. Thus it remains to be seen how these and other terms discussed here will be conventionally understood and used in the scientific literature.

1.4 DISCUSSION

The use of the definitions as proposed here redirects the study of interactions between biological systems and materials from efforts to obtain qualification to efforts at description and grading based upon the careful development of standard tests and the characterization of reproducible response relative to reference materials. This is the central idea that will guide our considerations.

I feel strongly that absolute qualification is not possible for an artificial or processed material in biological applications. It is necessary to establish minimum requirements for performance at various stages of materials development. Strategies for setting such levels are discussed in Chapter 18.

It has been suggested that no foreign material is biocompatible (old meaning), and that the best that can be expected is that the results of its use are "physiologically tolerable." This is a somewhat negative view that overlooks the benign responses elicited by many materials in living systems.

However, such a suggestion should attract our attention to another important point. Living systems differ most from machines in respect to the constant flux and change of their components, that is, in their physiology. Biological performance, particularly host response, ought not to be defined in terms of tissue structure and pathology but primarily in terms of physiology. Deviations from usual physiological conditions may lead to changes in the structure and function of living tissues. The key to understanding host response, and to a lesser degree material response, is knowledge of the participation of the material in the physiology of the host. Extrapolation of results obtained in tissue culture and animal models rests significantly on knowledge of how both normal and abnormal physiological processes in these systems differ from those in humans. So, in addition to concentrating on making *relative* rather than *absolute* determinations, we will also take care to attend to physiology (both normal and abnormal) and its interspecies variations.

Osborn (1979) attempted to take such physiological considerations into account by classifying biomaterials as either *biotolerant*, *bioinert* or *bioactive*, conveying respectively the sense of negative (but tolerable) local host response, absence of local host response, and positive (desired) local host response. It is possible to extend these ideas, taking into account as well the course of development of the *field* of biomaterials and eliminating the undesirable use of the prefix *bio-*. Examining the historical development of biomaterials, it is possible to define four phases or types of biomaterials, based upon changing concepts of host response:

Phase 1. Inert (biomaterials): Implantable materials which elicit little or no host response.

Phase 2. Interactive (biomaterials): Implantable materials which (since host response is inevitable) are designed to elicit specific, beneficial responses, such as ingrowth, adhesion, etc.

Phase 3. Viable (biomaterials): Implantable materials, possibly incorporating live cells at implantation, which are treated by the host as normal tissue matrices and are actively resorbed and/or remodeled.

Phase 4. Replant (biomaterials): Implantable materials consisting of native tissue, cultured in vitro from cells obtained previously from the specific implant patient.

We now recognize that searches for Phase 1 materials are pointless. Many biomaterials in present clinical use and ones in development are properly called Phase 2 materials. Preliminary research reports reveal great interest and promise in Phase 3 materials, and advances in control and manipulation of the genetic code in mammals suggest that no intellectual barrier exists to prevent the broad future realization of Phase 4 materials at both the tissue and organ level. In fact, a Phase 4 material (implantable, live tissue with the identical genetic code and immunological determinants of the recipient patient) represents the ultimate fulfillment of the original search for biocompatibility: implantable materials demonstrating harmonious interaction.

Medical practice today utilizes great numbers of artificial devices and implants. For any one application, a wide variety of similar designs exists with different degrees of efficacy. In contrast to this profusion of design, there is a paucity of materials choices. Perhaps no more than two or three dozen of the millions of available metal, polymer, and ceramic compositions have proven useful in medical devices and implants.

The limiting factor continues to be biological performance. Better understanding of biological performance and the factors affecting it will lead to a variety of useful new materials options. This in turn will lead to substantial expansion of the role that artificial devices can play in the prevention and treatment of human disability and disease. At the end of this progression, when the technology for preparation of Phase 4 materials is readily and widely available, artificial devices will be called upon to serve only as ''bridges'' to replantation and biomaterials will emerge in its rightful place as one of the healing arts.

REFERENCES

Osborn, J. F. (1979): Schw. Mschr. Zahnheilk. 89:1138.

Williams, D. F. (Ed.) (1987): *Definitions in Biomaterials: Proceedings of a Consensus Conference of the European Society for Biomaterials*, Chester, England, March 3–5, 1989. Elsevier, Amsterdam.

2
Introduction to the Biological Environment

2.1 GENERAL CONSIDERATIONS

The central idea developed in the previous chapter is that biological performance can be defined in terms of the interaction between materials and their operational setting, the biological environment. This is not qualitatively different from the normal consideration given during an engineering design process to the material aspects of performance and durability. However, there are two quantitative aspects that set biological performance apart and create the need for an independent study of material and host responses:

1. *High demand*: The biological environment, especially internal to living systems, is a remarkably aggressive one. It is a milieu of high

chemical activity combined with a highly variable spectrum of combined mechanical stresses.

2. *Invariant conditions*: Despite its aggressive aspects, the biological environment displays an extraordinary quality of constancy in both physical conditions and composition. Complex control systems exist to maintain that constancy; thus, deviations from established conditions attendant to the presence of materials may be expected to incite restoring responses.

The latter portions of this work deal with many aspects of this peculiar environment during examination of typical material and host responses. At this point it is advisable to examine general points that will serve as guides for discussion.

2.2 COMPARISON OF EXTERNAL AND INTERNAL CONDITIONS

The aggressive aspects of the biological environment may be understood if we examine the differences between conditions external and internal to living systems. Externally, we find the familiar aspects of the physical world. Most materials are inorganic and are partially or fully oxidized. While physical processes are interrelated, there is an absence of active environmental control systems. Time constants for change are long, determined by processes of chemical reaction and diffusion, and driven by sources that supply energy primarily through radiation, conduction, and convection. A wide variety of atomic species are present. There is a great variety in structure and chemical content and there is little evidence of compositional or structural optimization.

By contrast, the internal environment arises from a system in which materials (molecules and tissues) are largely organic and are partially or fully reduced. Most changes are mediated by active, energy-requiring control systems. In many cases, multiple parallel systems with different time constants and extensive intersystem interactions control a single transformation or process. Time constants are orders of magnitude shorter than for most inorganic reactions due to mediation by specialized organic catalysts (enzymes) and the derivation of energy from chemical sources through coupled reactions. While there is a great variety of chemical content and structure, a few elements, primarily carbon, oxygen, hydrogen and nitrogen, provide the vast majority of this complexity. Elements that are present are generally utilized and structures display a parsimonious efficiency, providing an overall impression of design optimization.

Whatever one's views on the origin and development of biological systems, one must be impressed by the complexity of these systems and their economy of action. They obtain their objectives by excluding, through accident, design, or active process, materials that are unnecessary or harmful to the function of their individual processes. These phenomena act to exclude all materials other than healthy, autologous (belonging to the same organism) tissue. Furthermore, the systems interact, both locally and on a regional and global scale. Thus, a constant aspect of the biological environment is that the introduction of a foreign material will elicit a host response, which may have local, systemic, and remote aspects.

2.3 PROBLEMS IN DEFINITION OF THE BIOLOGICAL ENVIRONMENT

It is difficult to define the actual environment in which a material or device is called upon to function. More will be said about this later. The difficulty arises from a lack of detailed knowledge of in vivo conditions and the local variations that can occur in the face of overall maintenance of conditions, termed homeostasis, necessary for life. Also, there is ambiguity in defining the region that is coupled with an implant. Implants in isolated locations can interact with the rest of the system through diffusion of ions and fluids, circulation of blood, and drainage of the lymphatics. Even the definition of absolute volumes of material in communication with the implant is difficult. As we will see in Section 14.2.2, the volume of water in which an implant is immersed in the human body may be taken as 10^{-15}, 8.4, or 1000 liters, depending upon the details of consideration.

A last general point has to do with the maintenance of homeostasis. In a particular location, temperature, pH, pO_2, equivalent electrical potential, hydrostatic and osmotic pressure, and tissue/fluid composition are carefully controlled. However, the observation of this active control should not lead us to unwarranted conclusions concerning its adaptability. The control systems most in evidence are those that control for the usual situation and small deviations. Superimposed upon these are "emergency" protection systems, such as coagulation, inflammation, and immune response. These initiate planned deviations that lead to restoration of normal conditions locally, as long as systemic integrity is maintained.

Taking a control systems viewpoint, we can foresee challenges that can overwhelm the restorative capabilities of control. Only challenges that occur within the design spectrum of the system can elicit satisfactory responses, except by chance. Thus, when viewing host response it is wise to recognize the limited environmental variations that occur in the absence of outside intrusion. It is also prudent to consider the qualitative and quantitative differences between chance intrusion and deliberate functional implantation.

Materials must be tested in vitro before implantation, even in animals. It is desirable to attempt, in large or small part, to replicate the operating environment which the material may encounter after implantation. Here it is useful to distinguish between four classes of exposure environments:

1. *Physiological*: Chemical (inorganic) and thermal conditions controlled.
2. *Biophysiological*: Physiological conditions with the addition of appropriate cell products (serum proteins, enzymes, etc.).
3. *Biological*: Biophysiological conditions with the addition of appropriate viable, active cells.
4. *Pericellular* (circumcellular): A special case of *biological*: the conditions in the immediate vicinity of appropriate viable, active cells.

These are termed "classes" of environments since the exact value of parameters within each depend upon the specific details of the location within tissue or organ and, in the case of materials (rather than device) testing, upon the design details and functional goals of the device.

In vitro testing is usually carried out under physiological or biophysiological conditions only. The problems associated with in vitro testing and its comparison to in vivo conditions will be discussed further in Chapters 16 and 17. In this chapter, "biological environment" is taken to mean that combination of conditions which an implanted material will encounter acutely and chronically in actual service: the combination of biological and pericellular conditions as well as the instantaneous requirements placed upon the design and function of the device in which it is incorporated. The combination of these intrinsic and extrinsic environmental effects with the overall patient requirements during the proposed period of implantation produces what is properly termed the *implant life history*: the total combination of requirements which the material must meet to be successful in its application.

Table 2.1 Macroscopic Parameters of the Reference Human[a]

Weight: 70 kg	Height (medium frame): 1.80 m
Surface area: 1.88 m^2	Volume: 0.065 m^3
Composition	Density
Water: 60% (42 liters)	Fat 0.9 g/cm^3
Solid: 40% (28 kg)	Whole body 1.07 g/cm^3
Distribution of tissue types	
(as percentages of body weight)	
Muscle 43	
Bone 30	
Internal organs	
Heart 0.4	
Liver 2	
Kidneys (2) 0.5	
Spleen 0.2	
Lungs 1.6	
Brain 2.3	
Viscera 5.6	
Skin 7	
Blood 7.2 (5 liters)	
Basal metaboli rate 37/kcal·m^2/hr	

[a]Values given for male individual, mid-thirties.
Source: Lentner, C. (1981).

2.4 ELEMENTS OF THE BIOLOGICAL ENVIRONMENT

We generally consider the human body in terms of a standard or reference configuration; this is the 70 kg man.* This standard has the macroscopic parameters given in Table 2.1.

Wide variations from these parameters exist. Age, level of activity, disease state, national origin, and genetic factors will also affect the absolute values. Furthermore, the values given here, as in the following tables in this chapter, are mean normative values for a mid-thirties male individual. They represent an average expectation and do not account for variations within physiological limits. It is common to describe such variations under the overall term "biological variation." We can also

---

*It is usual practice to speak of the "standard man" rather than the "standard human." Recent comments in the lay literature concerning the focus of Federal funding for biomedical research have drawn attention, once again, to differences between men and women. Thus, the values in Table 2.1 should be taken as guidelines and other sources, such as Lentner (1981), addressed for values applicable to specific applications.

Table 2.2 Physicochemical and Mechanical Conditions in Humans

	Value	Location
pH	1.0	Gastric contents
	4.5–6.0	Urine
	6.8	Intracellular
	7.0	Interstitial
	7.15–7.35	Blood
pO_2 (mm Hg)	2–40	Interstitial
	12	Intramedullary
	40	Venous
	100	Arterial
	160	Atmospheric
pCO_2 (mm Hg)	40	Alveolar
	2	Atmospheric
Temperature (°C)	37	Normal core
	20–42.5	Deviations in disease
	28	Normal skin
	0–45	Skin at extremities
Mechanical	Stress (MPa)	Tissues
	0–0.4	Cancellous bone
	0–4	Cortical bone
	4	Muscle (peak stress)
	40	Tendon (peak stress)
	80	Ligament (peak stress)
	Stress cycles (per year)	Activity
	3×10^5	Peristalsis
	3×10^6	Swallowing
	$0.5–4 \times 10^7$	Heart contraction
	$0.1–1 \times 10^6$	Finger joint motion
	2×10^6	Walking

define the physicochemical and mechanical conditions that are encountered in the body (Table 2.2).

When we consider the effects of release of material from the implant into the body, it is necessary to know the starting or nominal inorganic chemical composition of the body. While the concentrations of major elements have been known for some time, those of trace elements (those present in very low concentrations) are just beginning to be appreciated.

Table 2.3 Inorganic Composition of the Human Body[a]

			Total body burden	Concentration (average)
Basic elements[b]	Oxygen		43,000 g	61.4%
	Carbon		16,000 g	22.9%
	Hydrogen		7,000 g	10.0%
	Nitrogen		1,800 g	2.6%
		Total	67,800 g	96.9%
Physiologic elements[b]	Calcium		1,000 g	1.43%
	Phosphorus		780 g	1.11%
	Potassium		140 g	0.20%
	Sulfur		140 g	0.20%
	Sodium		100 g	0.14%
	Chlorine		95 g	0.14%
		Total	2,255 g	3.22%
Trace elements[b]	Magnesium		19 g	271 ppm
	Iron		4.2 g	61.4 ppm
	Zinc		2.3 g	33 ppm
	Iodine		130 mg	1.9 ppm
	Copper		72 mg	1.0 ppm
	Aluminum		61 mg	0.9 ppm
	Vanadium		18 mg	260 ppb
	Selenium		< 13 mg	< 190 ppb
	Manganese		12 mg	170 ppb
	Nickel		10 mg	140 ppb
	Molybdenum		< 9.5 mg	< 136 ppb
	Titanium		9 mg	130 ppb
	Chromium		< 6.6 mg	< 94 ppb
	Cobalt		< 1.5 mg	< 21 ppb
		Total	< 25.84 g	< 0.04%

[a]Data from Lentner (1981).
[b]Total body burden exceeds 70,000 g and 100% due to variety of primary sources and experimental error in individual values.

Table 2.3 presents nominal or reference human values; see Chapter 13 for a more detailed discussion.

These parameters, taken together, define the intrinsic physicochemical and mechanical parameters appropriate to generic biological environments. Exact details of a particular anatomical site may be required when designing tests for particular materials or devices.

There is an additional area where more detailed information is desirable. Blood is a delicate and pervasive tissue. It is essential to understand its makeup and normal values, especially for applications involving blood contact on a chronic basis. Information describing the composition and cellular distribution of blood is given in Table 2.4.

Beginning with the information given in Tables 2.1 to 2.4, it is possible to begin to develop a picture of the thermal, mechanical, and chemical environment that an implant will encounter when inplanted in a specific anatomical site. Some of the material responses during its implanted service life will be described in Chapters 3–7 of this work. This defined environment may be changed acutely and chronically by the presence of an implant. Chapters 8–15 deal with some variations in the biological environment that arise, locally and systemically, as a result of implantation, that is, the host response.

2.5 IMPLANT LIFE HISTORY

The thermal, mechanical, and chemical parameters described in previous sections are sufficient to predict, in general, the acute or instantaneous biological environment encountered by an implant. These acute values differ little from patient to patient; what differences there are have only small effects on acute host and material responses. Biological differences do exist that affect chronic host response to materials. These may only be discernible by clinical testing of a specific patient; analyses of body fluids and tissues are probably inadequate for a full understanding of individual differences. It is unfortunate that technology for determination of the functional behavior of implants and implant/patient interactions is weak compared with that available to biological scientists for the study of natural organs in situ.

Beyond these obvious similarities and possible individual biological differences, the demands and expectations of individuals vary considerably. Devising a total hip replacement prosthesis for a thirty-five-year-old head of a family presents a quite different engineering problem from such a design for a seventy-year-old nursing home resident. Accounting for these functional differences completes the description of the *service environment*; the full picture thus formed is termed the *implant life history*.

Implant life histories vary considerably from application to application and, of necessity, involve a good degree of estimation. Within given target (patient) groups, there will be a wide variety in the choice and intensity of work and leisure activities. As a result, implant life histories

Table 2.4 Components and Composition of Human Blood[a]

Blood
    Packed cell volume   38.5%
    Serum volume   61.5%

Serum composition (mean values)

Cations	mEq/l	Anions	mEq/l
Sodium	142	Chlorine	101
Potassium	4	Bicarbonate	27
Calcium	5	Phosphate	2
Magnesium	2	Sulfate	1
Total	153	Organic acids	6
		Proteins	16
		Total	153

Other elements

Iron	0.75–1.75 mg/l (=ppm)
Nickel	1.0–5.0 µg/l (=ppb)
Titanium	3.3 µg/l
Aluminum	2.0 µg/l
Copper	0.8–1.4 µg/l
Chromium	0.3 µg/l
Manganese	0.4–1.0 µg/l
Vanadium	<0.2 µg/l
Cobalt	0.15 µg/l

Serum proteins
    Total   65–80 g/l
Distribution (%)

Albumin	6.15
Globulins	34.5
α 8.2	
β 10.3	
δ 12.6	
Fibrinogen	4.0

Cellular distribution

Type		Typical dimension (µm)
Erythrocyte	4–$5.6 \times 10^6/\mu l$	8–9
Platelet	1.5–$3 \times 10^5/\mu l$	2–4
Leucocyte	2.8–$11.2 \times 10^3/\mu l$	
Leucocyte distribution (%)		
Neutrophils	59	10–15
Eosinophils	2.4	10–15
Basophils	0.6	10–15
Monocytes	6.5	12–20
Lymphocytes	31	7–18

*Data from Lenter (1981) and author's research.

Table 2.5 Implant Life History

Implant: Anterior cruciate ligament replacement
Type: Permanent
Patient indications: Post-traumatic replacement, age: 35–48
(est. mean life expectancy: 40 years)
ph = 7 ± 0.3 pO_2 = <40 mmHg pCO_2 = <40 mmHg
25°C ≤ T ≤ 37°C
Mechanical conditions:[a]
  Strain (range of maximum): 5–10%
  Loads: (moderate activity level, including recreational jogging)

Activity	Peak load (N)	Cycles/year	Total cycles
Stairs			
ascending	67	4.2×10^4	1.7×10^6
descending	133	3.5×10^4	1.4×10^6
Ramp walking			
ascending	107	3.7×10^3	1.5×10^5
descending	485	3.7×10^3	1.5×10^5
Sitting and arising	173	7.6×10^4	3.0×10^6
Undifferentiated	<210	9.1×10^5	3.6×10^7
Level walking	210	2.5×10^6	1.0×10^8
Jogging	630	6.4×10^5	2.6×10^7
Jolting	700	1.8×10^3	7.3×10^5
Totals:		4.2×10^6	2.9×10^8

[a]Adapted from Table III, Chen and Black (1980).

can only be regarded as *predictive guides* in the development, evaluation, and study of implantable biomaterials.

Table 2.5 gives an example of an implant life history, in this case for a permanent anterior cruciate ligament replacement. The supposed patient presents an example of an individual with moderate demands: presumably a full-time worker with evening and/or weekend physical recreational interests. If this individual was disabled in some manner, or was employed in a setting with unusual physical demands, such as mining or construction, or took part in other more demanding physical activities, such as wind surfing, mountain climbing or parachute jumping, these facts should be noted and the physical consequences accounted for, in terms of different estimated peak loads and numbers of repetitions.

2.6 PRE-IMPLANTATION HANDLING EFFECTS

We tend to think of the biological environment as one which the implant passes into after manufacture and storage. This assumption overlooks two intermediate processes common to all implant applications.

In the first place, the implant may become contaminated, accidentally or as a side effect of planned processing and handling, during manufacture, storage, and insertion. It is usually assumed that the implant surface is a pure, clean one with the composition of the bulk material. The truth may be far different. Organic films introduced during manufacture or by inadvertent handling may persist. Materials may be picked up from packaging used for storage or during sterilization. Contaminants may be transferred from surgical instruments.

For this reason, experimental studies of biological performance should include surface characterization of actual implant specimens, selected from the full group fabricated in a particular study, in the condition just before surgical insertion. Furthermore, care should be taken when materials are incorporated into devices for clinical trials and use, to see that the surface conditions are the same as those found during earlier developmental studies.

Secondly, all implants must be sterilized before use. Some may be supplied in sterile double-wrapped packages by the manufacturer, while others must be sterilized in the laboratory or hospital before use.

The common forms of sterilization used in implant practice are:

1. Cold solution
2. Dry heat
3. Moist heat (steam)
4. Gas
5. Irradiation

Some typical sterilization parameters for each of these common methods are listed in Table 2.6. The particular method and parameters used must be suited to the individual implant type to provide maximum safety with minimum cost and implant degradation. Newer methods include electron beam irradiation and radio-frequency plasma gas sterilization.

The process of sterilization, if overlooked, may affect our perception of both the material and the host response. It is possible for some sterilization processes, such as irradiation, to change material properties. This might be interpreted, in error, as a material response effect if detected after implantation, or might produce changes in host response that are secondary to the changes in the materials' properties. It is also possible

Table 2.6 Methods and Typical Parameters of Sterilization

Method	Temperature	Time	Notes
Cold solution	RT	1–3 hr	Commercial solutions; usually include formaldehyde or gluteraldehyde
Dry heat	160–175°C (max.)	0.5–2 hr	Time/temperature vary inversely
Moist heat	120–130°C (max.)	2–15 min	Time/temperature vary inversely
Gas	RT–55°C	1–24 hr	Gas used is usually ethylene oxide, 400–1200 mg/liter, 48 hr degassing required
Irradiation	RT	Not applicable	^{60}Co gamma radiation, 2–4 Mrad dose

for traces of liquid or gaseous sterilants to be carried into the implant site, thus modifying the host response. Finally, sterilization of an *unclean* implant may render it sterile but not pyrogen-free, thus affecting the host response (see Section 8.2.1).

Therefore, in any examination of material and host responses to implantation, it is necessary to pay close attention to actual surface conditions and sterilization effects as a prologue to exposure to the biological environment.

REFERENCES

Chen, E. H. and Black, J. (1980): J. Biomed. Mater. Res. 14:567.

Lentner, C. (Ed.) (1981): *Geigy Scientific Tables*. Ciba-Geigy, Basle.

Snyder, W. S. (Ed.) (1975): *Report of the Task Group on Reference Man*. International Commission on Radiological Protection, No. 23. Pergamon, Oxford.

BIBLIOGRAPHY

Altman, P. L. and Dittmer, D. S. (Eds.) (1964): *Biological Handbooks*: *Blood and Other Body Fluids*, 1961; *Biology Data Book*, 1964. Federation of American Societies for Experimental Biology (FASEB), Bethesda, Maryland.

Å strand, P.-O. and Rodahl, K. (1970): *Textbook of Work Physiology*. McGraw-Hill, New York.

Cooney, D. O. (1976): *Biomedical Engineering Principles*. Marcel Dekker, New York.

Frankel, V. H. and Nordin, M. (1980): *Basic Biomechanics of the Skeletal System*. Lea & Febiger, Philadelphia.

Ganong, W. F. (1989): *Review of Medical Physiology*. 14 edition. Appleton & Lange, Norwalk, Connecticut.

Gaughran, E. R. L. and Kereluk, K. (Eds.) (1977): *Sterilization of Medical Products*. Johnson & Johnson, New Brunswick, New Jersey.

LeVeau, B. (Ed.) (1977): *Williams and Lissner: Biomechanics of Human Motion*, 2nd edition. W. B. Saunders, Philadelphia.

Northip, J. W., Logan, G. A. and McKinney, W. C. (1974): *Introduction to Biomechanical Analysis of Sport*. Wm. C. Brown, Dubuque.

II

MATERIAL RESPONSE: FUNCTION AND DEGRADATION OF MATERIALS IN VIVO

3

Swelling and Leaching

3.1 INTRODUCTION

The simplest form of interaction between implant materials and the bio-logical environment is the transfer of material across the material-tissue interface in the absence of reaction. If the substance, primarily fluid, moves from the tissue into the biomaterial, the result in a fully dense material will be swelling due to conservation of volume. Even in the absence of fluid uptake, the biomaterial may absorb some component or solute from the surrounding fluid phase. If the fluid moves into the tissue, or if one component of the biomaterial dissolves in the fluid phase of the tissue, the resulting material porosity is said to be due to leaching. Both of these effects have profound influences on the behavior of materials despite the absence of externally applied mechanical stresses and obvious shape changes.

Swelling and leaching both result from the process of diffusion. Before considering the effects on materials' properties, we will examine the fundamentals of diffusion and the diffusion models appropriate to each situation.

3.2 FICK'S LAWS OF DIFFUSION

The fundamental relationship that governs diffusion in isotropic materials is

$$F = -D \frac{\partial C}{\partial x} \tag{3.1}$$

where

F = rate of transfer per unit area of cross section
D = diffusion constant
x = coordinate normal to cross section
C = concentration of diffusing material

Equation (3.1) is called Fick's first law, and diffusion that obeys this relationship is termed ''Fickian'' or Type I diffusion. In this simple case the diffusion constant, D, depends only upon the material diffusing (the solute) and the matrix through which it moves. Thus, D is independent of concentration, position, and time. However, D does depend upon the type of diffusion process taking place, which includes surface, grain boundary, and volume diffusion. This dependence is given by

$$D = D_0 e^{[-Q/kT]} \tag{3.2}$$

with dimensions of cm^2 sec^{-1}, where

Q = energy of activation of the diffusion process:
[$Q_{volume} > Q_{grain\ boundary} > Q_{surface}$ (4:3:2 or :1)]
D_0 = intrinsic diffusion constant:
[$D_{0(volume)} > D_{0(grain\ boundary)} > D_{0(surface)}$]

Thus it is easy to see that within a given material system (a substance diffusing through a biomaterial), $D_{surface} > D_{grain\ boundary} > D_{volume}$, and surface diffusion is favored at all temperatures.

When Eq. (3.1) is applied to the problem of one-directional flow in an infinite medium, a differential equation of this form can be obtained:

$$\frac{\partial C}{\partial t} = D\frac{\partial^2 C}{\partial x^2} \tag{3.3}$$

Equation (3.3) is usually called Fick's second law. The application of Eqs. (3.1) and (3.3) to the different geometries, and to the initial and boundary conditions, of specific situations is sufficient to determine the distribution and mass transfer rate of diffusing materials in all cases.

3.3 ABSORPTION

The simplest case that results in swelling is that of diffusion from a fluid with a fixed concentration, in the presence of perfect mixing, into an infinite medium. This is the case for the early period of absorption in any geometric arrangement; when the diffusing material is mostly near the surface, geometric factors have little effect. The arrangement, initial conditions, and change of concentration in the solid (biomaterial) phase with time are shown in Figure 3.1.

The exact solution for the concentration at a given point, as a function of time, is

$$C = C_0\left(\frac{x}{2(Dt)^{1/2}}\right) \tag{3.4}$$

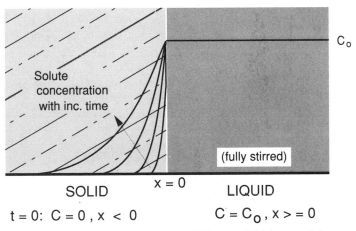

Figure 3.1 Absorption from liquid into solid biomaterial.

where

C_0 = external concentration

x = distance perpendicular to interface

$$\text{erfc}(a) = \left[\frac{2}{\sqrt{\pi}}\right] \int_a^\infty e^{-y^2}\, dy \qquad (3.5)$$

The function erfc(a) is related to the error function erf(a) and will be found tabulated in texts on diffusion and heat transfer.

Integration of Eq. (3.4) over distance for two values of time leads to the following relationship for the total mass transfer across the boundary M_t:

$$M_t = 2C_0 \left[\frac{Dt}{\pi}\right]^{1/2} \qquad (3.6)$$

The following conclusions follow directly from Eqs. (3.4) and (3.6):

1. The distance of penetration of any given concentration ("the diffusion front") increases in proportion to the square root of time.
2. The time required for any point to reach a given concentration is proportional to the square of the distance from the surface and is inversely proportional to the diffusion constant.
3. The total amount of diffusing material entering the biomaterial through a unit area of interface increases as the square root of time.

The situation shown in Figure 3.1 is correct for either volume or grain boundary diffusion, with appropriate values for the diffusion constant, when the liquid phase is well mixed, as in the case of fluid or solute uptake from arterial blood. If the liquid phase is not well mixed, as in the case of interstitial fluid surrounding a soft tissue implant site, then the situation is more complex. The simplest case occurs when a stagnation layer exists at the solid surface: this may be modeled as a third phase with a "resistance" to diffusion, often expressed by reducing the fluid phase concentration by a multiplier, k, which is less than unity. Thus, the adjusted concentration C_0' is given by:

$$C_0' = kC_0 \qquad (3.7)$$

A second complication arises when the solid phase is able to absorb water (or another solvent) as well as the solute under study. This produces a steadily thickening solvated surface layer in which the solute can diffuse more readily than in the unsolvated solid. This has the effect of increasing the 1/2 power exponent in Eqs. (3.4) and (3.6) to values closer to one (and similarly of altering conclusions 1–3 above). Such a process is termed "non-Fickian" or Type II diffusion.

3.4 EXAMPLES OF UNDESIRABLE ABSORPTION

Let us examine one special case, that of diffusional pickup by a sphere in a medium of constant concentration. The application is the problem of the "variant" heart valve poppet. The poppet-type of heart valve consists of a polymeric sphere (the "poppet") usually fabricated from a silicone elastomer, a constraining cage, and a valve seat with a sewing ring (Figure 3.2). Although this was an early and usually successful design for human aortic valve replacement, for a small percentage of patients the ball jammed in the cage or, in some cases, the ball actually escaped. Discoloration of such recovered balls, termed "variant," suggested that some material had been absorbed, leading to swelling.

The mathematical solution for this case is reached by transforming Eq. (3.3) into spherical coordinates, and solving for the appropriate boundary conditions. We find that M_t is proportional to Dt/a^2, where a is the radius of the sphere. That is, we can write:

$$\frac{M_t}{M_\infty} = k\left(\frac{Dt}{a^2}\right) \tag{3.8}$$

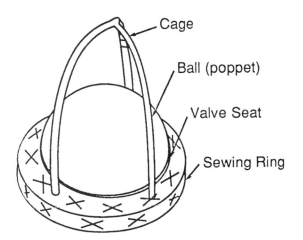

Figure 3.2 Diagram of a poppet type aortic heart valve.

Table 3.1 Weight Gain by
"Variant" Poppets

Weight gain (%)	Period of implantation (weeks)
0.1	1
0.1	2
0.9	11
2.1	18
4.15	34
3.4	39
2.5	48
3.1	52
3.3	52
5.5	80

Source: Adapted from Carmen and
Kahn (1968).

where

M_∞ = maximum amount of material absorbed

k = constant

The solution suggests that simple absorption should produce a weight gain that is *linear* with time. Data obtained by Carmen and Kahn (1968) from the study of 10 "variant" poppets are given in Table 3.1.

A linear regression ($r = 0.91$) yields a linear rate of weight gain of 0.27% per month in good functional agreement with our calculation. This rate presumably equals kD/a^2 from Eq. (3.8).

The absorption of materials from blood can have a variety of consequences. McHenry et al. (1970) studied poppets retrieved from five patients with clinical symptoms of aortic valve malfunction and reported discoloration (yellowing), fatty smell, and strut grooves. These poppets, which were implanted for longer times than those studied by Carmen and Kahn (1968), contained up to 16% simple and complex lipids by weight. The simple lipids (1.5–2% by weight) were composed of 60–65% cholesterol esters, 15–20% triglycerides, 5% fatty acids, and 10% cholesterol.

The material absorbed (in these two reports) is a portion of the "fatty" component of blood serum, usually called lipid on a collective basis. It can be presumed that the shape and physical property changes observed

Table 3.2 Lipid Content of Retrieved Finger Joint Prostheses

Duration of implantation (mon.)	Intact implants		Fractured implants			
	Cumulative[a]		Cumulative		Estimated[b]	
	No.	Lipid (wt.%)	No.	Lipid (wt.%)	No.	Lipid (wt.%)
12	15	0.475 ± 0.05	8	0.515 ± 0.2	8	0.515
24	24	0.461 ± 0.04	10	0.447 ± 0.2	2	0.175
36	37	0.469 ± 0.03	12	0.443 ± 0.1	2	0.423

[a]Cumulative: 24 month group includes data from 12 months; 36 month group includes data from 24 months.
[b]Estimated: 24 months = estimated lipid (wt.%) for implants failing between 12 and 24 months; 36 months = estimated lipid (wt.%) for implants failing between 24 and 36 months.
Source: Adapted from Swanson et al. (1973).

were a result of lipid absorption. Lipids are present in other body fluids and may be absorbed by implants that are not in the blood flow.

Swanson et al. (1973) reported on a study of 49 silicone elastomer finger joint prostheses that were retrieved, for various reasons, after implantation for up to 36 months in patients. All implants were examined for lipid content by chloroform extraction, the resulting weight change was determined, and direct analyses of the extractant fluid was performed. Of these implants, 12 had fractured in use. A summary of the findings is given in Table 3.2.

The question was raised as to whether lipid absorption had any relationship to the fracture in service. The authors concluded that there is no relationship between time of implantation, or lipid content, and risk of fracture.

It is surprising in the light of the Carmen and Kahn study that no relationship was seen between lipid content and time. However, two points must be made:

1. The boundary conditions for an implant within the tissue capsule in a finger joint are quite different from those present in flowing blood; C_0, C_s, and mass transfer are all different in this case. Thus, we should be little surprised that the previous solution does not predict the results.
2. The maximum weight gain reported in Table 3.2 is 0.515%. This is the value that would be expected after less than 2 months if we used the rate computed from Table 3.1 and is *one-tenth* of the maximum value reported for the heart valve poppets. So, an alternate explana-

tion for the failure to observe a linear increase in weight with time might be that the equilibrium level is far lower in the finger joint, and is reached before 12 months, the earliest summary date given in Table 3.2.

Examination of the data in Table 3.2, with an attempt to estimate mean lipid values of the failed implants within 12-month intervals, suggests another possible conclusion concerning the relationship between lipid content and device fracture. It is possible that there are two populations of failed implants: those that take up lipid with abnormal rapidity (represented by the failures in the first 12-month period) and those that take it up quite slowly (represented by the later failures). In the first case we might invoke weakening due to swelling, while in the second we might suggest decreased tear resistance associated with the absence of a plasticizing effect (see Chapter 6 for a more complete discussion of these phenomena).

We can summarize the effects of absorption in these series as change in color, change in volume (swelling), and possible change in mechanical properties (as evidenced by the presence of strut grooves). Mechanical property changes secondary to absorption, in the general case, can include reduction in modulus of elasticity, increase in internal viscosity, increase in ductility (or possible decrease in the presence of reaction with the absorbed species), change in the coefficient of friction, and reduction of wear resistance.

3.5 OSMOTIC EQUILIBRIUM

Problems arise even in the absence of obvious property changes. The increase in volume due to absorption results in the application of a negative internal hydrostatic stress to the material. We can understand this effect through the following calculation. If a material is immersed in a solution such that the solute is soluble in both the liquid phase (solvent) and the solid phase (material), there will be an initial interfacial pressure:

$$\Pi = MRT \tag{3.9}$$

where

M = molarity of solute in liquid phase

R = gas constant

   $= 8.31 \times 10^7$ erg $°K^{-1}$ mol^{-1}

T = absolute temperature

This is van't Hoff's law, and the pressure Π is called the osmotic pressure. It is an expression of the fact that the activity of the solute in the liquid phase is greater than that in the solid phase.

Driven by the osmotic pressure, the solid phase will absorb solute. When this happens, the solid phase expands as if subject to a *negative* hydrostatic pressure. This pressure P is called the swelling pressure and

$$P = \Pi - p \qquad (3.10)$$

where p = hydrostatic restoring force of the swollen material. That is, as the material swells, an elastic restoring force is generated through the strain of the material that counteracts the osmotic pressure. As the material swells, this restoring force rises, the activity of the solute in the solid phase rises, and the swelling pressure approaches zero as equilibrium conditions are obtained. Obviously, changes of solute concentration in the liquid phase disturb this equilibrium and can produce further expansion or contraction of the solid phase.

Advantage is taken of this phenomenon in the production of hydrophilic gels for biomedical applications such as contact lenses. These materials have low elastic moduli and high affinity for water. Since water has a molarity of 55.6, high osmotic pressure coupled with low negative hydrostatic restoring forces permit the fabrication of compositions with an equilibrium water content approaching 98%, while volume expansions may exceed 5000% (Refojo, 1976)!

3.6 LEACHING

The simplest case that results in leaching is that of removal of diffusing material from the surface at a constant rate. This is approximately the case of elution by moving blood; it is closely parallel to the in vitro situation of evaporation from a surface. The arrangement, initial conditions, and change of concentration of the solute in the solid (biomaterial) phase with time are shown in Figure 3.3.

There is, in most real situations, an additional condition. If the fluid medium is in motion but not fully stirred (as shown), we must assume some rate of transfer. The simplest case is to take this rate as proportional to the surface concentration at any moment. In particular, this rate is taken to be linearly dependent upon the difference between a surface concentration C_s and bulk concentration C_0. Thus, the boundary condition is:

$$-D \frac{\partial C}{\partial x} = \alpha(C_0 - C_s) \text{ at } x = 0 \qquad (3.10)$$

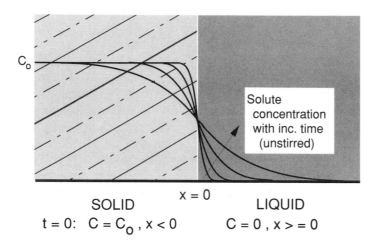

C_o

Solute
concentration
with inc. time
(unstirred)

SOLID   x = 0   LIQUID

t = 0: $C = C_o$, x < 0   C = 0 , x > = 0

Figure 3.3   Leaching from solid biomaterial into a liquid.

If $h = \alpha/D$, Eq. (3.10) leads to the following solution:

$$\frac{C - C_s}{C_0 - C_s} = \mathrm{erfc}\left(\frac{x}{2\sqrt{Dt}}\right) (-e^{[hx + h^2 Dt]})\ \mathrm{erfc}\left(\frac{x}{2\sqrt{Dt}} + h\sqrt{Dt}\right) \quad (3.11)$$

We find that for a particular value of h, M_t is proportional to $(Dt)^{1/2}$. Other conditions such as concentration-dependent diffusion constants, surface and bulk reactions, etc., can affect the former of these relationships.

3.7 EXAMPLE OF PLANNED LEACHING: DRUG RELEASE

When drugs are administered by intramuscular injection, their release into other compartments of the body is governed by diffusion. The effect of this may be seen in Figure 3.4.

On the left hand portion of this figure, the saw-toothed trace shows the effective drug concentration, at some point distant from the point of administration, resulting from three divided doses. After the first dose, the concentration at that point rises through the sub-optimal into the optimal therapeutic range, then begins to decline rapidly due to the combined effects of storage in other tissues, metabolic degradation, and excretion. The declining dashed line segment shows the continued concentration decline which would occur if the second (and third) dose was not

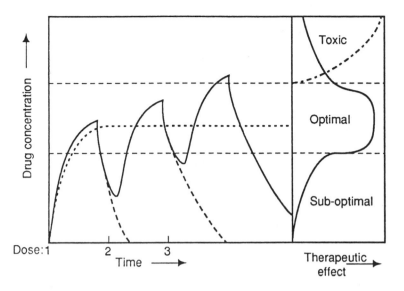

Figure 3.4 Drug release strategies. Source: Adapted from Chien (1989).

administered. Unfortunately, the selected temporal spacing of the doses, for their size, was too short: the concentration after the third dose is sufficient to produce toxic side effects. The relationship between drug concentration in the target tissue or organ and its effect is shown on the right side of Figure 3.4: this relationship is referred to as a "dose-response" curve, sometimes called a Bertrand curve.

The curvilinear dashed line in the left hand portion of Figure 3.4 represents an optimum strategy: the drug concentration rises smoothly and rapidly to an optimal level and then is maintained constant. With care, such an effect can be obtained by continuous intravenous injection. However, it is possible to utilize diffusional effects to produce an implant which can produce the same results.

Figure 3.5 schematically depicts a drug release system that would accomplish this goal. It consists of two elements: a barrier membrane (B) and a drug reservoir or filler (F). The drug is suspended as a solute in the filler. If the constants for the drug in question are such that the condition

$$D_B << D_F, D_{tissue} \tag{3.12}$$

is met, then there will be a near constant release rate until M_t becomes a significant fraction of M_∞. This secondary condition can be met by making the drug volume of the filler phase large compared with the desired dose (and removing the device when its function has been served) or by

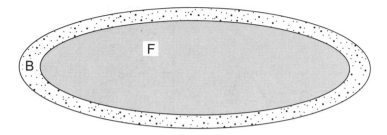

Figure 3.5 Sustained (reservoir) drug release device.

repeated injection of drug into the device. Such an arrangement is termed a reservoir or "sustained" release device.

This is only a simple example of the use of diffusional leaching to produce drug release. Adjustment of geometries and the use of different forms of drugs can produce a wide range of release rate (and resultant tissue concentration) profiles with time. Combining the effects of absorption (which may alter diffusion rates by changing osmotic pressure) with changes in the composition of the solvent of the filler phase produces still more options for the drug release device designer.

3.8 EFFECTS OF SWELLING AND LEACHING

There are adverse aspects to large deformations that may be caused by swelling and leaching. It is possible to exceed the creep stress in the material. This will produce continual deformation and absorption, rather than the attainment of an equilibrium solute concentration. Even if this does not happen, swelling reduces the elastic limit of a material (that is, the available strain to the elastic limit) and may lead to a mode of failure termed static fatigue or "crazing," especially in brittle materials. Crazing is the development of microcracks that merge and can eventually result in fracture.

Leaching, the reverse of swelling, usually has a less pronounced effect on properties. The primary undesirable aspects of unplanned leaching are the local and systemic biological reactions to the released products. Excessive leaching (for instance, intergranular leaching in metals) can result in a reduction in fracture strength. The defects produced by leaching can coalesce into macroscopic voids. If these begin to constitute a significant percentage of the volume of rigid materials, the elastic modulus will decline. The decrease is proportional to the second power of the volume percent of voids (see Section 6.3.2).

REFERENCES

Carmen, R. and Kahn, P. (1968): J. Biomed. Mater. Res. 2:457.

Chien, Y. W. (1989): Med. Prog. Technol. 15:21.

McHenry, M. M., Smeloff, E. A., Fong, W. Y., Miller, G. E., Jr. and Ryan, P. M. (1970): J. Thorac. Cardiovasc. Surg. 59:413.

Refojo, M. F. (1976): In: *Hydrogels for Medical and Related Applications*. J. D. Andrade (Ed.). American Chemical Society, Washington, D.C., pp. 37ff.

Swanson, A. B., Meester, W. D., Swanson, G. deG., Rangaswamy, L. and Schut, G. E. D. (1973): Orthop. Clin. N.A. 4(4):1097.

BIBLIOGRAPHY

Brophy, J. H., Rose, R. M. and Wulff, J. (1964): *Thermodynamics of Structure*, Vol. II of: *The Structure and Properties of Materials*. J. Wulff (Ed.). John Wiley, New York, pp. 75ff.

Crank, J. (1957): *The Mathematics of Diffusion*. Oxford University Press, London.

4

Corrosion and Dissolution

4.1 CHEMISTRY OF CORROSION

All of us have an intuitive understanding of corrosion. The purpose of this chapter is to consider the principles of chemistry that underlie this understanding, to characterize and classify the types of corrosion that may occur, and to see how these considerations apply to the use of metals as implants.

The layman tends to equate corrosion with the "rusting" of iron. However, the term has a far broader application. The word corrosion comes down to us from the Latin, *rodere*, to gnaw. We have come to use it to describe any chemical attack on solid materials, especially metals. Similar processes attacking glass and polymers resemble corrosion in their physical results; however, their chemistry is not as well understood at this time. The chemical processes that contribute to the phenomenon of corrosion can be termed processes of reaction and/or dissolution in the presence of water.

In the case of metals, the following four generic reactions are the most usual (*note*: valence states and changes given are typical; other values are possible):

Ionization: The direct formation of metallic cations, generally under acidic or reducing (oxygen-poor) conditions:

$$\underset{0 \quad +1}{M \rightarrow M^+ + e^-} \tag{4.1}$$

Oxidation: The direct reaction of metal with oxygen, either gaseous or dissolved, without the participation of water. In the extreme violent examples we recognize oxidation as burning:

$$\underset{0 \qquad\quad +4}{M + O_2 \rightarrow MO_2} \tag{4.2}$$

Hydroxylation: The reaction of metal with water under alkaline (basic) or oxidizing conditions to yield a hydrated oxide or hydroxide. Since most hydroxides are only sparingly soluble in alkaline conditions, this process often leads to the formation of a passivating film, of which more will be said later:

$$\underset{0 \qquad\qquad\qquad\qquad +2}{2M + O_2(\text{diss.}) + 2H_2O \rightarrow 2M(OH)_2} \tag{4.3}$$

Reaction: The combination of metal or metallic ions with other cations and anions; this is often termed *complex formation*:

$$\underset{+2 \qquad\qquad +2}{MO_2^{2-} + HCl \rightarrow MOCl^- + OH^-} \tag{4.4}$$

Each of these processes has the effect of decreasing the amount of pure metal present and of producing metal-bearing ions and compounds. Consideration of the effects of corrosion must take note of both the attack on the parent metallic component and the formation of reaction products.

4.2 CLASSIFICATION OF REACTIONS

It is necessary to make some order out of these various chemical processes which contribute to corrosion so that we can approach the prediction of corrosive attack in a systematic manner. All of these and other processes involved in corrosion can be classified by answering two questions:

1. Does the reaction depend upon pH?
2. Does the reaction depend upon local electrical potential?

For the purpose of discussion, we will designate the answers to these questions with a '' + '' or a '' − .'' Thus, a reaction such as dissolution of gas, which is independent of both pH and potential, would be an example of a '' − '' reaction. There may be as many as 30 possible reactions just for the interaction of an elemental metal with pure water in the presence of oxygen.

There is a further observation that simplifies our efforts of organization. For any particular combination of pH and potential, there will be a single *dominant* reaction for a specific metal in a specific solution; that is, of all of the possible reactions, one will be the maximum contributor to the degradation of a metallic part and to the concentration of metal in solution.

4.3 THE POURBAIX DIAGRAM

4.3.1 Reactions of Chromium in Pure Water

Classification of all possible reactions between a metallic element and water (and its constituents) by their pH and potential dependence, combined with the determination, from various types of experiments, of those combinations of pH and potential that favor particular reactions, provides the information needed for a graphic representation of the overall reaction system, such as Figure 4.1.

This is called a *pH-potential* or, more usually, a *Pourbaix* (poor BAY) diagram, after Marcel Pourbaix, who popularized its use. This particular

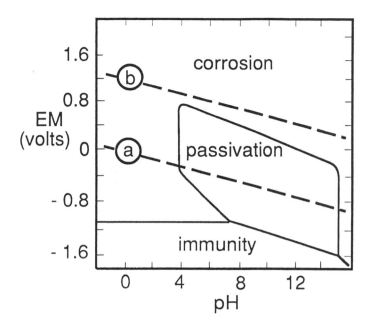

Figure 4.1 Pourbaix diagram for chromium in pure water.

diagram is that for chromium in pure water and summarises the reactions among five primary species (Cr, O_2, H_2, H^+ and OH^-).

There are three major regions in a Pourbaix diagram. Each of these represents combinations of ranges of pH and potential for which a reaction or a related group of reactions are dominant; each region corresponds to one of the three fundamental conditions that describe the response of metals to aqueous solutions:

1. *Immunity*: In this region, the dominant reaction is ionization. However, throughout this region the resulting equilibrium concentration of chromium in solution, in all ionic forms, is less than 10^{-6} M. This concentration, 10^{-6} M, is generally taken as being the boundary between corrosion and immunity. If reaction processes yield a total equilibrium concentration (of all metal-bearing ions) of less than this value, it is engineering practice to speak of the metal as being immune from corrosion, or more simply *immune*, under that particular set of conditions. Since the level of 10^{-6} M (typically 50 ppb) greatly exceeds normal physiological concentrations for most less common ions (especially those containing trace metals), it is better practice in

considering implant applications to assume that metal is released in some concentrations for *all* combinations of pH and potential.

2. *Passivation*: In this region, the dominant reactions lead to the formation of oxides and hydroxides. Since these products for chromium are largely insoluble at a pH above 4, they cling to the interface between the metal and the solution, reducing and eventually preventing further reaction. This renders the chromium *passive*. Throughout this region, the solubility of the oxides and hydroxides is low enough that the total concentration of chromium in solution is, as it was in the immune region, less than 10^{-6} M.

3. *Corrosion*: In this region, a variety of processes can attack either metallic chromium (at low or high values of pH) or the passive coating (if present, at intermediate values of pH). The result throughout the region is a total equilibrium concentration of chromium in solution that equals or exceeds 10^{-6} M; thus, the chromium is said to *corrode*.

Two additional features are of interest. The diagonal dotted lines in Figure 4.1 define the reactions of gaseous oxygen and hydrogen with water. The upper line, b, is that for oxygen; the lower line, a, is that for hydrogen. Both of these reactions are of the '' ++ '' type, and so the lines slope. The region between the lines is that in which water is stable. Above the oxygen line, oxygen is released, while below the hydrogen line, hydrogen is released.

Dominant reactions of the '' + − '' type produce vertical region boundary segments, while those of the '' − + '' type produce horizontal segments. Reactions of the '' − '' type do not produce regions of dominance on this type of diagram.

In biological systems that are pH controlled by buffering, the local oxygen or hydrogen partial pressure can then define an effective local potential. Thus, tissues that are perfused with arterial blood and are maintained at pH 7.37 have an equivalent potential of 0.782 V. This last fact makes it possible to apply the Pourbaix approach to the prediction of metallic corrosion in vivo.

4.3.2 Reactions of Chromium in the Presence of Aqueous Chloride Ion

Figure 4.2 is again the Pourbaix diagram for chromium, but with two important changes:

1. The solution is now water with 1.0 N chloride ion to better simulate the situation in vivo. The principal effect of this is to radically shrink

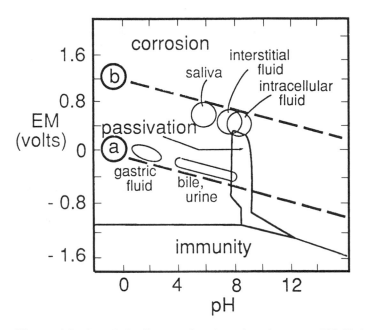

Figure 4.2 Pourbaix diagram for chromium in water (1N Cl$^-$).

the passive region. This results from reaction of the chloride anion with both free metal ions and the passive layer to form soluble complexes, thus raising the effective solubility of chromium.

2. The areas of pH and potential (as defined by pO$_2$) for various body fluids have been superimposed. Note that the areas for interstitial and intracellular fluids actually lie closer to pH 7.0; they are plotted as more alkaline for clarity. From this we can see that *pure* chromium would perform satisfactorily in neutral conditions in the bile duct or urinary tract, but would be unsatisfactory in the stomach, where pH may approach 1. General tissue applications would be marginal since they lie on the boundary between the passivation and corrosion regions.

It should be clear that a Pourbaix diagram is useful in predicting corrosion in only a general way. The following limitations should be realized:

1. The local microconditions, as determined by local ionic conductivity and by limitations in the diffusion of oxygen, hydroxide, and hydrogen ions, dictate the exact equivalent potential that may be expected. Thus, the regions shown on this diagram for different physiological

situations may be considered as reflecting average conditions. In particular, pericellular conditions may be radically different.

2. The areas of dominance and other details of the diagrams are those which hold after all reactions have come to equilibrium. Reactions may be very slow, as is the case for many involving chromium compounds, leading to prolonged nonequilibrium conditions.

3. Reactions and their kinetics depend upon the history of the metal to some degree. Thus a bare piece of chromium placed under conditions that lie within the passive region of the Pourbaix diagram will undergo reactions leading to formation of hydroxides; however, if its surface is pretreated to produce a passive (oxidized) layer, then the dominant reactions, under the same set of conditions, will be hydration and dissolution of this surface layer. Such reactions may be very slow; in the case of pre-passivated chromium-containing alloys they are so slow that such passive films are termed *metastable* and the materials may be used for some applications with conditions within the corrosive region.

4. Pourbaix diagrams are available for most elemental metals in pure water but they do not exist for the vast majority of alloys or for other aqueous solutions. However, since the corrosion resistance of chromium-containing stainless steels and cobalt-based ''super alloys'' depends to a great degree on the presence of chromium hydroxide passivation films, the diagram for chromium in the presence of chloride (Figure 4.2) is quite useful in understanding the chemical aspects of their material response in vivo.

5. Finally, as will be discussed in Section 4.10, the presence of active cell products in vivo may modify both the rate of reactions and the nature of their products.

4.4 THE ELECTROCHEMICAL SERIES

4.4.1 Ideal Series

It is possible to plot a complete Pourbaix diagram for any real metal or alloy in a defined solution. However, a further simplification may be made if one is only interested in the *relative* corrosion resistance of metals. We can begin by noting that the boundary between the immunity region and the corrosion region for acid and neutral pH is horizontal. The left hand intercept (or more correctly, the potential for pH = 0) is a single potential value. This potential will be different for each metal alloy. On the left side of Table 4.1 a number of metals are ranked by this potential in an *ideal* (sometimes termed *absolute*) electrochemical series.

Table 4.1   Ideal and practical electrochemical series

Potential	Ideal	Practical
	Noble or cathodic	
	↑ Gold	Platinum
	Platinum	Gold
	Silver	Stainless steel (passive)
	Copper	Titanium
$E = 0$	Hydrogen	Silver
	Lead	Nickel
	Tin	Stainless steel (unpassivated)
	Nickel	Copper
	Cobalt	Tin
	Iron	Lead
	Chromium	Cast iron
	Aluminum	Wrought iron
	Titanium	Aluminum
	↓ Magnesium	Magnesium
	Base or anodic	

This electrochemical series is arranged with the most noble or cathodic potentials (with respect to the H/H^+ half cell reaction) at the top and increasingly base or anodic potentials as one proceeds down the list. Note that the apparent sign of an electrode depends upon whether it is self-polarized (as in corrosion) or externally polarized (as in electroplating). A self-polarized anode is negative while an externally polarized one is positive and vice versa for the cathode. The oxidation-reduction nature of the reactions identical in both situations: reduction takes place at the cathode and oxidation at the anode.

Despite the use of potentials obtained under acidic, oxygenated conditions, the ideal series is a reasonable measure of the relative corrosion resistance of metals under a variety of conditions in pure water. The higher the place in the list (the more cathodic the potential), the more noble or corrosion resistant the metal.

We shall also see (Section 4.7.2) that the relative position in an electrochemical series determines which of a coupled pair of dissimilar metals may undergo corrosion.

4.4.2   Practical Series

In the rightmost column of Table 4.1, many of the same metals, and some alloys, are listed in a *practical* electrochemical series. The use of a practi-

cal series, in this case for the exposure of these metals to seawater, begins
to take into account the situation peculiar to a specific application. Seawater exposure, particularly in tropical climates, is the engineering condition that most closely simulates the biological environment.

Comparison of the practical to the ideal series demonstrates some
interesting differences related to the differences in environment:

1. The two most noble metals in both series, platinum and gold, change
    their relative positions. This reflects the fact that, while neither is
    strongly attacked by seawater, gold forms chlorides more readily
    than does platinum.
2. Titanium moves well up the list, reflecting the highly insoluble nature
    of most of its compounds, particularly its TiO_2 passivation layer that
    forms spontaneously in air.
3. Unpassivated stainless steel, an alloy of iron, nickel, and chromium
    (as well as other minor elements), is not particularly high on this list.
    The choice of stainless steel as an implant material depends primarily
    upon its mechanical properties and machinability in the presence of
    an *acceptable* level of corrosion resistance (when passivated before
    use) and moderate local host response to its corrosion products.

A practical series begins to take factors of corrosion other than equilibrium thermodynamics into account. That is, it reflects not only the possibility of corrosion, but also details of actual attack in specific environments. Pourbaix diagrams and electrochemical series tell us something
about the *likelihood* of corrosion. They define equilibrium conditions and
imply rates of corrosion as being proportional to deviations from equilibrium. It is important to know something about the *rates* of corrosion.
These rates will help to determine, for instance, the rate of release of
metallic ions from an implant.

4.5 CORROSION RATE

In engineering applications, corrosion rates are expressed as rates of
surface recession per year. The common unit is the *mpy* or mil per year.
This unit is far too big to be used to examine corrosion rates in the
biological environment for the same reasons that an equilibrium concentration of 10^{-6} M is too high to be considered indicative of immunity.

A more direct measure may be obtained by examining Eqs. (4.1)
through (4.3). These are characteristic of metallic corrosion. Note that in
each case the valence of the metal is reduced, with a required transfer of
charge or electrons to another species. Since the sites of reduction (cath-

ode) and oxidation (anode) are separated in space, an equivalent current must flow.

Corrosion (at the anode) of one molecular weight of metal, with an accompanying valence change of $+1$ (for instance, from 0 to $+1$), will result in the transfer of 1 Faraday (F) of charge. Thus, for a corroding anode one may determine the corrosion rate by measuring the *net* current flow and dividing by the area. The unit of corrosion, defined in this way, is then amp/cm^2.

4.5 POTENTIAL-CURRENT RELATIONSHIPS IN CORROSION

Let us look briefly at an ideal case of corrosion. We shall take the reaction at the anode to be Eq. (4.1).

$$M \rightarrow M^+ + e^-$$

At the cathode we shall assume that the reaction is the reduction of dissolved oxygen:

$$O_2(d) + 2H_2O + 4e^- \rightarrow 4OH^- \tag{4.5}$$

An alternative reaction, the reduction of hydrogen ion with the release of gaseous hydrogen, is also possible:

$$2H^+ + 2e^- \rightarrow H_2(g) \tag{4.6}$$

The latter reaction will occur preferentially if either the oxygen potential is very low or the metal is extremely active (base or non-noble). In this case, we will consider Eq. (4.5) to be the cathodic reaction.

Let us now look at Figure 4.3. The initial potentials at the anode and cathode are E_{A_0} and E_{C_0}. Due to the differences in potential, current begins to flow (corrosion takes place). This may be represented by moving to the right of the diagram. Note that, due to the familiar Ohm's law relationship between current, potential difference, and resistance, the effective potential of the cathode drops and that of the anode rises. If nothing happens to intervene, the potentials become equal. This mixed potential E_M is then maintained, and the current i_3 is defined by

$$\frac{E_M}{\text{Total resistance}} = i_3 \tag{4.7}$$

The current (i_3) divided by the area of the anode yields the corrosion rate.

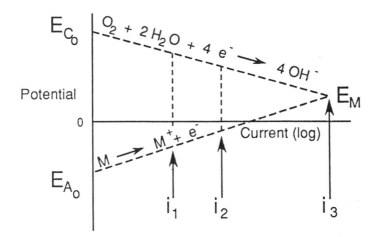

Figure 4.3 Potential-current (log) relationships in corrosion.

If an external resistance, R_{ex}, that is large compared to the previous resistance, is inserted between anode and cathode, then the resulting current is i_2 and is defined by

$$\frac{E_{C_0} - E_{A_0} - i_2 R_{ex}}{\text{Total resistance}} = i_2 \tag{4.8}$$

The current i_2 is less than i_3; thus, less corrosion takes place in a given period of time. This is the situation when a passivation or insulating layer can be maintained on a metal in a pH-potential region which would normally promote corrosion.

If the supply of oxygen is limited by diffusion, for instance, then the potentials of anode and cathode may remain more widely separated, and a still smaller current i_1, resulting in less corrosion, may flow.

In either case ($i = i_1$ or i_2), the effective potential will be somewhere between E_{C_0} and E_{A_0}, depending upon the relative areas of the cathode and anode.

Thus, it should be clear that the actual rate of corrosion may vary widely for a given set of equilibrium conditions. Local oxygen supply, conductivity of the bathing electrolyte, and the extent of the electrode surfaces, as well as the presence of various inhibitors and enhancers of corrosion, may affect the result.

Corrosion in real environments is not usually detected by measurement of potentials and currents, or of concentrations of ions in solutions. Rather, it is recognized by the evidence of attack on the bulk material, the chemical gnawing away of the fabricated part.

4.7 FORMS OF CORROSION

I am indebted to Mars Fontana and Norbert Greene (Fontana, 1986) who have collected many diverse descriptions of the physical appearance of corrosion and grouped them into eight categories or forms, depending upon mechanism and common features of the result of attack. The eight forms of corrosion will be briefly examined and some comments made on their mechanisms.

4.7.1 Uniform Attack

Uniform attack, or general overall corrosion, is a self-explanatory term. This is the process that is taking place in the corrosion region and, by oxide/hydroxide dissolution, in the passivation region of the Pourbaix diagram. It is the most common form of corrosion. In the absence of equilibrium concentrations of their constituent ions in the bathing solution, it will occur for all metals. Even in the immunity region, uniform attack will result in a slow removal of metal from implants. Thus, it is fair to state that, due to uniform attack, all metals have a finite corrosion rate in vivo. However, uniform attack may not be noticed until, or unless, significant amounts of metal are lost. Since all metals currently used in implants are relatively highly resistant to uniform attack, little *evidence* of such attack is ever seen in implant applications.

Uniform attack is usually measured in terms of surface recession with units of mils (10^{-3} in) per year, abbreviated *mpy*. An approximation to this rate may be obtained from Equation (4.9) if the surface area of an implant (A, in^2), the density of the alloy used (D, g/cm^3), the exposure time (T, hours), and the total weight loss (w, mg) are known:

$$\text{mpy} = \frac{534\ w}{DAT} \tag{4.9}$$

For a typical implant alloy, with a density near 8 g/cm^3, one mpy $\approx$ 0.7 mg/cm^2/day. In vivo uniform corrosion rates for well-passivated alloys are thought to be about 1/100 of this (Steineman, 1980).

4.7.2 Galvanic Corrosion

Galvanic (or two-metal) corrosion takes place when two different metals are in physical (electronic) contact and are immersed in an ionic conducting fluid medium such as serum or interstitial fluid. This is also referred to as *couple* corrosion. An example of a situation which may lead to this is

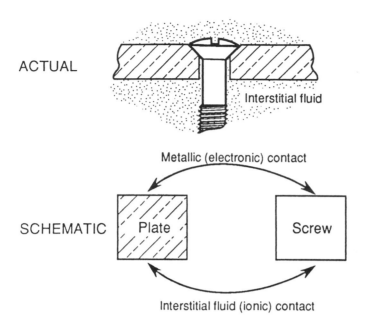

Figure 4.4   Conditions for galvanic corrosion.

shown in Figure 4.4. The difference between the screw and plate, responsible for the galvanic process, may be due to dfiferent compositions (major and/or minor constituents) or processing.

For a particular combination of pH and potential (defined by pO_2), the two metals will have different electrochemical potentials. The one which is less noble than the other, that is, below it on a suitable practical electrochemical series, becomes *anodic*. The surface of the less noble metal that is in contact with the solution will experience attack, of a uniform nature, with a release of metallic ions. The other metal becomes *cathodic*. Electrons move to it through the physical connection, driven by the intermetallic potential difference. This may then reduce either dissolved oxygen or hydrogen ions, depending upon local conditions. Although the less noble metal of the pair *may not* corrode, due to its being in a *passive* region in its respective Pourbaix diagram, the more noble metal *cannot* corrode under any conditions. Thus, being cathodic, it is said to have *cathodic protection*.

The actual details of galvanic corrosion depend upon a large number of complicating factors including the relative size of the areas of electronic and ionic contact, as well as on the actual metal pair involved. However, it is safe to assume that some galvanic corrosion will occur in any metal pair in acid pH. Note that all three conditions must be met for galvanic corrosion to take place; thus the presence of two or more different compositions of metallic implants within an animal or patient will not produce galvanic effects unless the implants are in actual contact, such that an electron current may flow between them.

4.7.3 Crevice Corrosion

Crevice corrosion is one of a number of forms of corrosion that are related to structural details. The basic requirement for the occurrence of this process is the presence of a crevice, a narrow, deep crack: either an interface between parts of a device, such as between plate and screw head, or a defect such as an incomplete fatigue crack. The details of the initiation of crevice corrosion are not yet clear. Once begun, however, it is characterized by oxygen depletion in the crevice, anodic metallic corrosion along the crevice faces, and cathodic protective conditions on the metal surface around its mouth. Static nonflowing conditions in the solution seem to favor crevice corrosion, perhaps because of the formation of a metallic ion concentration gradient away from the open end of the crevice. Since the areas of attack are concentrated, evidence of crevice corrosion can easily be seen in the mating areas in multipart devices, such as between screw and plate in retrieved fracture fixation devices (Colangelo and Greene, 1969). This is a serious effect; the majority of multipart fracture fixation devices retrieved from patients show either crevice corrosion and/or pitting corrosion (see Section 4.7.4) (Cook et al., 1985). In conjunction with stress corrosion, crevice corrosion may change the behavior of metals subjected to cyclic loading (see Section 6.5.2).

4.7.4 Pitting Corrosion

Pitting corrosion is a special case of crevice corrosion. It is a more isolated, symmetric form of attack; inclusions, scratches, or handling damage may initiate it. Pitting corrosion proceeds through processes similar to those for crevice corrosion, although static conditions and reduced oxygen potential seem less important. Pits often occur in large numbers, like freckles, and grow down in the direction of gravity in unstirred solutions.

Pits are a hazard in highly stressed implants as they constitute points of stress concentration and may serve as the starting points for mechanical cracks to develop. Like the effects of crevice corrosion, they are easy to see. When very small they change the surface finish, often producing a "frosted" or matte appearance. Larger, more developed pits often have accumulations of colored corrosion products in them and show up as green, brown, or black spots against the otherwise polished surface of the implant. For this reason it is inadvisable to clean implants vigorously after removal before they have been examined for evidence of corrosion. Multiple part implants often show evidence of crevice, pitting, and fatigue corrosion (Cohen and Lindenbaum, 1968). In practice it may be difficult to distinguish between pitting and crevice corrosion at early stages of attack.

4.7.5 Intergranular Corrosion

Intergranular corrosion is somewhat related to crevice corrosion but has different origins and produces different effects. It is more common in devices that are made by casting than those made by other processes. During the cooling of metal in the mold, crystals or grains form and grow. As each crystal touches the ones adjacent to it, disordered regions are produced called grain boundaries. These extend and persist since the orientation of the grains, as they form and grow, is random. Eventually, when the entire piece is solid, the last remaining fluid to freeze must do so along these boundaries. The boundaries may contain impurities, oxides, or compounds such as ferrites and carbides. As a result the chemistry of a grain boundary will be different from that of the grains on either side. If so, it will probably have a different electrochemical potential. If this material is noble with respect to the rest of the metallic part, very little will happen except that the uniform attack rate may go up slightly. However, if the material in the grain boundaries is base or anodic with respect to the grains, a galvanic cell will be produced. Since the volume of the grain boundaries in a solid is a very small part of the total volume, the grain boundaries can corrode rapidly. Obviously, once this process starts, crevice corrosion in the resulting defect can also proceed. The total effect, in engineering applications, is to produce a part that suddenly crumbles into grains under a mechanical stress, although appearing essentially normal up to that point. A less radical effect is sometimes seen in brass doorknobs in old houses. Because of the perspiration left on the knob by generations of hands, the intergranular corrosion of zinc precipitates "etches" the surface and makes the individual grains visible.

Intergranular corrosion is obviously more common in alloys than in pure metals and is favored by high levels of impurities and inclusions. Stainless steel, if not properly heat treated, may corrode by this mechanism due to a relative depletion of chromium from the grain boundaries. Welding of alloys that results in local melting and resolidification can also lead to a variant of this, which is called *knife edge* attack. The name derives from the appearance of the failure, a straight crack through the metal parallel to and near the weld. Again, proper heat treatment after welding will restore the right compositional distribution and reduce or prevent this type of attack.

4.7.6 Leaching

Leaching is similar to intergranular corrosion. However, in this case the components of a particular alloy are sufficiently weakly bound to each other, and differ enough in chemical reactivity, that there is a large difference in the rate of loss of the alloy component by uniform attack. Thus, leaching as a form of corrosion is a special case of leaching, with a reaction, and was discussed in Chapter 3. Attack of this kind will remove metal with a regular periodic variation of effect at a microscopic level. It is peculiar to certain alloy systems but can be induced by two conditions:

1. The introduction into the solution around the metal of an agent that attacks one component of the alloy in preference to another. For instance, F^- ion will selectively remove aluminum from copper-aluminum alloys.
2. The presence of more than one phase in the alloy. Usually all of the grains in an alloy have the same composition. The alloy is then said to have a single phase. However, it is possible for there to be grains of two or more different, discrete compositions. Such an alloy is said to possess multiple phases. Since electrochemical potential varies as chemical composition, these phases may have a different susceptibility to various forms of corrosive attack. Note that heat treatment to reduce the size of the grains will not change this situation. For this reason, multiphase alloys are not usually used in corrosive applications. Thus, there was considerable academic concern when ASTM F562, a multiphase alloy of cobalt, nickel, chromium, and molybdenum containing 35% nickel, was introduced for use in implants. Experience suggests, however, that, despite the differences in the phase compositions, all of the phases are sufficiently passive under the conditions experienced in soft and hard tissue implant sites that leaching does not occur in this alloy.

Leaching produces surface appearances that are similar to those produced by pitting and intergranular attack. It can only be verified by analysis of either the corrosion products or the surface concentration of the corroded metal part. The composition of the products or the surfaces that differ significantly from the bulk composition of the alloy are definite proof of leaching.

4.7.7   Erosion Corrosion

Erosion corrosion is a rare form of corrosion. This is an acceleration of attack on a metal because of the relative movement between the surrounding fluid and the metallic surface. It is not a unique process, but it serves to increase the rate of attack by several other mechanisms. This happens since many corrosion processes tend to be self-limiting. That is, the accumulation of the products of corrosion at the interface between metal and solution tends to reduce the rate of reaction. Flowing solution will sweep away these corrosion end products as well as providing new amounts of dissolved reactants, such as chloride ion and oxygen. In extreme cases, the solution may physically erode the passive layer in regions of passivity. The reformation of this layer and its removal by continued flow produces progressive attack on the metal and renders that region of the pH-potential diagram corrosive rather than passive as predicted.

The physical damage resembles pitting, except that these pits are elongated in the direction of flow and are generally larger and less symmetric than those seen in pitting corrosion under static conditions. The peculiarities of flow, especially if it is a stable pattern, will often result in etching a clear picture of the course of the flow on the surface of the metal.

4.7.8   Stress and Fatigue Corrosion

Stress corrosion is the last of the eight forms of corrosion. Simply stated, tensile stress increases the chemical activity of metals. A flexed metal plate will sustain a tensile stress on one side and a compressive stress on the other side. This produces a difference in electrochemical potential that renders the convex surface anodic with respect to the concave. Corrosion, as an acceleration of uniform attack, or perhaps secondary to tensile rupture of the passive film, will then attack the convex surface.

Since the formation of even a small crack in a static, loaded structure, such as a flexed plate, will concentrate stress, this attack tends to initiate cracks that grow rapidly. The result can be rapid structural failure. The cracks extend in between grains, but this process can be differentiated from intergranular cracking due to the small number and relative isolation of stress corrosion cracks.

Many metals display an endurance limit to cyclic loading. In the presence of crevice corrosion, either in physical cracks or in cracklike defects in a passive surface layer produced by single or repeated cyclic loading, this limit may be abolished. That is, the maximum stress that can be reached without failure continuously decreases as the number of load cycles increases. This phenomenon, which is a dynamic form of stress corrosion, is termed *fatigue* corrosion and may be an important limit on the life of metallic implants undergoing cyclic mechanical deformation.

4.8 CORROSION IN IMPLANT APPLICATIONS

Which of these eight forms of corrosive attack are important in implant applications? As a general rule, corrosive attack is more common on multipart implants than single part devices. Some studies indicate that a majority of multipart orthopedic fracture fixation devices show evidence of corrosion after recovery at the end of treatment (Cook et al., 1985). Uniform attack occurs on these as on all other implants. The primary physical evidence suggests that crevice and pitting corrosion are the next most important forms (Cohen and Lindenbaum, 1968). Crevice corrosion occurs in the gap between the screw and the plate in screw-plate assemblies. The attack is most often seen on the plate, within the hole but near the longitudinal surfaces. Occasionally, crevice attack will be noted on the portions of the screw opposite these areas. These areas have a high stress concentration due to the reduction in the cross section of the plate at the hole. Frank mechanical fracture of the plate through a screw hole can often be associated with microscopic evidence of crevice corrosion.

Pitting most often occurs on the underside of screw heads. Despite the characteristic freckle appearance of the pits, they may be hard to distinguish from mechanical scoring of the screw head and shaft during insertion and/or removal. Such scoring may result from rubbing against a burr in the hole in the plate, or against a fragment of bone caught between plate and screw during removal.

Galvanic corrosion may also occur between plates and screws. There is a slight tendency for this naturally since the plates and screws are fabricated by different processes and, thus, if not properly heat treated, may have slightly different electrochemical potentials. Mixing screws and plates from different manufacturers may also produce galvanic effects as each manufacturer uses a slightly different heat treatment schedule. Obviously, using screws of a different composition from the plate will also cause problems.

Attack of this sort is often discovered through reports of persistent operative site pain. It may occur, however, as judged by frequent observations of tissue discoloration during routine device removal procedures, without any apparent sensation. Galvanic corrosion may leave a discoloration with a ''burned'' or sooty appearance on either the screw or the plate in the area of contact.

Stress corrosion is also possible but extremely rare. Intergranular corrosion, leaching, and erosion do not occur to any real extent in modern multipart devices in orthopedic applications. A solid-state version of erosion corrosion may occur if the plate is loose or fixation is poor. Relative motion between plate and screws may result in physical removal of material or *fretting*. This may disrupt the passive film and produce accelerated corrosion in much the same way that erosion corrosion takes place. This phenomenon is difficult to distinguish from simple wear and is called fretting corrosion.

In single part devices such as cranial plates, intramedullary rods, endoprostheses, pins, and cerclage wire, the effects are rather more limited. Uniform attack does occur, as previously noted. Stress corrosion, or more generally, stress enhancement of fatigue failure (fatigue corrosion), is probably the most common destructive form. Although rare in prostheses, its incidence is very high in the highly stressed cerclage wire used for uniting bone fragments. Intergranular corrosion does occur occasionally and is probably most often associated with surface inclusions or casting defects in cast prosthetic sections. It is rarely active enough to lead to mechanical failure in the absence of cyclic loading.

Corrosion in blood contact areas is much more complex. The abundant supply of oxygen and the continued flow of electrolytes render most processes highly active. In addition, the presence of many small organic molecules influences rates. Sulfur-bearing molecules such as cystine appear to accelerate corrosion, while neutral molecules such as alanine may inhibit corrosion (Svare et al., 1970) in much the same way as rust inhibitors in engineering applications. Furthermore, corrosion may profoundly affect surface properties and thus influence thromobogenic behavior; this will be discussed in Section 9.3.2.

We generally consider corrosion to be undesirable. However, there are applications where the response to local concentrations of corrosion products is necessary to the successful function of an implant. For instance, the copper IUD (intrauterine device) depends for its contraceptive properties on the release of copper ions by a corrosion.

It may also be desirable to accept a higher corrosion rate because of other more critical properties of an alloy. In a surgical setting, the stainless steel spring clips used for the repair of large cranial aneurysmal defects have been deliberately fabricated from type 301, 416, and 420 steel alloys (McFadden, 1969). These alloys corrode at a more rapid rate than the more usual grade of stainless steel used in implants, 316L (ASTM F 138). However, these alloys are superior to 316L for spring fabrication. One feature of local response to these products is undoubtedly an increased fibroplasia, perhaps producing a more rapid and mechanically sound scarring process, but otherwise, they appear to be adequately tolerated.

4.9 ENGINEERING VARIABLES AFFECTING CORROSION RATES

Despite the prevalence of corrosive attack on implanted devices, the rate of failure, either of structure or function, is quite small. Why is it that a particular device can perform well in 99 patients, and then cause problems in the 100th? The answer is not simple.

In addition to all of the normal biological variables of human health and disease, we must add at least four "engineering" variables:

1. Composition of the implants, in particular, variations within the implant, and extremes of implant-to-implant variation can affect corrosion rates in many ways, as has been discussed.
2. Manufacturing variables, including casting conditions, metal purity, amount of cold work, and the degree and type of heat treatment, have a profound effect on corrosion rates (Sutow et al., 1976).
3. Handling, both in delivery and insertion, can affect results. Often, corrosion initiation can be traced to unintended physical damage.
4. Anatomical location and small differences in it may have an important effect. Positioning of the implant will affect the stresses upon it as well as the local environment it experiences.

4.10 CORROSION FACTORS PECULIAR TO THE BIOLOGICAL ENVIRONMENT

In addition to the factors just discussed, it is apparent that organic molecules present in implant sites may affect corrosion rates. These are thought to act in four ways:

1. *Formation of organometallic complexes.* If conditions at any point in the pH-potential region of interest favor the formation of organometallic complexes, the metallic content of these represents an addition to the corrosion rate.
2. *Alteration of charge of corrosion products.* Many organic molecules are potent oxidizing agents, producing the possibility of different ionic valences than predicted by pH-potential considerations. For example, in the presence of serum proteins, it is believed possible that significant amounts of chromium may be released from alloys as Cr(VI) rather than Cr(III) (Rogers, 1984).

3. *Modification of the passive layer.* Combination of organic species with the passive layer, either in the passive region or in adjacent regions of metastability (low passive film-dissolution rate), may alter the nature of the passive film. These may act to stabilize or destabilize the film, or to change its electrical conductivity. Svare et al. (1970) showed that, while alanine and albumin had little effect on the corrosion rate of copper, physiologic concentrations of cystine (17.6 mg/liter) in a solution of normal osmolarity reduced the corrosion rate by two orders of magnitude. On the other hand, nickel that was previously passivated by formation of a $NiCO_3$ surface film experienced a complete loss of passivity in this solution (at a pH and pO_2 appropriate for passivation) and an attendant rise in corrosion rate of two orders of magnitude.
4. *Changes in wear conditions.* While the presence of serum proteins generally elevates corrosion rates, in in vitro experiments it markedly reduces fretting corrosion rates of stainless steel (Brown and Merritt, 1981).

4.11 FINAL REMARKS

The one certain thing that we can say about corrosion is that it results in the release of cations from all metallic implants. Some of these, such as the ferric and ferrous ions, are familiar parts of the internal environment. Some are trace elements with known biological roles, such as trivalent chromium ions. Others are rare enough in nature that they do not have known metabolic roles and are released in the body, even in the absence of abnormal corrosion processes, at concentrations that are orders of magnitude above their normal in vivo occurrence. The consequences of such release will be considered at length in later chapters.

REFERENCES

Brown, S. A. and Merritt, K. (1981): J. Biomed. Mater. Res. 15:479.

Cohen, J. and Lindenbaum, B. (1968): Clin. Orthop. 61:167.

Colangelo, V. J. and Greene, N. D. (1969): J. Biomed. Mater. Res. 3:247.

Cook, S. D., Renz, E. A., Barrack, R. L., Thomas, K. A., Harding, A. F., Haddad, R. J. and Milicic, M. (1985): Clin. Orthop. 194:236.

Deltombe, E., de Zoubov, N. and Pourbaix, M. (1966): In: *Atlas of Electrochemical Equilibria in Aqueous Solutions*. M. Pourbaix (Ed.). Pergamon, Oxford, pp. 256ff.

Fontana, M. G. (1986): *Corrosion Engineering*, 3rd edition. McGraw-Hill, New York.

McFadden, J. T. (1969): Jour. Neurosurg. 31(4):373.

Rogers, G. T. (1984): Biomaterials 5:244.

Sutow, E. J., Pollack, S. R. and Korostoff, E. (1976): J. Biomed. Mater. Res. 10:671.

Svare, C. W., Belton, G. and Korostoff, E. (1970): J. Biomed. Mater. Res. 4:457.

BIBLIOGRAPHY

Fraker, A. C. and Griffin, C. D. (Eds.) (1985): *Corrosion and Degradation of Implant Materials: Second Symposium*. STP 859. American Society for Testing and Materials, Philadelphia.

Luckey, H. A. and Kubli, F., Jr. (Eds.) (1983): *Titanium Alloys in Surgical Implants*. STP 796. American Society for Testing and Materials, Philadelphia.

Pohler, O. E. M. (1983): In: *Biomaterials in Reconstructive Surgery*. L. R. Rubin (Ed.). C. V. Mosby, St. Louis, pp. 158ff.

Pourbaix, M. (1984): Biomaterials 5:122.

Scully, J. C. (1990): *The Fundamentals of Corrosion*, 3rd edition. Pergamon Press, Oxford.

Steinemann, S. G. (1980): In: *Evaluation of Biomaterials*. G. D. Winter, J. L. Leray, and K. deGroot (Eds.). John Wiley, Chichester, pp. 1ff.

Syrett, B. C. and Acharya, A. (Eds.) (1979): *Corrosion and Degradation of Implant Materials*. STP 684. American Society for Testing and Materials, Philadelphia.

Vermilyea, D. A. (1976): Physics Today 29:23.

Williams, D. F. (1976): Ann. Rev. Mater. Sci. 6:237.

5

Reactions of Biological Molecules with Biomaterial Surfaces

5.1   INTRODUCTION

Strictly speaking, in a biomaterials-tissue system there are no surfaces: as Andrade (1973) has pointed out, there are only interfaces. In this chapter the solid-liquid interface produced by the contact of the solid biomaterial with body fluids will be considered briefly.

The solid-liquid interface can affect dissolved species in the surrounding fluid at two levels of characteristic dimension:

1.   The molecular level (3–15 Å); these effects are essentially *chemical*.

2. The macromolecular level (15–500 Å); these effects are more of a *mechanical* nature.

Chemical effects depend upon the detailed chemistry and ionic charge distribution of the surface. We shall consider the effects that local chemistry can have on some of the events of coagulation (Section 9.3) and on carcinogenesis (Section 12.2). These effects can be either undesirable side aspects of the biomaterial selected for other properties (as in the general blood conduit problem), or deliberately induced effects required to mediate a cellular response. Examples of induced effects are common in the results of various surface treatments used to reduce or eliminate thrombus formation on the surface of blood contact materials. A less common induced effect is the production of surface activity (in the chemical sense) to directly stimulate *cellular* response.

An example of a putative cell-stimulating material is "Bioglass™." This is a system of glassy materials developed by Hench (1984). These glasses are soft and easily attacked by physiological fluids. Their virtue is that, when placed in a suitable tissue site a tenacious bond to hard or soft tissue is formed by cellular activity. This is a special case of the general tissue adhesion phenomenon reviewed by Manly (1970).

A typical composition (by weight %) in this system (type 45S5) is: SiO_2, 45%; P_2O_5, 6%; CaO, 24.5%; Na_2O, 24.5% (for a fuller discussion of such materials, see Section 10.3.4.3). The effect of the interaction of this glass with physiological solutions in vivo is to overwhelm the local buffering capacity and produce a local pH in the range of 9–10. Whatever direct effect this may have on cellular activity, this will produce a variety of physiochemical changes in proteins, including dissociation and denaturation. The observed cellular response may be secondary to these physiochemical changes in proteins.

Dissociation is understood in the general chemical sense as the separation of ions from molecular species. In addition, it refers to the disaggregation of multimolecular organic complexes, such as enzyme-cofactor complexes. These associations, like all chemical bonding processes, depend upon free energy considerations and may be affected by local pH.

5.2 DENATURATION

Denaturation is a problem peculiar to large organic molecules such as proteins. We recognize four levels of structure in these molecules:

1° The chemical composition as defined by atomic content and primary bonds between atoms.

2°   The spatial arrangement of portions of a molecule as defined by the requirements of bond angulation at each atomic center and by weak intramolecular bonds other than main chain bonds.

3°   The spatial arrangements determined by strong intramolecular bonds and secondary folding to produce stable domains.

4°   The aggregation of 3° structures by weak associative bonding (hydrogen or Van der Waals bonds) constitutes the final level of structure.

The 1° level and, to some degree, 2° level of structure are determined during synthesis. The 3° level may be produced by a self-assembly process or occur secondary to a, usually extracellular, one-time cleavage of a portion of the synthesized molecule. Thus, if these structures are disturbed by heat or local chemical activity, the molecule may not be able to revert to its original or native structure. Such a molecule is said to be *denatured* and may arouse a variety of biological responses despite a normal or near normal *chemical* composition [1° level (or order) structure].

Finally, the 4° level of structure, since it is dependent upon *weak* bonds, is quite sensitive to pH and concentration changes. The results of these changes may be as simple as slight alterations in configuration or as profound as the dissociation of multimolecular structures, such as those formed by enzymes and cofactors.

5.3 ORGANOMETALLIC COMPOUNDS

5.3.1 Definitions

Beyond the effects on structure, surfaces may obviously be reactive. As a class, metals are the most reactive implant materials. They may provoke a host response due to the formation of corrosion products (discussed in Section 8.2.5). However, many of these corrosion products are organometallic complexes or compounds and, as such, have a special behavior of biological importance.

As the name implies, organometallic compounds have two components: one is an organic moiety, while the other is the metallic moiety. The association between the two can be a weak interaction or, at the other end of the spectrum, a very strong interaction, as in the case of the Fe-hemoglobin complex. There also can be a degree of specificity of structure, again as seen in the heme molecule. If another metallic ion were to replace the iron, the function of oxygen transport would be impaired. This is the case when elevated chromium levels are present in heme synthesis sites (Smith, 1982).

Three terms are useful in discussions of organometallic compounds:

Chelation: A type of interaction between an organic compound (having two or more points at which it may coordinate with a metal) and the metal, to form a ring-type structure.

Coordination: The joining of an ion or molecule to a metal ion by a non-ionic valence bond to form a complex ion or molecule.

Ligand: Any ion or molecule that, by donating one or more pairs of electrons to a central metal ion, is coordinated with it to form a complex ion or molecule, as in the cobalt complex $[CoCl(NH_3)_5]Cl_2$, in which Cl and NH_3 in the bracketed portion are ligands coordinated with Co.

5.3.2 Stability

In a free metal cation all of the five d orbitals have the same energy level. However, in a chelate or complex, some of the filled orbitals are oriented toward the chelating atoms. Repulsion between non-bonding electrons in a d orbital and those of the chelating atom causes electrons in these orbitals to be less stable with respect to the other orbitals. In addition, bonding can preferentially stabilize one orbital with respect to the others. The theory dealing with repulsion from the field produced by the chelating atoms is called crystal field theory; the total effects are dealt with in ligand field theory.

By preferentially filling low-energy orbitals in organic ions, the metallic d orbitals can stabilize the molecular system. For example, if three orbitals have a low energy and two have a higher energy, as in an octahedral complex, the configuration would be much more stable with six electrons occupying the low-energy levels than with the electrons spread throughout all five orbitals. The gain in bonding energy achieved in this manner is referred to as the *crystal field stabilization energy* (CFSE).

Complexes with more of the electrons in the lower energy levels are more stable than those with all the d orbitals equally filled. Trivalent chromium and cobalt, for example, with 3 and 6 electrons, respectively, will form very stable complexes or ions. Consequently, the tendency of these ions to form complexes is very great. A complex of univalent copper, on the other hand, has zero CFSE because the five orbitals are completely filled. As a result, cuprous complexes will be less stable than those of trivalent cobalt or trivalent chromium.

Crystal field effects are important in predicting the rates and mechanisms of reactions of coordination compounds. The essential feature here is that ions which are strongly crystal-field stabilized will be slow to react, while nonstabilized ions will be more liable (reactive). This explains why Co^{3+}, which has considerable CFSE, is so nonreactive. In order for Co^{3+} to react, the octahedral configuration, which creates the large CSFE, must first be disrupted.

5.3.3 Production

The production of the organometallic compounds is controlled by a dynamic equilibrium which is established after implantation of a metallic specimen or device. This equilibrium is established between the alloy and the intermediate organometallic compound, as well as between the alloy and the more traditional inorganic ions. The rate of corrosion will then depend on the removal of the intermediate compound. If more is removed by deposition in tissues, then corrosion could proceed at an increased rate. Because the removal of the intermediate is the rate-limiting step, differences seen between biological response to powder and bulk implants probably reflect different surface (interface) reaction conditions.

The equilibrium will also depend upon two other factors:

1. The organometallic complex may be formed either on the surface or in solution. If it is formed on the surface, then the ratio of the implant surface area to the fluid volume available for equilibrium (SA/FLV) will govern both the formation rate and the equilibrium concentration. This is apparently the case for complexes formed between chromium or nickel and serum proteins (Woodman et al., 1984). If the complex forms preferentially in solution, then SA/FLV affects only the rate of formation, as is the case for cobalt.
2. The chemical composition of the surface may affect both the strength of the initial association and the rate of loss (desorption) of complexes which form at the interface. Table 5.1 presents data for absorption and desorption of albumin, the most common serum protein, by a variety of materials. A single monolayer corresponds to 0.2–0.7 $\mu g \cdot cm^{-2}$, depending upon packing of molecules. ''Desorption'' in this context is actually exchange, since release in the absence of proteins in solution (a highly unphysiological condition!) may be different. In particular, under these conditions, polyethylene releases no measurable albumin into an albumin-free solution (Brash et al., 1974). Fluid flow near the interface also has an effect, with release/exchange rates increasing with increasing flow rate. The situation at the implant-tissue interface is more complex than this, due to competition between proteins (see Section 5.5).

Table 5.1 Albumin Absorption and Desorption from Surfaces

Material	Absorption ($mg \cdot cm^{-2}/24$ hr) (Conc. 2 mg/ml)	Desorption (%/24 hr) (Conc. 1 mg/ml)
Metals:		
Silver	2.01 ± 0.22	23
Vanadium	0.13 ± 0.06	73
Titanium	0.05 ± 0.02	86
Oxides:		
TiO_2	0.15 ± 0.02	70
Al_2O_3	0.06 ± 0.01	83
Polymers:	(Conc. 3.7 mg/ml)	(Conc. 0.1 mg/ml)
Polyethylene	0.28	42
	(Conc. 1 mg/ml)	(Conc. 0.2 mg/ml)
Cuprophane	0.28 ± 0.05	90 +
Polyurethane	1.0–2.8	undetermined

Metals and oxides: Absorption/desorption in 0.01 M citrate/phosphate buffered saline (pH = 7.4) at 37°C using ^{125}I labeled human albumin.
Source: Williams et al. (1988).
Polymers: Absorption/desorption in Tyrode's solution (pH = 7.4) at 25°C using ^{125}I labeled human albumin.
Source: Brash et al. (1974).

5.4 MECHANICAL ASPECTS OF INTERFACES

Far more important than these chemical effects that take place at atomic dimensions are the mechanical effects that occur on a larger scale. These are primarily associated with the fact that the solid-liquid interface is a *phase* boundary. As such it has an interfacial energy associated with it. Let us consider the following situation (Figure 5.1). A spherical molecule P is approaching a solid surface S in liquid L. Suppose that it will stick or adhere to the surface. The work of adhesion is then (for adhesion of phase A to B)

$$W_{AB} = \gamma_A + \gamma_B - \gamma_{AB} \qquad (5.1)$$

where

γ_A, γ_B = "free" surface tensions

γ_{AB} = interfacial surface tension.

For this situation, we can write:

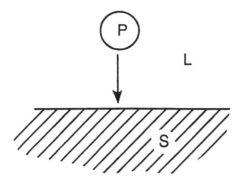

Figure 5.1   Molecule approaching an interface.

$$W_{SP} = \gamma_{SL} + \gamma_{PL} - \gamma_{PS} \tag{5.2}$$

Remember that the surface tensions are negative, and that W_{SP} must be negative for adhesion to take place. For example, suppose:

$$\gamma_{SL} = 70 \text{ dyne/cm}$$
$$\gamma_{PL} = 40 \text{ dyne/cm}$$
$$\gamma_{PS} = 50 \text{ dyne/cm}$$

then

$$W_{SP} = -70 - 40 + 50$$
$$= -60 \text{ (a change from } -110 \rightarrow -50) \tag{5.3}$$

Thus, adhesion would result.

However, let us now look more closely at the interface between the molecule and the solid (Figure 5.2). The normal interfacial equilibrium condition (the Young-Dupree equation) must be satisfied:

$$\gamma_{SL} = \gamma_{PS} + \gamma_{PL} \cos \theta \tag{5.4}$$

Note that for any degree of adhesion ($W_{SP} \leq 0$), θ will be less than 180°. If we initially consider the molecule to be a sphere, this condition can only be achieved by deformation of the natural (free) shape. A careful analysis, combining the Young-Dupree equation with the restoration forces resulting from this molecular deformation, would permit a more exact calculation of θ. However, the simple form can be taken as an estimator of the deformation. Thus, the larger the value of θ, the smaller the deforming force and the smaller the likelihood of mechanical damage to the molecule.

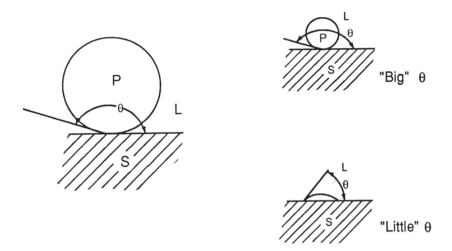

Figure 5.2 Conditions for molecular adhesion at an interface.

There is an interesting point concerning the opposite extreme, i.e., when θ goes to zero ($\cos \theta = 1$). This is called full wetting and requires that:

$$\gamma_{PS} = \gamma_{PL}; \; \gamma_{SL} = 2\gamma_{PL} \tag{5.5}$$

This value of surface tension is called the critical surface tension (γ_C). It can be determined by measuring θ for a variety of structurally related liquids and extrapolating the data to determine the limiting value of surface tension (γ_C) as θ approaches zero. A plot of $\cos \theta$ versus γ_{LV} is called a *Zisman* plot. The critical surface tension, γ_C, is determined by the intercept at $\cos \theta = 1$. A schematic result is shown in Figure 5.3 for a material surface with $\gamma_C = 28$ dyne/cm. [See Section 9.3.3 for a discussion of the supposed role of γ_C in cell-surface interactions; DePalma et al. (1972) presents interesting examples of Zisman plots obtained on metallic implants before and after blood contact.]

5.5 RESULTS OF INTERFACIAL ADHESION OF MOLECULES

A few of the effects that can result from molecular adhesion to biomaterials or tissues are:

1. Enzyme activity and rate constants are closely dependent upon details of the 3° and 4° molecular structure of enzymes. Accidental or delib-

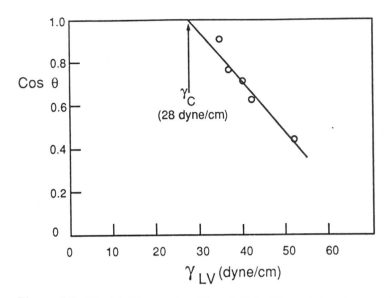

Figure 5.3 Model Zisman plot (five fluids). (Note: 1 dyne/cm = 1 erg/cm^2.)

erate enzyme adhesion at interfaces can be expected to modify synthetic behavior significantly.

2. Organic substrate response to the enzymatic action is also structure specific. Many substrate molecules have a degree of orientational freedom, possessing features such as saturated bonds, about which free rotation is possible. Association of this type of substrate with a surface could hinder or prevent such rotation. Depending upon the configuration that the molecule is "frozen" in, enzymatic attack might be either accelerated or inhibited.

3. Many molecules, as synthesized, have a "tail" portion that serves to inactivate them. In the normal course of events, enzymatic processes act to strip this small segment and release the molecule into an active substrate pool. Contact with surfaces, and the forces resulting from adhesion, may cause premature activation. Some molecules are, apparently, designed specifically to become activated in this manner by contact with foreign surfaces. An example is fibrinogen which is reduced slightly in molecular weight and converted to the active protein, fibrin, by surface contact.

4. Immunological response to proteins is also strongly dependent upon 2°, 3°, and 4° order structure. Contact with surface by native proteins produces unnatural configurations of the following types:
    a. Conversion or activation of molecules as mentioned in 3 (above).

b. Transient deformations during surface contact that are restored upon subsequent desorption.

c. Partial or total denaturation due to surface adhesion forces.

There is considerable evidence that molecular deformations of each of these three types can excite antibody production and trigger a variety of immune responses, either directly or on a subsequent challenge.

In an effort to study the possible immunological results of the surface denaturation of proteins, Stern et al. (1972) exposed a series of polymers, including epoxies, silicones, and polacrylamide, to fresh rabbit serum. The serum was then injected into the host animals, and the production of antibodies was investigated. Unless the in vitro exposure included exposure to macrophages as well as serum, no antibody titers were developed. However, in the presence of peritoneal macrophages, a number of these materials produced positive titers. This is evidence of a cell-mediated response to denatured serum proteins, recognized as foreign bodies (see Section 11.4.1). The absence of effect (antibody production) when the serum was directly injected suggests that the denaturation was reversible and present only when the serum proteins were adsorbed to the test surfaces.

This experiment reminds us of a further complication. The data in Table 5.1 were obtained from pure albumin (one protein) solutions. The actual exposure in the biological environment involves many proteins, as in Stern's use of serum in vitro. In such a situation, proteins encounter the surface depending upon the product of their concentration and their self-diffusion velocity, which is approximately inversely related to the square root of their molecular weight. (Additional factors, such as molecular shape, also affect self-diffusion rates.)

Thus we should think of protein-surface interactions in vivo (or in vitro from mixed solutions) as a succession of events, with early arrivers (low molecular weight/high concentration) being potentially displaced by late arrivers (high molecular weight/low concentration). This process, first recognized by Leo Vroman in the blood coagulation process (see Section 9.2) and termed, by others, the *Vroman effect*, is shown schematically in Figure 5.4.

Here we see that even after the total surface concentration of protein (solid line) reaches a steady state value, the composition of the film continues to change, as molecules of B displace those of A and, in turn, are displaced by molecules of C. Remember also that these are equilibrium surface concentrations; the data of Table 5.1 suggest that continuing exchange of each species may take place.

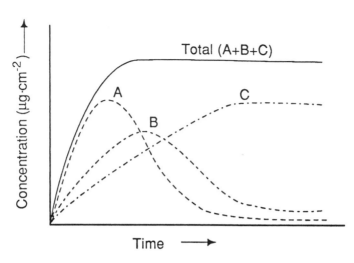

Figure 5.4 Vroman effect (schematic).

5.6 EFFECTS OF CHARGED INTERFACES AND IONS

The discussion so far has focused upon interfaces that are considered to be electrically neutral. That is, the deforming force on molecules results simply from interfacial free energy and the equilibrium requirements for adhesion. If there is a net surface charge, then a potential gradient will exist in the vicinity of the surface. This has three major effects:

1. Uncharged molecules will suffer deformation in structure due to the interaction of their internal dipoles (primarily associated with covalent bonds) with the electrical field.
2. Charged molecules and zwitterions (molecules with no net charge but with equal amounts of negative and positive charge) will undergo an additional set of constraints due to attraction or repulsion of their charge centers. These additional forces will produce additional structural deformation.
3. Charged molecules and ions will also move along the electrical field gradient, attracted to surfaces of opposite charge. This motion, called electrophoresis, is utilized in analytical processes to separate ions with different ratios of charge to ionic mobility. Electrophoresis can act in vivo to change the concentration of ions as well as pH, thus potentially altering 3° and 4° structure.
4. Finally, charged molecules and ions may interact with magnetic fields. Moving charges experience a transverse force from constant

magnetic fields, while time varying magnetic fields can produce oscillatory rotation of zwitterions and local current flow through motion of ions and molecules with net charge.

One should also remember that net (nonzero) surface potentials may arise (1) by electrochemical equilibria; (2) from differences in dielectric constant across the interface; (3) from external sources such as direct potential imposition [as suggested by the experiments of Sawyer et al. (1965) in the reduction of thrombogenic behavior by changing surface potential (see Section 9.3.2)]; or (4) from the other member of a galvanic (corrosion) couple (see Section 4.7.2).

REFERENCES

Andrade, J. D. (1973): Medical Instrumen. 7:110.

Andrade, J. D., Hlady, V., Herron, J. and Lin, J.-N. (1988): *Proc. 2nd Intern. Symp. Bioelectronic and Molecular Electronic Devices*, Dec. 12–14, 1988, Fujiyoshida, Japan (unpaginated).

Bernabeu, P. and Caprani, A. (1990): Biomaterials 11:258.

Brash, J. L., Uniyal, S. and Samak, Q. (1974): Trans. Amer. Soc. Artif. Int. Organs 20:69.

DePalma, V. A., Baier, R. E., Ford, J. W., Gott, V. L. and Furuse, A. (1972): J. Biomed. Mater. Symp. 3:37.

Hench, L. L. and Wilson, J. (1984): Science 226:630.

Manly, R. S. (Ed.) (1970): *Adhesion in Biological Systems*. Academic Press, New York.

Sawyer, P. N., Wu, K. T., Wesolowski, S. A., Brattain, W. H. and Boddy, P. J. (1965): Proc. Natl. Acad. Sci. 53:294.

Smith, G. K. (1982): *Systemic Transport and Distribution of Iron and Chromium from 316L Stainless Steel Implants*. Ph.D. Thesis, University of Pennsylvania, Philadelphia.

Stern, I. J., Kapsalis, A. A., DeLuca, B. L., and Pieczynski, W. (1972): Nature 238:151.

Williams, R. L. and Williams, D. F. (1988): Biomaterials 9:206.

Woodman, J. L., Black, J. and Jiminez, S. A. (1984): J. Biomed. Mater. Res. 18:99.

BIBLIOGRAPHY

Adamson, A. W. (1990): *Physical Chemistry of Surfaces*, 5th edition. John Wiley, New York.

Friedberg, F. (1974): Quarterly Rev. Biophys. 7(1):1.

Gabler, R. (1978): *Electrical Interactions in Molecular Biophysics.* Academic Press, New York.

Ivarsson, B. and Lundström, I. (1986): Crit. Rev. Biocompat. 2(1):1.

Tanford, C. (1961): *Physical Chemistry of Macromolecules.* John Wiley, New York.

Zangwill, A. (1988): *Physics at Surfaces.* Cambridge University Press, Cambridge.

6

Mechanics of Materials: Deformation and Failure

6.1 INTRODUCTION

Mechanical integrity is a nearly universal requirement for implant materials. All materials must cohere or ''hold together'' if they are to be expected to stay in one shape, in one location, and to perform their designed function. The requirement may be only that they withstand the various stresses that exist in the implant site. A more rigorous requirement exists if part of the intended function for the implant is a mechanical one, such

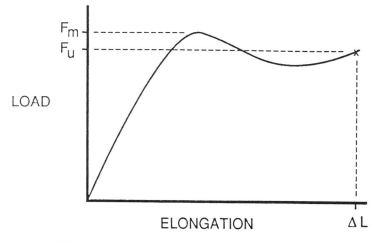

Figure 6.1   Load-elongation curve.

as a heart valve replacement or a fracture fixation device. Then the application may require the preservation of a minimum value of a property, such as withstanding permanent deformation, or of a design (mean) value of another one, such as possessing a particular spring constant. These are extrinsic behaviors but they depend in part on intrinsic properties: in this case, on yield strength and elastic modulus, respectively.

Unfortunately, the environmental exposure of materials alters their mechanical properties in a variety of ways. As discussed in Section 2.2, the chemical and physical environment of the human body is different from external engineering environments and is, by comparison, quite severe.

In this chapter, we will briefly consider the origin of intrinsic mechanical properties of materials. We will also consider how materials fail in mechanical applications, and how these properties and types of failure are affected by the biological environment.

6.2 MECHANICS OF MATERIALS

The simplest experiment that can be performed to characterize the mechanical properties of a solid material is to machine a specimen with well-defined dimensions (a "standard" specimen) and load it to failure in tension. Figure 6.1 shows the result of such a model experiment, obtained by plotting the applied load directly against the resulting elongation. This curve is characterized by the following points:

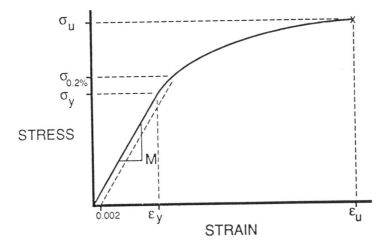

Figure 6.2 Stress-strain curve.

F_m: Maximum load that can be sustained

ΔL: Elongation to failure

F_u: Load at failure

Unfortunately, these numbers are *extrinsic* (dependent upon the specific dimensions of the specimen). It is common practice to transform the load-elongation curve into a stress-strain curve.

Stress, σ, is given by the force divided by the cross-sectional area perpendicular to the line of force application:

$$\sigma = \frac{F}{A} \tag{6.1}$$

Strain, ε, is given by the ratio of the change in length to the original length, L_o:

$$\varepsilon = \frac{\Delta L}{L_o} \tag{6.2}$$

These conversions produce *intrinsic* values that are independent of specimen dimensions, given that certain basic rules are obeyed in the design of the specimen. In particular, it is necessary to assure that the specimen be uniform in dimensions in the region in which L_o, ΔL, and A are measured. Figure 6.2 shows a stress-strain curve that might be obtained by conversion of the model load-elongation curve of Figure 6.2.

This curve is characterized by several intrinsic parameters (Table 6.1).

There are, of course, other intrinsic parameters that characterize materials, and these may be determined in forms of load application other than pure tension. However, in this chapter we will concentrate on tensile behavior and, in particular, the elastic modulus E, the yield stress σ_y, and the ultimate stress σ_u. These three parameters tend to dominate mechanical design.

The relationship of these parameters to mechanical design may be summarized as follows:

Elastic modulus: This is the intrinsic "spring constant" of the material; thus it specifies the proportional deformation as a result of stress within the limit of *recoverable* deformations.

Yield stress: This sets the upper stress limit for the design of a body fabricated from a plastically deformable material that must not undergo permanent deformation from its original shape.

Ultimate stress: This defines the stress that produces fracture and thus sets the maximum strength of the material.

These three parameters taken together provide a measure of stiffness, deformability, and strength of a material.

6.3 ELASTIC MODULUS

6.3.1 Fundamental Aspects

The elastic behavior of materials has its origin in the basic bond structure at the atomic level. That is, for deformations below the yield point, the major effect is the elastic (recoverable) deformation of interatomic bonds. Each of the four major materials classes, metals, ceramics, polymers, and composites, possesses its own characteristic bond structures.

The metals are characterized by the looseness of binding of their valence electrons. Thus, we think of solid metals as aggregates of positively charged ions with a neutralizing negative electron cloud. The resulting cohesion is high but the individual bonds lack strong directionality. In real metals there are regions of high order with almost exclusively metallic bonding (grains) separated by zones of disorder (grain boundaries) that contain impurities and other forms of bonding.

Ceramic materials, on the other hand, are primarily ionically bonded. They are made up of geometric arrays of cations and anions with strong ordering and a resulting high directionality of bonds. Again, real (nonideal) ceramics contain grains (ordered) and grain boundaries (disordered) in combination as do metals. In this case, ionic bonding is possible

Table 6.1 Intrinsic Parameters from a Stress-Strain Curve

Symbol	Name	Units	Definition
σ_y	Yield stress	MPa	Stress to start permanent (plastic) deformation
ε_y	Yield strain	MPa	Strain at the moment of yielding
$\sigma_{0.2\%}$	0.2% offset stress	MPa	Stress to produce 0.002 strain
σ_u	Ultimate stress	MPa	Stress to produce fracture
ε_u	Ultimate strain	none	Total strain to fracture
M	Modulus	GPa	Ratio of stress divided by strain (slope of line in elastic (proportional) region ($=\sigma_y/\varepsilon_y$))
E	Young's modulus	GPa	Modulus determined in tension
—	Work of fracture	J/m^3	Area under the stress-strain curve

in the boundary regions but is usually weaker than within the grains because of inclusions, mismatches between adjacent grains, and disorder effects.

Polymers exhibit the third type of bond, the covalent bond. This bond, formed by orbital sharing of electrons between atoms, while not particularly strong, is highly directional. Engineering polymers consist of long-chain molecules with covalently bonded "backbones." The chains may be ordered in a regular array in regions, forming crystals, or may be uniformly amorphous; the more usual structure is again a combination of order and disorder. The overall structure is stabilized by occasional inter-chain covalent bonds or ionic bonds between charged side groups (crosslinks) and by diffuse attraction of hydrogen, oxygen, and nitrogen atoms to —OH groups (Van der Waals bonds).

In order of strength, these bonds may be classified as follows:

Ionic > metallic > covalent > Van der Waals

Thus, it should come as no surprise that we can rank elastic moduli as:

Ceramic > metallic > polymeric

Composite materials do not appear in this range since they can be particle- or fiber-reinforced metals or polymers and thus display a wide range of elastic moduli, depending upon the contributions of their components. A further complication is introduced by the nature of the bonding between the matrix and the reinforcing material. This bonding depends on the chemistry and microstructure of the interface and affects yield and failure but has little effect upon elastic behavior. The ability to adjust the modulus (and other properties) to meet the requirements of a specific application is, of course, one of the great attractions of composites.

6.3.2 Environmental Effects

What effects does environment have upon the elastic moduli of these material types? Practically speaking, the elastic moduli of metals and ceramics are unaffected by exposure to biological environments. This is due both to the great strength of internal bonding in these materials and to the relative simplicity of their structure when compared with polymers and composites.

Polymers may experience profound changes in elastic moduli in response to the internal environment. Table 6.2 summarizes the principal mechanisms and their effects.

We have already discussed absorption and leaching in Sections 3.3 and 3.4. The principal effect of absorption of low molecular weight species is to swell the matrix, moving the crystalline "islands" further apart and thus weakening the already weak bonds between them. This permits easier deformation, in the same way that lubrication makes it easier for surfaces to move over each other. However, many polymers already contain plasticizers in the form of low-molecular-weight fragments of the basic polymer, deliberately added low-molecular-weight agents, and water. Thus, the *loss* of these by leaching would be expected to reverse the effect of absorption and increase the elastic modulus. In real applications there is competition. However, since biomedical polymers tend to be simple (low additive) materials due to host response considerations, and tend to have high molecular weight due to strength considerations, the usual effect of exposure to physiological fluids is to *lower* the effective elastic modulus. Highly crystalline or highly crosslinked polymers should be less sensitive than amorphous, low-molecular-weight ones.

An illustration of both plasticizing and antiplasticizing effects can be seen in data on the elastic moduli of polymers in compression in Table 6.3. In this study (Jacobs, 1974), standard compression test cylinders were made of a commercial polymethyl methacrylate (PMMA) surgical cement, a duplicate formulation compounded in the laboratory, and a commercial medical grade of ultrahigh-molecular-weight polyethylene (UHMWPE). These were tested as fabricated (except for UHMWPE, for which the fabrication date was unknown), and after 120 days exposure to a variety of environments, including subcutaneous implantation in the rabbit. The slope of the stress-strain curve at 1% strain, $E_{1\%}$, was used for comparison since, in common with most other polymers, these polymers do not possess a single well-defined elastic modulus in the elastic region

Table 6.2 Environmental Effects on Mechanical Properties of Polymers

	Effects on	
Phenomenon	Modulus (E)	Yield stress (σ_y)
Absorption	Decrease ("plasticizing")	Increase
Leaching	Increase ("antiplasticizing")	Decrease
Chain scission	Decrease	Decrease
Cross linking	Increase	Increase

of the stress-strain curve. Exposure of both PMMA formulations to high humidity or saline solutions that duplicated the ionic concentration of serum at 37°C produced a reduction of 30% in $E_{1\%}$ when compared with dry, room-temperature storage. This illustrates the plasticizing effect of absorbed water. Implantation, while producing a similar reduction, was much less damaging. This could be interpreted in one of two ways:

1. Implantation might have prevented the loss of residual monomer by leaching. An effect similar to leaching, loss of monomer by evapora-

Table 6.3 Variation of Compressive Moduli of Polymers with Environmental Exposure

Test Condition	PMMA (raw materials) $E_{1\%}(\times 10^5$ psi)	PMMA (commercial)[a] $E_{1\%}(\times 10^5$ psi)	UHMWPE (commercial)[b] $E_{1\%}(\times 10^5$ psi)
As fabricated	3.0 ± 0.2[c]	3.4 ± 0.2	na
Post laboratory storage (24°C/120 days)	3.9 ± 0.2	3.8 ± 0.5	0.83 ± 0.06
Post humid storage (97%RH/37°C/120 days)	2.7 ± 0.3[d]	2.6 ± 0.2[d]	0.81 ± 0.05
Post saline storage (0.9% NaCl/37°C/120 days)	2.7 ± 0.4[d]	2.8 ± 0.2[d]	0.80 ± 0.06
Post implantation (rabbit, subcutaneous/120 days)	3.2 ± 0.2[d]	3.1 ± 0.4	0.80 ± 0.05

[a]Simplex-P™ (North Hills Plastics, Ltd.)
[b]Zimmer-USA
[c]±95% confidence interval
[d]Different from "post laboratory storage" ($p < 0.05$)
na = not available
Note: $E_{1\%}$ = tangent modulus at 1% strain.
Source: Adapted from Jacobs (1974).

tion, is probably responsible for the increase of $E_{1\%}$ due to dry storage when compared with the as-fabricated value. Residual monomer would serve as a plasticizer but, since it is hydrophobic, might exclude the more efficient plasticizer, water.

2. A crosslinking agent or an antiplasticizer might be absorbed from serum in the animal, counteracting the plasticizing effects of water absorption.

On the other hand, UHMWPE, with its more crystalline nature and far higher average molecular weight ($\approx 2 \times 10^6$ vs. 2×10^4), is unaffected by the environmental exposures used in this experiment.

Chain scission is the polymeric equivalent of the processes of corrosion and dissolution of metals discussed in Sections 4.1 and 4.2. The principal mechanisms are intrinsic scission (no external chemical species involved), oxidation, hydrolysis, or chemical attack. Figure 6.3 summarizes these mechanisms and provides some generic examples.

Chain scission reduces the elastic moduli of polymers through three routes:

1. The scission reaction may release a very small molecular fragment, such as a water molecule, that can act as a plasticizer.
2. Although the principal resistance to small deformations in polymers is due to stretching and/or disruption of weak bonds, there is some contribution due to "tangling" of long molecules. In much the same way that long strands of spaghetti tend to trap each other, this tangling forces an elongation of a portion of the strong, covalently bonded molecules, even at modest deformations. Thus, scission of molecules, by reducing average molecular weight, releases these trapped molecules and permits greater strain before covalent bond stretching can contribute significantly to the elastic modulus.
3. The disorder associated with shorter chain length may reduce crystallinity and thus reduce the average strength of bonding, leading to lower moduli.

Crosslinking is the reverse of scission. The formation of new bonds between chains increases the effective molecular weight, further tangles and traps molecules, and may reduce the effective concentration of plasticizers by chemical combination. A common mechanism for crosslinking polymers in the laboratory is exposure to ionizing radiation. This produces active free radicals, as in chain scission, that link with free radicals in neighboring chains, forming covalent crosslinks. While clinical doses of x-radiation do not produce measurable changes in the properties of polymeric implants in patients (Eftekhar and Thurston, 1975), high dose

INTERNAL MECHANISMS

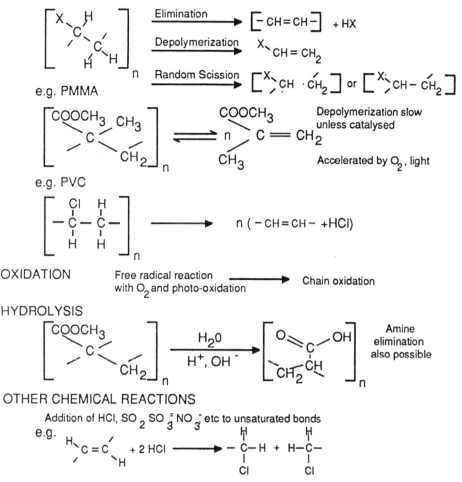

Figure 6.3 Mechanisms of chain scission in polymers. Source: Adapted from Allara (1975).

radiation in the laboratory is a convenient device for studying the mechanical consequences of crosslinking. Irradiation of simple pure polymers such as polyethylene suggests a linear increase of modulus with the number of crosslinks (Grobbelaar et al., 1978), and a more pronounced effect may occur if a number of low-molecular-weight agents that can be incorporated are present during irradiation.

The environmental effects on the elasticity of composites is more difficult to generalize. In an ideal model, the elastic modulus of a randomly oriented composite, E_C, made of materials A(matrix) and B(reinforcing or filler phase), can be calculated from

$$E_C = E_A V_{fA} + E_B V_{fB} \qquad (6.3)$$

where V_{fi} = volume fraction of material (phase) i. Then, any effect on the modulus of either material is seen as a proportional effect of the modulus of the composite. A special case would be the formation of voids in a material, either by leaching of a second phase or by aggregation of internal defects. Since the modulus of a void is zero, the modulus of a porous material, for small pore volumes, would be given by

$$E = E_0(1 - V_{fP}) \qquad (6.4)$$

where E_0 = elastic modulus of fully dense material. Thus, we would expect the modulus to decrease linearly with increasing volume fraction of pores.

In real materials the effect is somewhat greater at small void volume fractions but becomes less pronounced for more porous materials. Equation (6.5) was derived for rigid ceramics (MacKenzie, 1950) and has been shown experimentally to describe effects in materials with Poisson ratios near 0.3:

$$E = E_0(1 - 1.9V_{fP} + 0.9V_{fP}^2) \qquad (6.5)$$

There is a further problem in describing the effects of environment on the elastic moduli of composites. Equation (6.3) is based upon an assumption that there is a perfect bond between the phases so that each phase experiences an equal internal strain for a given macrostrain of the composite material. Real composites rarely display such perfect bonding, and the bond itself is often the weak point for environmental attack. The consequences of this are unpredictable but the usual effect is a reduction in modulus.

6.4 YIELD STRENGTH

6.4.1 Fundamental Aspects

The yield strength is defined by the stress that is necessary to produce unrecoverable deformation in a material. Deformation at lower stresses may be linear in the case of a simple solid, or increasingly nonlinear as strain increases, as in the case of many polymers. Recovery may be rapid at lower strains and become slower as peak strain increases. Finally, at the yield stress, conditions of deformation are such that a residual unrecoverable strain remains, even after long times at zero stress.

Within crystals, unrecoverable strain is produced by migration and aggregation of defects and by the slippage of material along defect planes. However, in complex materials and composites, slip, leading to unrecoverable deformation, may occur preferentially along grain and phase boundaries.

6.4.2 Environmental Effects

At room and body temperature, the processes leading to either crystalline deformation or grain boundary slip in both metals and ceramics are little affected by environmental exposure because of the relatively high bonding energies. However, the situation for polymers and polymer-based composites is different, as noted in the earlier discussion of elastic modulus. The effects are summarized in Table 6.1.

Absorption and leaching produce what appear to be paradoxical effects on yield strength. That is, we might expect a lower modulus, as results from absorption of a plasticizer, to accompany a lower yield strength. In general, however, the yield strength is raised. While motion along a particular grain boundary may become easier, this may lead to increased load sharing with adjacent material and, in fact, may produce modest elevations of yield stress in inhomogeneous materials. A similar but inverse effect is seen when plasticizers are leached from the material.

On the other hand, chain scission produces an overall reduction in molecular weight, making plastic deformation more dependent upon the interruption of weak bonds and, thus, reducing the yield stress. Crosslinking increases the tangling effect of long molecules and thus substitutes strong covalent bonds for weaker bonds, raising the yield stress.

The situation in composites is more complex, and no generalizations can be made. This is the case since the environment may effect not only the matrix but the matrix-filler bond as well. The results depend upon the details of the composite material in question and its exposure.

It is possible for materials to undergo unrecoverable deformation under constant load at stress below the yield stress. This is the familiar creep process. Creep rates are generally very slow for temperatures below one-half the melting temperature of the material. However, for temperatures above one-half the melting temperature, or in the presence of plasticizers, creep can be significant. Creep is possible in many biomedical polymers.

Creep is characterized by an initial or primary creep phase in which the creep rate diminishes rapidly. This is followed by a long secondary creep phase with a strain rate that is essentially constant in logarithmic time. In this secondary creep phase, the Dorn-Weertman equation can be used to describe the creep rate:

$$\dot{\varepsilon} = A\sigma^{n}e^{-Q/RT} \qquad\qquad (6.6)$$

where

$\dot{\varepsilon}$ = creep rate

σ = stress

n = experimentally fitted parameter ($\cong 5$)

Q = activation energy

The activation energy (Q) is usually taken to be the activation energy for self-diffusion but may be considered more generally as an intrinsic activation energy for creep (Parsons and Black, 1977). Thus, environmental effects on the creep rate can be interpreted in terms of changes in the activation requirements of the creep process.

There is a final or tertiary process of creep, characterized by a rapidly increasing strain rate leading to fracture. Little is known about the mechanism of this process or about environmental effects on it.

It should also be clear from this discussion that creep in biomedical applications is primarily observed in polymers and polymer-based composites. In general, secondary creep rates decrease with increasing yield stress at a given temperature, but the relationship is weak. However, they increase with increasing temperature and with the presence of plasticizers. This latter effect may dominate in polymer matrix composites, producing significant increases in creep rate (Soltész, 1986).

6.5 FRACTURE STRENGTH

6.5.1 Fundamental Aspects

Fracture occurs when the cohesive strength of a material is exceeded. It represents an accentuation and final stage of the processes that earlier led to yielding, if that is possible in a particular material. However, it is generally observed that ultimate strengths, such as the ultimate tensile stress, are small compared with those expected, based upon cohesive energy calculations.

Typical calculations of cohesive energy or theoretical maximum strength lead to values of σ_{u} equal to $E/10$. This would predict a tensile strength of 12.7 GPa for Ti6Al4V, a common alloy useful in implant applications. The actual value of σ_{u} is typically 0.9 GPa; that is, $\approx E/140$.

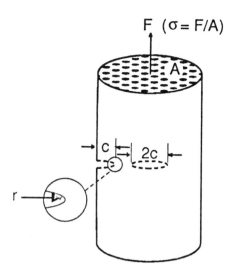

Figure 6.4 The ideal Griffith crack.

This is a relatively strong material; weaker materials such as stainless steel have σ_u in the range of $E/250$ to $E/350$. This observation was first explained for brittle materials (those that fail without significant unrecoverable strain) by Griffith (Guy, 1971). He suggested that defects exist in the surface and the body of real materials as seen in Figure 6.4. He calculated that the presence of an elliptical crack in brittle materials produces a stress concentration at the "point" of crack, as given by Eq. (6.7):

$$\sigma_m \cong 2\sigma \left(\frac{c}{r}\right)^{1/2} \tag{6.7}$$

where

σ = apparent or "macro" stress

σ_m = elevated stress at "point" of crack

s = overall (macro) stress

c = 1/2 major diameter of internal elliptical crack

= major width of surface crack

r = radius of curvature at "point" of crack ($r \ll c$)

Thus, Griffith suggested that, while the macrostress might be well below the true ultimate stress, the elevated local stress, σ_m, near a defect might exceed the ultimate strength, and a crack would propagate. By considering the energy required to form the crack (the difference between elastic strain energy released in the material near the newly formed crack and the increase of interfacial surface energy due to formation of the new material-environment interfacial area along the crack), he also calculated the minimum stress required to propagate the crack:

$$\sigma = \left(\frac{2\gamma E}{\pi c}\right)^{1/2} \tag{6.8}$$

where

γ = surface tension (material-environment)

E = elastic modulus

Fortunately, most materials undergo plastic deformation before fracture. Thus, as stresses about a defect are increased, as predicted by Eq. (6.7), plastic deformation will take place before fracture, even if the macrostress is below the yield stress. Orowan (Guy, 1971) dealt with this problem by replacing the term γ in Eq. (6.8) with the quantity $(\gamma + p)$, where p = the work of plastic deformation at the "point" of the propagating fracture. Since p is typically 1000 times γ in magnitude, Eq. (6.6) then becomes approximately,

$$\sigma \cong \left(\frac{Ep}{c}\right)^{1/2} \tag{6.9}$$

and we can recognize that such materials will be proportionally stronger since they will require far higher stresses to propagate existing defects into fracture surfaces.

A special case of Eq. (6.7) occurs for spherical pores, the situation discussed previously with respect to the reduction of elastic modulus by pores. This has been studied empirically, and the usual relationship [parallel to Eq. (6.5)], derived by Ryskewitsch (Kingery, 1960), is

$$\sigma_u' = \sigma_u e^{(-nV_{fP})} \tag{6.10}$$

where σ_u' is the actual fracture strength for a material with pore volume fraction V_{fP} and n is an empirically fitted constant between 4 and 7.

The difference between the stresses predicted by Eqs. (6.8) and (6.9) results in the classification of materials as those that fail in a brittle mode and those that fail in a ductile mode. Brittle failure, characteristic of ceramics and of polymers at low temperatures, occurs without significant residual deformation and is governed by relations of the form of Eq. (6.8). Materials with yield stresses well below ultimate stresses tend to be

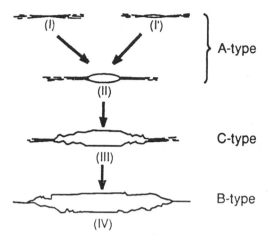

Figure 6.5 Plausible crack discs in section (vertical dimension exaggerated). Source: Adapted from Takahashi (1973).

ductile and fail in a manner governed by Eq. (6.9). Because yielding occurs during failure, they also exhibit significant unrecoverable strain.

The existence of Griffith defects has been shown repeatedly. An elegant example is the work of Takahashi (1973) in polymethyl methacrylate, as shown in Figure 6.5. The A-type cracks are those seen after modest stresses, and B- and C-types after higher stresses, presumably exceeding the limit imposed by Eq. (6.9).

Given that Griffith defects exist, there are two conclusions that can be drawn from this analysis:

1. Equation (6.7) predicts that the magnitude of the stress concentration near a defect will vary *inversely* as the minimum radius of curvature of the defect. Thus, a sharp crack is a more extreme stress riser than a semicircular notch. This effect is the basis of the commercial practice of drilling a hole at the advancing tip of a slowly propagating crack, as in a bridge strut, to prevent further propagation by reducing σ_m.
2. As the defect major diameter ($=2c$) increases, stress concentration increases [Eq. (6.7)], and the stress necessary to propagate brittle fracture [Eq. (6.8)] or ductile fracture [Eq. (6.9)] decreases. Thus, there is a critical minimum size for a Griffith defect to contribute to fracture, given that r remains constant. In practice, larger defects decrease strength to a limit, beyond which the process is reversed since r begins to increase.

Calculations of minimum crack lengths in composites are much more complex. Marom (1975) has shown that the minimum or critical defect size in polymer-based composites is much greater than that in the unfilled resin and is dependent upon orientation of stresses with respect to the reinforcing fiber.

6.5.2 Environmental Effects

The biological environment can have significant effects upon the details of crack propagation and, thus, upon the strength of materials. These effects will now be briefly discussed.

Any form of chemical attack, whether corrosion, dissolution, or leaching, that can increase the size of pre-existing defects or produce new defects by preferential attack clearly weakens a material. Such an attack may take place preferentially near defects in stressed materials and is termed "stress-enhanced attack" or "stress corrosion" in the case of metals. Such effects are well recognized in metals and have been demonstrated in silicone rubber (Rose et al., 1973).

Crazing due to swelling can also produce Griffith defects where none previously existed or can expand preexisting ones.

In brittle materials, the fact that $\gamma_{SL} < \gamma_{SA}$ for all but hydrophobic materials reduces the required propagation stress predicted by Eq. (6.8).

Thus, any of the degradative phenomena discussed in Chapters 3 and 4 can be expected to reduce the ultimate strength of biomaterials, and all classes of materials are susceptible.

Table 6.4 summarizes behavior for a range of typical implant materials, non-absorbable sutures, during a 24-month experiment in rabbits (Postlethwait, 1970). The data given in the table are the ultimate tensile loads normalized by dividing by the strength of materials retrieved after 1 week of implantation (in the abdominal wall) to remove the effect of differing diameters of specimens.

Absorbable materials, such as gut (natural) or polygalactic acid (synthetic) sutures, will show more pronounced and rapid loss of strength. However, the rate for an individual material and surgical situation is hard to predict. The loss of strength is affected by the material composition, fabrication, post-fabrication handling (production of surface defects) and, because of possible pH dependence of degradation by hydrolysis (Chu, 1982) and/or enzymatic attack (Salthouse et al., 1969).

In addition to failure by fracture at stresses that exceed ultimate strength, materials may fail by fatigue. Fatigue fracture, fracture at stresses below ultimate after a number of cyclic deformations, is recognized as one of the major sources of mechanical failure of implants.

Table 6.4 Degradation of Tensile Strength of Sutures In Vivo

	Suture type				
	Multifilament		Monofilament		
Period of implantation	Silk	Cotton	Polyamide (Nylon)	Polypropylene	Polyester (Dacron)
2 weeks	0.87	0.98	0.98	1.05	0.92
4 weeks	0.58	1.07	0.97	1.14	1.03
3 months	0.20	0.67	0.88	1.03	0.91
6 months	0.36	0.52	0.79	0.93	0.70
12 months	0.58	0.50	0.89	0.97	0.90
24 months	dis.	0.58	0.72	0.99	0.96
Comment	Slow dissolution	Separated	Swollen	No visible change	No visible change

Note: Data normalized by 1 week value of ultimate tensile stress.
Source: Adapted from Postlethwait (1970).

Fatigue failure is characterized by the construction of an S-N (stress vs. number of cycles) curve as shown in Figure 6.6. The ultimate (fracture stress) decreases with cyclic loading in air until an apparent limit, termed the "endurance limit," is reached. In this example, the endurance limit is reached between 10^6 and 10^7 cycles. However, in an aqueous or corrosive environment where stress corrosion is possible during the high stress part of each cycle, this endurance limit is apparently abolished, and the fracture strength continues to decrease with cyclic loading.

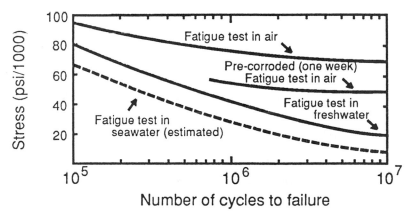

Figure 6.6 S-N curves for a Cr-V steel. Source: Dumbleton and Black (1975).

In polymers and polymer-based composites, this effect is also seen, probably secondary to plasticization and/or "decoupling" (secondary to bond failure) of the reinforcing phase from the matrix. In composites, a reduction in modulus is also observed as internal damage accumulates with an increasing number of load cycles; the decrease is pronounced if the aqueous environment is such that the fiber-matrix interface is wetted preferentially (is hydrophilic).

Ceramics display an additional fatigue problem called *static* fatigue, which is a sudden failure by brittle fracture at stresses well below ultimate when subjected to steady noncyclic loads. In Al_2O_3 (Krainess and Knapp, 1978) and in glasses (Adams and McMillan, 1977), this effect is believed to be due to the formation of weak bonds by absorbed water displacing stronger ionic bonds.

Thus, it should be clearly understood that biological environments, as encountered by implants, can be expected to produce a wide range of significant changes in the mechanical properties of materials.

REFERENCES

Adams, R. and McMillan, P. W. (1977): J. Mat. Sci. 12:643.

Allara, D. L. (1975): Environ. Health Perspec. 11:29.

Chu, C. C. (1982): J. Biomed. Mater. Res. 16:117.

Dumbleton, J. H. and Black, J. (1975): *An Introduction to Orthopaedic Materials.* C. C. Thomas, Springfield, Illinois.

Eftekhar, N. S. and Thurston, C. W. (1975): J. Biomech. 8:53.

Grobbelaar, C. J., duPlessis, T. A. and Marais, F. (1978): J. Bone Joint Surg. 60B:370.

Guy, A. G. (1971): *Introduction to Materials Science.* McGraw-Hill, New York.

Jacobs, M. L. (1974): *Evaluation of Three Polymer Resins for Use in Polymer Based Composites for Hard Tissue Prostheses.* M.S. Thesis, University of Pennsylvania, Philadelphia.

Kingery, W. D. (1960): *Introduction to Ceramics.* John Wiley, New York.

Krainess, F. E. and Knapp, W. J. (1978): J. Biomed. Mater. Res. 12:241.

MacKenzie, J. K. (1950): Proc. Phys. Soc. (London) B63:2.

Marom, G. (1975): Int. J. Frac. 11:534.

Parsons, J. R and Black, J. (1977): Trans SFB 1:78.

Postlethwait, R. W. (1970): Ann. Surg. 171:892.

Rose, R. M., Paul, I. L., Weightman, B., Simon, S. R. and Radin, E. L. (1973): J. Biomed. Mater. Res. Symp. 4:401.

Salthouse, T. N., Williams, J. A. and Willigan, D. A. (1969): Surg., Gyn. Obstet. 129:691.

Soltész, U. (1986): In: *Materials Sciences and Implant Orthopedic Surgery.* R. Kossowsky and N. Kossovsky (Eds.). Martinus Nijhoff, Dordrecht, pp. 355ff.

Takahashi, K. (1973): J. Macromol. Sci.-Phys. B8(3-4):673.

BIBLIOGRAPHY

Chu, C. C. (1983): In: *Biocompatible Polymers, Metals, and Composites.* M. Szycher (Ed.). Technomic, Lancaster, Pennsylvania, pp. 477ff.

Ducheyne, P. and Lemons, J. E. (Eds.) (1988): Annals NY Acad. Sci. 523:1.

Hayden, W., Moffatt, W. G. and Wulff, J. (1965): *Mechanical Behavior,* Vol. III of *The Structure and Properties of Materials.* J. Wulff (Ed.). John Wiley, New York.

Hull, D. (1981): *An Introduction to Composite Materials.* Cambridge University Press, Cambridge.

Jones, R. M. (1975): *Mechanics of Composite Materials.* McGraw-Hill, New York.

Kronenthal, R. L. (1975): In: *Polymers in Medicine and Surgery.* R. L. Kronenthal, Z. Oser and E. Martin (Eds.). Plenum, New York, pp. 119ff.

Ward, I. M. (1971): *Mechanical Properties of Solid Polymers.* Wiley-Interscience, London.

7
Friction and Wear

7.1 INTRODUCTION

In the previous chapter we considered the mechanical behavior of materials under stress. The areas dealt with were those concerning the properties of singular parts or components. When devices contain more than one component, or are able by design or chance to move against natural tissue, another class of mechanical effects must be considered.

There is a general resistance to the motion of one material body over another which is termed *friction*. When static friction is overcome and relative motion takes place, it is accompanied by a modification of the interface by a variety of processes that are collectively known as *wear*. Introduction of surface treatments or interposed materials to make relative motion easier is called, collectively, *lubrication*. The study of these three phenomena (friction, wear, and lubrication) is the science of *tribology*. In this chapter we shall consider these phenomena and their presence in and alterations by biological environments.

7.2 FRICTION

If an attempt is made to move one body over the surface of another, a restraining force oriented to resist motion is produced. This restraining or friction force, F_f, is given by

$$F_f = \mu F_\perp \tag{7.1}$$

where

$F_\perp$ = force perpendicular to interface

μ = coefficient of friction

The force perpendicular to the surface, $F_\perp$, may be generated by compressive or gravity forces. The coefficient of friction, μ, is a ratio or unitless number, with values usually between 0 and 1, which describes the relationship of the frictional restraining force to this perpendicular force. It is characteristic of the interface, depending upon the composition and finish of the pair of materials involved, and is affected by lubrication. Furthermore, the coefficient is greater just before surfaces begin to move (initial conditions $\rightarrow \mu_i$) than when the surfaces are in continuing or steady motion (sliding conditions $\rightarrow \mu_s$). Table 7.1 gives some typical values of μ_i and μ_s.

Frictional behavior arises from the physical situation of the surfaces having a relatively small area of contact due to microsurface roughness. The small size of this area, perhaps as little as 1% of the geometric

Table 7.1 Initial and Sliding Coefficients of Friction

Materials combinations	Lubricant	μ_i	μ_s
Rubber tire/concrete	None (dry)	1.0	0.7
Rubber tire/concrete	Water	0.7	0.5
Leather/wood	None (dry)	0.5	0.4
Steel/steel	None (dry)	—	0.5
Steel/polyethylene	None (dry)	—	0.1
Steel/ice	Water	0.03	0.01
Cartilage/cartilage (hip)	Synovial fluid	—	0.002
	Ringer's	—	0.01–0.005
CoCr/CoCr (hip prosthesis)[a]	None (dry)	—	0.55
	Veronate buffer	—	0.22
	Serum	—	0.13
	Synovial fluid	—	0.12
	Albumin (sol.)	—	0.11
CoCr/PE(UHMW)[a]	Serum	—	0.08
Al_2O_3/Al_2O_3[b]	Ringer's	—	0.1–0.05

[a]Weightman et al. (1972).
[b]Dörre et al. (1975).

interface area, leads to local yielding and bonding due to high stresses at the points of actual contact. Thus, relative motion results only when these bonded areas can be disrupted and moved relative to one another. This disruption produces the frictional restraining force and also clearly leads to the wear process.

Frictional restraining forces are complex, but a number of generalizations can be made:

1. The coefficients of friction, μ_i and μ_s, are essentially independent of $F_\perp$ [they may be affected, however, by tangential (lateral) forces].
2. For a given $F_\perp$, coefficients of friction are independent of stress, that is, of the apparent or geometric interfacial surface area.
3. Coefficients of friction depend upon surface texture, the material pair, and the lubricant involved. However, in general, coefficients are lower for a pair of unlike materials of the same roughness than for identical materials and are lower for a given material pair in the presence of lubricating agents than in their absence.
4. Static and dynamic coefficients of friction are not closely related to wear rates (Galante and Rostoker, 1973). In particular, low frictional coefficients do not lead necessarily to low wear rates.

Table 7.2 Composition of Synovial Fluid in Comparison to Serum

Component	Synovial fluid (g/L)	Serum (g/L)	Synovial/serum
Protein (total)	18	70	0.26
Albumin	11.3	34.3	0.33
α_1-globulin	1.26	4.2	0.30
α_2-globulin	1.26	8.4	0.15
β-globulin	1.62	11.9	0.14
γ-globulin	3.06	11.2	0.27
Lipid (total)	2.4	7.0	0.34
Phospholipids	0.8	2.0	0.40
Urate	0.016	0.018	0.88
Glucose	0.66	0.91	0.73
Hyaluronate	2–4	4.2×10^{-5}	$\sim 7 \times 10^4$

Sources: Proteins—Lentner, C. (1981), Vol. 1. Other—Levick, J. R. (1987).

It is clear from Table 7.1 that artificial material pairs do not possess coefficients of friction which closely approach those possible in natural joints, particularly at the low velocities at which joints operate. There is little that can be done about this situation as long as body fluids are depended upon for lubrication. However, it is important from a design point of view to know what actual coefficients of friction may be expected.

Table 7.1 suggests the importance of an appropriate lubricant in laboratory evaluations. For materials in contact with blood, such as heart valve components, the appropriate lubricant is fresh serum. For device components in soft tissue locations, a 50 : 50 mixture of serum and normal saline approximates the intracellular exudate. For joint replacement components, the appropriate lubricant is synovial fluid. It has been shown (Woodman et al., 1977) that the synovial tissue remaining in the vicinity of a joint produces essentially normal synovial fluid that is available for lubrication of the artificial joint replacement. [Differences in composition between synovial fluid and serum (Table 7.2) suggest that dilute serum:saline solutions are superior to saline to simulate synovial fluid but may be inadequate replacements.]

7.3 LUBRICATION

The principle of lubrication is to provide a film or layer to separate two surfaces during relative motion in order to reduce both frictional restraining forces and wear. Lubrication modes or processes are classified by the nature and the magnitude of the average surface separation characteristic for each type.

7.3.1 Hydrodynamic

Hydrodynamic lubrication is perhaps the most common process and occurs when the motion of one body relative to the other draws a continuous film of lubricant into the contact area. The characteristic surface separation for typical lubricants and engineering finishes is between 10^{-3} and 10^{-4} cm. In this mode, all of the work of friction is dissipated by viscous shear of the lubricant.

7.3.2 Elastohydrodynamic

Elastohydrodynamic lubrication occurs at smaller separations, between 10^{-4} and 10^{-5} cm. In this case, the motion of one body of the pair is able to transmit force through the lubricant to generate sufficient stress for transient elastic deformation of the other body. While this may be satisfactory in the short term, in the long term it may lead to localized fatigue failure of one or the other surface, with an accompanying increase in wear rate.

7.3.3 Squeeze Film

Squeeze film lubrication occurs in either hydrodynamic or elastohydrodynamic conditions if the lubricant is sufficiently viscous to respond elastically (rather than by increased flow) to temporarily increased normal loads. Thus, a squeeze film lubricant, although highly viscous, may reduce wear in situations where transient overloads occur.

7.3.4 Boundary

Boundary lubrication occurs when the lubricant coats the opposing surfaces rather than acting as a low-shear interface. This coating acts to modify the frictional character of the surfaces to reduce both frictional restraining forces and wear. Characteristic mean surface separations depend sensitively on the nature of the lubricant but are usually less than 10^{-5} cm.

7.3.5 Mixed

Mixed lubrication occurs when a fluid lubricant operating in hydrodynamic or elastohydrodynamic mode is able to coat the surfaces, by an

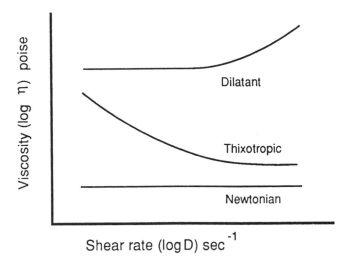

Figure 7.1 Types of lubricant behavior. Source: Adapted from Dintenfass (1963).

adhesive process, thus providing additional protection at high loads through bonding lubrication.

Natural joints in the skeletal system probably demonstrate a combination of these lubrication modes (Wright, 1969):

Boundary lubrication during motion initiation
Elastohydrodynamic lubrication during motion
Squeeze film lubrication during high load events

This combination of behavior results both from the structure of the joint and from peculiarities in the nature of the lubricant, synovial fluid. The typical composition of normal synovial fluid is summarized and compared to that of serum in Table 7.2. It resembles serum in inorganic species but contains 30–50% the amount of protein and lipids with very significantly greater amounts of hyaluronate. The hyaluronate is largely responsible for synovial fluid's response to shear (see Section 7.3.6) and a portion of the lipid phase is thought to be a surfactant (= avidly attracted to surfaces) which produces low coefficients of friction in the boundary lubrication regime (Hills and Butler, 1984).

7.3.6 Types of Lubricant Behavior in Response to Shear

In general, lubricants display three types of relationships between apparent viscosity and shear rate (Dintenfass, 1963) as shown in Figure 7.1. Conventional lubricants have a viscosity that is independent of shear rate.

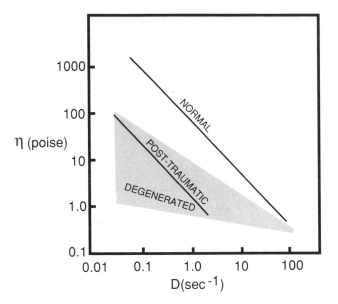

Figure 7.2 Viscosity-shear rate relationships for synovial fluid. Source: Adapted from Dintenfass (1963).

Thus, surface separations as a function of relative velocity may be determined fairly easily, taking the lubricant properties as a constant. Under these conditions the lubrication process may remain constant over a wide range of velocities.

However, some lubricants are thixotropic, that is, they become reversibly less viscous as shear rates increase. An everyday example of such a fluid (although not a lubricant) is nondrip ceiling paint. This appears nearly solid in the can but becomes quite thin as it is brushed on the wall.

The inverse of thixotropic is also possible; such a material would be called dilatant and would become reversibly more viscous with increasing shear rate. Such a material would not contribute to easy relative motion but might act to reduce wear at high relative velocities.

Either thixotropic or dilatant lubricants can produce a change in lubrication mode as a function of relative velocity of the surfaces. For instance, a material pair with a thixotropic lubricant can display hydrodynamic lubrication at intermediate pressures and velocities and boundary lubrication at high pressures and velocities.

Synovial fluid, which is an ultrafiltrate of serum with the addition of a long chain polysaccharide complex, hyaluronic acid, is a highly thixotropic lubricant (Figure 7.2). Trauma resulting in joint effusion may reduce the osmolarity of synovial fluid, producing a generally less viscous but still thixotropic lubricant. However, in the presence of a persistent joint disease such as rheumatoid arthritis, the fluid both thins and tends to lose its thixotropic property. This permits closer approach of the joint surfaces and may produce increased wear as a contributing factor to joint degeneration.

7.4 WEAR

7.4.1 Introduction

Wear is a more pronounced problem than frictional restraint for two reasons:

1. Wear produces biologically "active" particles that can excite an inflammatory response (see Chapter 8).
2. Wear produces shape changes that can affect function.

There are several mechanisms of wear. Probably the most important mechanism in biomedical applications is adhesive wear. This arises from the junction making and breaking process previously described. The rate of production of wear debris, expressed as a volume, is given most generally by Eq. (7.2):

$$V = \frac{kF_\perp x}{3p} \tag{7.2}$$

where

V = volume of wear debris
k = Archard's coefficient
$F_\perp$ = perpendicular force
p = surface hardness
x = total sliding distance

Figure 7.3 shows the range of k values of typical engineering situations. For the situation of a polymer on a metal, in vivo, values for k should lie between 10^{-5} and 10^{-7} with conditions described by the lower right-hand corner of the diagram. Note that the ordinates are labeled differently. The left ordinate refers to transfer film formation (see Section 7.4.2), while the right ordinate refers to the production of loose particles.

It is interesting to note that, while values of μ lie within a small range (between 0 and 1), k, and thus wear rates, vary over many orders of

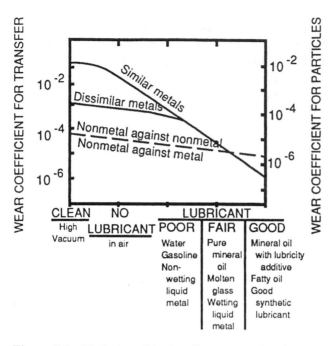

Figure 7.3 Variation of Archard's constant, k, with wear and lubrication conditions. Source: Adapted from Rabinowicz (1976).

magnitude. This further reinforces the observation that friction forces and wear rates have little direct relationship.

7.4.2 The Transfer Film

Figure 7.3 suggests that there are two major wear processes possible for any combination of materials: particle production as described above, or the formation of a "transfer film." A *transfer film* is produced when a hard material, such as a metal, moves on a softer material (for instance, a polymer) and shears off and picks up a coating of polymer, as shown in Figure 7.4. This film bridges across the asperities on the surface of the metal, replacing metal-polymer contact with polymer-polymer contact and, by increasing the actual contact area (as a function of the apparent contact area), reduces local stresses.

The formation of a transfer film may lead to one of two circumstances:

1. If the film is stable, then wear rates may be reduced after an initial high-wear interval during film formation.

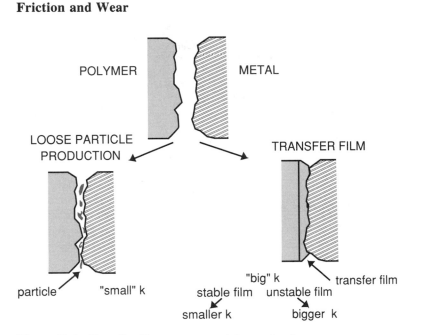

Figure 7.4 Transfer film versus particle production.

2. If the film is unstable, it may peel and increase wear by either abrasive or "three-body" wear.

McKellop et al. (1978) reported some interesting results involving transfer films (Table 7.3). The material pair is F138 type steel [316L (vacuum melted)] and ultra-high-molecular-weight polyethylene (UHMWPE), with serum, distilled water, or saline as lubricants. Wear rates are reported as a calculated linear recession of the UHMWPE surface, based upon measurement of the volume of wear debris. Thus, the wear rate does not include a creep component. Distilled water apparently permitted the formation of a transfer film and produced the lowest wear rate. Wear in saline was some 65 times greater (than in distilled water) and was apparently accompanied by crevice corrosion of the stainless steel in the interface between substrate and transfer film as indicated by orange discoloration. This leads to occasional release of the film and very high wear rates: apparently case 2 cited above.

Serum produced an intermediate wear rate with only occasional transfer. This suggests that, although serum does not completely reproduce synovial fluid, active molecules present in vivo can coat metal surfaces in metal/polymer wear combinations and, although preventing the formation

Table 7.3   Wear of UHMWPE[a] in Different Lubricants In Vitro

Lubricant	Number of specimens	Average wear rate (μm/10^6 cycles) (range)	Observations: (μ_s = dynamic coefficient of friction)
Serum (bovine)	4	0.65 ($\pm$17%)	μ_s = 0.07–0.12 normally, μ_s = 0.35 during temporary high friction. Polymer transfer onto metal counterfaces occurred only during high friction phase.
Distilled water	3	0.08 ($\pm$60%)	μ_s = 0.07–0.13 at start. A heavy polymer transfer layer formed by 3 × 10^5 cycles; μ_s then ranged from 0.14 to 0.18. The transfer layer remained intact for the duration of the test.[b]
0.9% saline (Ringer's solution)	3	5.2 ($\pm$17%)	μ_s = 0.07–0.10 at start. Heavy, orange-colored transfer layers formed as μ_s increased to 0.27. These layers occasionally broke up and μ_s dropped to the initial level.

[a]Against 316L (VM) stainless steel counterface at 3.45 MPa (500 psi) nominal contact stress, 10^6 sliding cycles @ 60 cpm [travel: 5 × 10^4 m (est.)]
[b]McKellop et al. (1978) also report transfer layer to be unstable at 690 MPa.
Source: Adapted from McKellop et al. (1978).

of stable transfer films, the molecules reduce wear through surface lubrication.

7.4.3   Other Wear Mechanisms

There are three other mechanisms of wear that are of concern to us. The first is abrasive wear. This is wear of a soft surface produced by a "ploughing" of the surface by large asperities in the harder counter surface. This clearly occurs in vivo, as will be seen in the later discussion of the size of wear debris.

Corrosive wear of metals occurs secondary to the physical removal of a passive or protective surface layer. The exposed surface may be more susceptible to wear, perhaps being softer and/or more chemically reactive, or wear may be accelerated by the repetition of cycles of passive film formation and mechanical removal. Figure 7.5 suggests this possibility graphically. In this experiment, a rod of passivated F75 cobalt-base alloy was pressed against a dimple in a block of UHMWPE (Jablonski et al., 1986). After the equilibrium potential has been established [in an aerated 0.9% saline solution, reference: standard calomel electrode (S.C.E.)], a half sinusoidal stress, with a peak value of 3.4 MPa, was applied during 40% of a 60° back and forth rocking cycle, at 36 cpm (cycles per minute). These conditions replicate those thought to exist in the human hip joint prosthesis during slow walking. The potential (vs. S.C.E.) became more negative, indicative of the flow of an increased corrosion current. When motion ceased, the potential became less negative, paralleling the (presumed) re-establishment of the passive layer. The time to 50% of the maximum potential change ($t_{1/2}$max) decreased as the peak stress and the rate of rocking were increased. Similar results have been reported in all metal total joint replacement prostheses (Thull, 1977).

The third and final type of wear to be considered is a surface fracture or fatigue wear. There are three routes to this process:

1. Stresses produced by asperities may not exceed yield stress but may exceed the endurance limit for the softer of the material pair (generally a polymer). This will lead to local fatigue failure of the softer surface and local fracture, "mud-caking," or spalling.

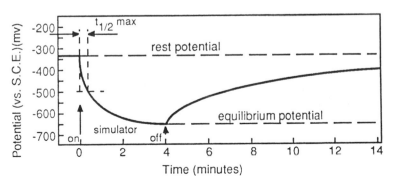

Figure 7.5 Effect of articulation on corrosive wear. Source: Stone (Jablonski et al. 1986), unpublished data.

2. Design of devices may produce a rigid support to the soft component. If the soft component is too thin in comparison to the magnitude of the contact stress and its apparent contact area, then local stress elevation occurs (Bartel et al., 1985), contributing to fatigue failure, as observed above. Dowling et al. (1978), although suggesting alternative mechanisms, have observed such polymer surface failure in UHMWPE/ metal hip prostheses after 8 years or more of implantation.

3. Free particles, perhaps adventitial (e.g., bone or PMMA fragments), or fragments of incomplete or shed transfer films may roll between the moving surfaces. If these particles are relatively undeformable, they can easily generate stresses above the ultimate tensile strength of the surface, leading directly to surface cracking. As cracks join up, wear debris is released. This process is usually called "three-body" wear.

7.4.4 Evidence of Wear In Vivo

Wear is a considerable problem in vivo. Of the five heart valve poppets studied and reported in Table 3.2, four showed signs of wear as evidenced by strut grooves. These would be expected to have a profound effect on local hemodynamics and on valve closure. All 21 hip cups that were recovered after periods of 14–159 months and studied by Dowling et al. (1978) showed signs of either adhesive (early) or fatigue (late) wear. Of these, nine showed evidence of the formation of a secondary socket or bearing area, with resultant theoretical effects on joint function (Dumbleton et al., 1984).

However, of more importance may be the formation of wear debris. Particulate materials usually elicit different host responses than bulk materials. The observed cellular response is frequently different in nature (see Section 8.2.3) and far more vigorous than in comparable bulk materials.

7.4.5 Size of Wear Debris

Rabinowicz (1976) discussed a criterion for predicting the diameter of wear particles produced by adhesive wear processes:

$$d = \frac{6 \times 10^4 \ W_{12}}{p} \tag{7.3}$$

where

d = diameter of wear particle

W_{12} = surface energy of adhesion between materials 1 and 2

p = hardness of wearing surface

Similar relationships for abrasive wear and for stress concentration phenomena would suggest that hardness is the key parameter in determining typical wear debris particle size. Since the elastic modulus, E, is a good estimator of the hardness, p, we would expect particle diameter, d, to vary inversely with E.

Although Rabinowicz suggests that equation 7.3 correlates well with observation (presumably in predominantly adhesive wear situations, in vitro), in biomedical applications broad ranges of wear debris particle sizes are seen. Table 7.4 surveys the findings of wear debris particles in a number of joint replacement situations. Several comments seem in order.

The finding of wear debris particles, either in implant sites or in regional lymph nodes where they were deposited by phagocytosis and transport, seems quite variable. This suggests that several wear processes are taking place simultaneously, and that the mechanical and environmental details of each application influence the type of particles produced. Polymer particles seem to be produced by fatigue, abrasion, and, possibly, by shredding of transfer films, resulting in either fibers, plates, or grains. Metals, on the other hand, are released by less apparent processes, probably including adhesion, third-body fatigue, and local corrosion. Metal particles are usually smaller than polymer particles, as expected from their relatively higher elastic moduli. The relative absence of expected metallic "fines" ($d < 1$ μm) in many clinical cases, in comparison to results of in vitro wear experiments (Table 7.4), may reflect the extremely high dissolution rates of such particles due to their high specific surface areas or cellular transport from the site of production.

Finally, it should be pointed out that no animal or clinical study to date has shown a good correlation between wear loss of an articulating implant component and a volumetric estimate of wear debris in the tissues of the experimental subject. This is due to a number of factors:

1. Systemic distribution of particles (previously referred to)
2. Dissolution of metallic debris
3. Deformational changes in polymeric components (creep, fluid absorption, etc.)

Thus, estimates of wear from examination of implants probably provide only upper bounds, while examination of tissues provides unrealistic lower bounds on wear rates. Carefully conducted in vitro wear testing with appropriate mechanical conditions and lubricants is required to measure actual wear rates with any degree of certainty.

Table 7.4 Characterization of Biomaterial Wear Debris Particles

Source	Polymer debris	Metal debris
Recovered human tissues (during THR revision) (Walker et al., 1973)	PMMA: "slivers": 2–3 μm → 20 μm long. UHMWPE: particles: 1 μm, "shreds": 3–12 μm long	CoCr: round or rod shaped particles, 1–2 μm (typ.)
Recovered human tissues (during THR revision) (Mirra et al., 1976)	PMMA: "chips": 100 μm–1 mm (typ.). UHMWPE: fibers with smooth borders: 5–40 μm long, 2–5 μm wide	SS, CoCr: "small" spheres or splinters: 1–4 μm (typ.)
Recovered human tissues (during THR revision or at autopsy) (Vernon-Roberts et al., 1976)	PMMA: spheres, "ovoids", splinters: 0.5–80 μm (typ.) UHMWPE: spears, splinters and sheets: 0.5–50 μm	SS, CoCr: granules, rods and needles: 0.1–3 μm (typ.)
Aspirated synovial fluid (after THR or TKR) (Mears et al., 1978)	PMMA: "irregular chunks": 1 μm–1 mm (typ.) UHMWPE: "shreds": 1–10 μm (diameter), up to 300 μm long	0.25 μm–1 mm
Total joint replacement (in Ringer's solution)—in vitro simulator. (Freeman et al., 1969)		CoCr: irregular particles: 0.1–1 μm (typ.)
Metal/metal/water; metal/polymer/liquid N_2—in vitro (Wagner et al., 1976)	UHMWPE: ellipsoids: 1 (minor axis)–100 μm (major axis)	SS, CoCr: irregular grains, 0.1–10 μm (typ.), 50% <1 μm

CoCr: Cobalt-chromium alloy; SS: stainless steel; PMMA: poly methyl methacrylate, THR: total hip replacement; TKR: total knee replacement; UHMWPE: ultra-high-molecular-weight polyethylene.
typ.: typical dimension.
Source: Dumbleton et al. (1984) (see for primary citations).

7.4.6 Anomalous Wear

One final comment is in order. We normally think of wear in terms of a harder material wearing away a softer material. However, there is ample evidence that the reverse can occur. In natural tissues it is observed that soft tissues, such as tendons, moving over bone will rapidly form grooves while appearing unchanged. This may reflect dynamic remodeling but may also involve a direct wear process of the harder bone by the softer tendon. Gent and Pulford (1979) have shown a similar situation in the wear, in vitro, of steel and bronze alloys by a variety of elastomers. Obviously, no dynamic remodeling is possible here. The mechanism proposed is the formation of active radicals on the elastomer surface by mechanical cleavage, followed by chemical attack of the metal surface by these active sites. The enhancement of the effect by the exclusion of oxygen seems to support such a mechanism. The evidence discussed in Section 4.10 for the participation of amino acids in corrosion processes certainly suggests that we should consider the possibility of such anomalous wear processes when implant materials move in contact with biological materials.

7.5 CONCLUSIONS

Frictional restraint to relative motion of device components and the associated wear, releasing particulate debris, are inherent in the nature of materials. Intelligent material selection and component design can minimize but not eliminate these phenomena in current biomedical devices fabricated from present biomaterials. Increased knowledge of the mechanistic details of friction and wear phenomena in these applications can be expected to produce improved performance. However, it may be the case that accumulation of wear debris, and the host response to such accumulation, will prove to be the ultimate limitation on the useful lifetime of articulating biomedical devices, such as total joint replacements.

With this in mind, we will now turn, in the next section, to consideration of host response to biomaterials and their degradation products.

REFERENCES

Bartel, D. L., Burstein, A. H., Toda, M. D. and Edwards, D. L. (1985): J. Biomech. Eng. 107:193.

Dintenfass, L. (1963): J. Bone Joint Surg. 45A:1241.

Dörre, E., Beutler, H. and Geduldig, D. (1975): Arch. Orthop. Unfall.-Chir. 83:269.

Dowling, J. M., Atkinson, J. R., Dowson, D. and Charnley, J. (1978): J. Bone Joint Surg. 60B:375.

Dumbleton, J. H. and Black, J. (1984): In: *Complications of Total Hip Replacement*. R. S. M. Ling (Ed.). Churchill Livingstone, Edinburgh, pp. 212ff.

Freeman, M. A. R., Swanson, S. A. V. and Heath, J. C. (1969): Ann. Rheum. Dis. (Suppl.) 28:29.

Galante, J. O. and Rostoker, W. (1973): Acta Orthop. Scand. (Suppl. 145).

Gent, A. N. and Pulford, C. T. R. (1979): J. Mater. Sci. 14:1301.

Hills, B. A. and Butler, B. D. (1984): Ann. Rheu. Dis. 43:641.

Jablonski, J. E., Stone, J. and Black, J. (1986): Trans. SFB 9:196.

Lentner, C. (Ed.) (1981): *Geigy Scientific Tables*, Vol. 1. Ciba-Geigy, Basle.

Levick, J. R. (1987): In: *Joint Loading*. H. J. Helminen, I. Kiviranta, A.-M. Säämänen, M. Tammi, K. Paukkonen, and J. Jurvelin (Eds.). Wright, Bristol, pp. 149ff.

McKellop, H., Clarke, I. C., Markolf, K. L. and Amstutz, H. C. (1978): J. Biomed. Mater. Res. 12:895.

Mears, D. C., Hanley, E. N., Jr., Rutkowski, R. and Westcott, V. C. (1978): J. Biomed. Mater. Res. 12:867.

Rabinowicz, E. (1976): Mater. Sci. Eng. 25:23.

Thull, R. (1977): Med. Prog. Tech. 5:103.

Walker, P. S. and Bullough, P. G. (1973): Orth. Clin. N. A. 4(2):275.

Weightman, B., Simon, S., Paul, I., Rose, R. and Radin, E. (1972): J. Lubric. Tech. 94:131.

Woodman, J. L., Miller, E. H., Dumbleton, J. H. and Kuhn, W. E. (1977): Trans. ORS 2:13.

Wright, V. (Ed.) (1969): *Lubrication and Wear in Joints*. J. B. Lippincott Company, Philadelphia.

BIBLIOGRAPHY

Bowden, F. P. and Tabor, D. (1986): *The Friction and Lubrication of Solids*. Oxford University Press, Oxford.

Davies, D. V. (1966): Fed. Proc. 25:1069.

Dumbleton, J. H. (1981): *Tribology of Natural and Artificial Joints.* Elsevier, Amsterdam.

Lee, L.-H. (Ed.) (1985): *Polymer Wear and Its Control.* ACS Symposium Series 287. American Chemical Society, Washington, D.C.

McCutchen, C. W. (1978): In: *The Joints and Synovial Fluid*, Vol. I. L. Sokoloff (Ed.). Academic Press, New York, pp. 437ff.

Weightman, B. (1977): In: *The Scientific Basis of Joint Replacement.* S. A. V. Swanson and M. A. R. Freeman (Eds.). John Wiley, New York, pp. 46ff.

Interpart 1
Implant Materials: Properties

I1.1 INTRODUCTION

The preceding discussions concerning material response have necessarily been generic; that is, they have largely treated biomaterials by classes related to their predominant chemical bond type (and resulting physical properties) rather than dealing with the response of a particular, specific composition of material to the biological environment. This is how it should be since the details of such response depend upon the composition (including impurities) of a specific material, the methods by which it is fabricated and finished, the device application in which it is used, the animal and/or clinical model in which it is evaluated, etc. The professional literature on material response is broad, so much of the information needed to make a preliminary selection based upon material response is usually available.

However, before such a selection may be made, a more fundamental question must be answered: does the material meet the physical requirements of the design? Again, this is a complex problem, requiring a well-structured design process for successful solution. This and related issues of design are dealt with in Chapter 20. In this Interpart, tables of basic materials' properties are provided as an introduction to the broader issue of materials selection.

The reader must be warned that most properties tabulated here are *nominal* or *typical* properties. Materials specifications permit a range of compositions and processing conditions; in addition, the choice of starting materials (feed stocks) may also affect final properties. Therefore, reference should be made to original sources (as provided in the reference and bibliography sections or elsewhere) for exact details of composition, fabrication, and test conditions producing the values or ranges cited.

The properties tabulated are density and typical hardness, when known, and those mechanical parameters which may be obtained from a conventional stress-strain curve. Standards, such as those produced by the ASTM (American Society for Testing and Materials), BSI (British Standards Institute), and ISO (International Standards Organization) are cited when relevant. However, standards, whether consensus or regulatory in nature, most often describe minimum requirements; properties of actual manufactured materials frequently exceed these values. Other properties, such as fatigue endurance limit, coefficients of friction (for material pairs), workability parameters, etc., may be of equal or greater importance in material selection and should be sought out during the design process.

Finally, it should be remembered that high performance requirement materials applications, such as in medical and surgical devices and implants, require a high degree of confidence in design and performance parameters. Thus, some form of property verification, up to and possibly including 100% nondestructive testing, may be required in medical and surgical applications to assure safe and effective material response in use.

I1.2 METALS

Biomaterials in common use are drawn from the stainless steels (Table I1.1), the cobalt-base "superalloys" (Table I1.2), and from the titanium/titanium-base alloy system (Table I1.3). Several refractory and precious

Table I1.1 Stainless Steels

Material	F55 type 2	F138 type 2	F1314	F55 type 2	F138 type 2	High N$_2$	F138 type 2	22-13-5
Condition	AN	AN	AN	CW	CW	CW	HF	HF
Source	[1,2]	[1,3]	[1,4]	[1,2]	[1,3]	[1,4]	[1,3]	[5]
Density (gr/cm^3)	7.9	7.9	7.98	7.9	7.9	–	7.98	–
E (tensile) (GPa)	200	200	200	200	200	200	200	–
Hardness (Hv)	183	–	205	320	350	365	–	–
$\sigma_{0.2\%}$ (MPa)	170	170	460	690	750	975	240	785
σ_{UTS} (MPa)	465	480	770	850	950	1090	550	945
Elong. (min. %)	40	40	35	12	10	14.5	55	34

Key: AN: annealed; CW: cold worked; HF: hot forged; –: unavailable.
Sources: [1]: BSI 3531 (Part 2, Sec. 2)(Amend. 2, 1983)
    [2]: ASTM F55-82
    [3]: ASTM F138-82
    [4]: ASTM F1314-90
    [5]: Anonymous pamphlet, Zimmer (1985)

Table II.2 Cobalt-Base Alloys

Material	Cast CoCrMo	Wrought CoCrMo	Wrought CoNiCrMo	Wrought CoNiCr-MoWFe	Wrought CoCrMo	Wrought CoNiCrMo	Wrought CoNiCr-MoWFe
Condition	AN	AN	AN	AN	CW	CW	CW
Source	[1,2]	[1,3]	[4]	[5]	[1,3]	[4]	[5]
Density (gr/cm^3)	7.8	9.15	–	–	9.15	–	–
E (tensile) (GPa)	200	230	–	–	230	–	–
Hardness (Hv)	300	240	–	–	450	–	–
$\sigma_{0.2\%}$ (MPa)	455	390	240–450	275	1000	1585	825–1310
σ_{UTS} (MPa)	665	880	795–1000	600	1500	1795	1000–1585
Elong. (min.%)	10	30	50	50	9	8	8–18

Key: AN: annealed; CW: cold worked; —: unavailable.
Sources: [1]: BSI 3531 (Part 2, Sec. 4-5) (Amend. 2, 1983)
[2]: ASTM F75-82
[3]: ASTM F90-82
[4]: ASTM F562-84
[5]: ASTM F563-83

Table I1.3   Titanium and Titanium-Base Alloys

Material	Ti type 4	Ti6Al4V	Ti5Al2.5Fe	Ti6Al4V	Ti6Al7Nb	Ti5Al2.5Fe
Condition	AN	AN	AN	HF	HF	HF
Source	[1,2]	[1,3]	[7]	[6]	[4,5]	[7]
Density (gr/cm^3)	4.5	4.4	4.45	4.4	4.52	4.45
E (tensile) (GPa)	127	127	—	127	105	—
Hardness (Hv)	240–280	310–350	—	—*	400	—
$\sigma_{0.2\%}$(MPa)	430–465	830	815	—*	800–900	900
σ_{UTS}(MPa)	550–575	900	965	—*	900–1000	985
Elong. (min.%)	15	8	16	—*	10–12	13

Key: AN: annealed; HF: hot forged; *: specified by manufacturer [cf Donachie (1988)]; —: unavailable.

Sources: [1]: BSI 3531 (Part 2, Sec. 6) (Amend. 2, 1983)
        [2]: ASTM F67-83
        [3]: ASTM F136-79
        [4]: Anonymous pamphlet, IMI Titanium Ltd, 1989
        [5]: Semlitsch et al. (1985)
        [6]: ASTM F620-79
        [7]: Borowy and Kramer (1985)

Table I1.4   Other Metals and Alloys

Material	Ta	Ta	Pt	Pt10Rh	Pt10Rh	W
Condition	AN	CW	AN	AN	75%CW	SN
Source	[1,2]	[1,2]	[1]	[1]	[1]	[1]
Density (gr/cm^3)	16.6	16.6	21.5	20	20	19.3
E (tensile) (GPa)	186	186	147	—	—	345
Hardness (Hv)	—	—	38–40	90*	165*	225
σ_y(MPa)	140	345	—	—	—	—
σ_{UTS}(MPa)	205	515	135–165	310	620	125–140
Elong. (min.%)	20–30	2	35–40	35	2	~0

Key: AN: annealed; CW: cold worked; SN: sintered bar, *: Brinell hardness; —: unavailable.
Sources: [1]: Metals Handbook, 8th ed., Vol. 1 (ASM, Metals Park, 1961)
        [2]: ASTM F560-78

metals which have seen limited use in biomedical applications are covered
in Table I1.4.

I1.3 POLYMERS

The cautionary note previously sounded concerning the reliability of tabulated materials properties applies especially to polymers. In the case of
this material class, three additional problems interfere with interpretation
of data:

1. All polymers are viscoelastic; therefore, mechanical property measurements depend upon the strain *rate* used in evaluation. Since viscoelastic materials generally become stiffer and less ductile as strain
   rates increase, testing rates should equal or exceed those expected to
   be encountered in service.
2. Properties of engineering polymers are closely related to average
   molecular weight and molecular weight distribution as well as to
   curing conditions and time (thermosets) and fabrication temperatures
   and post-fabrication heat treatment (thermoplastics).
3. Sterilization, either by ethylene oxide or ^{60}Co irradiation, may alter
   final properties.

The data are presented in two tables: Table I1.5 for thermosets and
Table I1.6 for thermoplastics. Both types of systems have important roles
as biomaterials, although thermoplastics tend to be preferred due to the
relatively greater ease in fabricating them without low-molecular weight
leachable components.

Finally, fatigue behavior of biomedical polymers, especially of
PMMA-type "bone-cements," is a sufficiently controversial subject that
no data are provided here. The reader is referred to the professional
literature.

I1.4 CERAMICS

At room and body temperature, ceramic materials suitable for biomedical
applications possess negligible ductility; thus no tensile or elongation data
are included in Table I1.7. Data are provided only for some of the most
common structural ceramics; no information is provided on resorbable or
so-called "bioactive" ceramics due to their variety and complexity [the
reader is referred to deGroot (1983) and Hench (1984)].

Table I1.5 Thermoset Resins

Material	EP	PMMA 10BaSO$_4$	PMMA	PEU	PSU	SR	SR (HP)
Condition	CR	24hrCR	24hrCR	CR	CR	HV	HV
Source	[1]	[2,3]	[3]	[4]	[5]	[5,6]	[6]
Density (gr/cm^3)	1.11–1.40	1.183	1.088	1.1	1.2	1.12–1.23	1.15
E (tensile) (GPa)	2.4	1.31	2.4–3.1	5.9*	3.7*	<1.4*	2.4
Hardness (Shore A)	—	—	—	75	88	25–75	52
$\sigma_{y(C)}$(MPa)	—	—	15.8	—	—	—	—
σ_{UCS}(MPa)	100–170	70	69–125	—	—	—	—
σ_{UTS}(MPa)	28–90	28–46	9.7–32	45	40	5.9–8.3	8.3–10.3
Elong. (min.%)	3–6	4.6	2.4–5.4	750	540	350–600	700

Key: CR: room temperature cured; EP: epoxy; HV: heat vulcanized; PMMA: polymethyl methacrylate; SR: silicone rubber; HP: high performance; *: MPa; —: unavailable.

Sources: [1]: *Modern Plastics Encyclopedia* (McGraw-Hill, New York, 1990); Note: typical values; not specific medical grades.

[2]: ASTM F451-76

[3]: Lautenschlager et al. (1984)

[4]: Boretos et al. (1968)

[5]: Braley (1970); see also ASTM F604-78

[6]: Frisch (1984)

Table I1.6 Thermoplastic Resins

Material	PE (UHMW)	PE (UHMW)	PE (UHMW)	PE (UHMW)	PLA (STCP)	PMMA	PSF
Condition	MM	EX	CM	HC	CM	CM	IM
Source	[1]	[1,2]	[1-3]	[3]	[4]	[5]	[6,7]
Density (gr/cm^3)	.93-.944	.93-.944	.93-.944	—	—	1.186	1.23-1.25
E (tensile) (GPa)	—	1.24	1.36	2.17	4-5	2.6-3.2	2.3-2.48
Hardness (Shore D)	—	—	62	66	—	—	—
$\sigma_{y(C)}$(MPA)	21	21-28	19-29	28	—	—	65-96
σ_{UCS}(MPA)	—	—	—	—	—	80-125	—
σ_{UTS}(MPA)	34	34-47	27-40	—	50-60	50-75	106*
Elong. (min.%)	300	200-250	350	230	2-3	2-10	20-75

Key: IM: injection molded; MM: molded, machined; EX: extruded; CM: compression molded; HC: high crystallinity; PE(UHMW): ultra high molecular weight polyethylene; PLA(STCP): polylactic acid stereo co-polymer; PMMA: polymethyl methacrylate; PSF: polysulfone; *: flexural strength; —: unavailable.

Sources: [1]: ASTM 648-83
[2]: Roe et al. (1981)
[3]: Anonymous pamphlet, Depuy (1989)
[4]: Christel et al. (1985); note: resorbable; properties depend upon L/D ratio.
[5]: Lautenschlager et al. (1984)
[6]: Dunkle (1988)
[7]: ASTM 702-81

Table I1.7   Ceramic Materials

Material	Al_2O_3	C	C	C	ZrO_2
Condition	HP	LTI	VT	ULTI	SHP
Source	[1,2]	[3]	[3]	[3]	[4]
Density (gr/cm³)	3.93	1.7–2.2	1.4–1.6	1.5–2.2	6.1
Grain size (μm)	3–4	30–40*	10–40*	8–15*	<0.5
E (tensile) (GPa)	380	18–28	24–31	14–21	200
Hardness (Hv)	23,000	150–250	150–200	150–250	1300
σ_{UFS}(MPa)	550	280–560	70–210	350–700	1200
σ_{UCS}(MPa)	4,500	—	—	—	—

Key: HP: high purity; LTI: low temperature isotropic; SHP: sintered, hot isostatic pressed; ULTI: ultra low temperature isotropic; VT: vitreous (glassy); *: angstroms; —: unavailable.
Sources: [1]: Boutin et al. (1988)
         [2]: ASTM F560-78
         [3]: various; compilation in anonymous pamphlet, Intermedics Orthopedics (1983)
         [4]: Christel et al. (1989)

I1.5 COMPOSITES

Composites, or more properly composite materials, is a term which has come to describe a wide range of engineered or designed materials. Mechanical properties of composite materials depend upon both the properties of the phases (matrix and strengthening, or reinforcing, phase or phases) as well as on the nature of the phase interfaces, their volume fractions, and the local and global arrangement of the reinforcing phase(s).

This complexity dictates an economy of action here, since an entire volume could be (and has frequently been) devoted to the subject. Thus, as a brief guide, Table I1.8 presents properties of some more common reinforcing phases useful in composite biomaterials design, and Table I1.9 presents properties of a few representative composites which have been evaluated to some degree as biomaterials.

Table I1.8 Reinforcing Phases

Material	E-glass	S-glass	C-glass	C[a] (Low E)	C[a] (High E)	PA (K29)[b]	PA (K49)[b]	PA (K149)[b]
Condition	AN,CF	AN,CF	AN,CF	HT,CF	HT,CF	CF	CF	CF
Source	[1,2]	[1,2]	[1,2]	[1]	[1]	[1]	[1]	[1]
Density (gr/cm³)	2.62	2.50	2.56	1.76	1.9	1.44	1.44	1.47
Diameter (μm)	3–20	3–20	3–20	7–8	7	12	12	12
E (tensile) (GPa)	72–81	85–89	69	230	390	83	131	286
σ_{UTS}(MPa)	3450	4580	3000–5300	3300	2400	2800–3600	3600–4100	3400
Elong. (min.%)	4.9	5.7	4.8	1.4	0.6	4.0	2.8	2.0

Material	Al_2O_3	Al_2O_3-48SiO_2	SiO_2	βSiC	αSiC	βSiC	BN	B_4C
Condition	CF	DF	CF	CF	WH	WH	CF	WH
Source	[1]	[1]	[1]	[1]	[1]	[1]	[3]	[3]
Density (gr/cm³)	3.95	2.73	2.2	2.55	3.2	3.19	1.91	2.52
Diameter (μm)	20	2–3	9	10–15	0.6	0.1–0.5	7	—
E (tensile) (GPa)	379	100	69	180–200	690	400–700	90	483
σ_{UTS}(MPa)	1380	1900	3450	2500–3200	6900	3000–14000	1380	13800

Key: AN: annealed; CF: continuous fiber; DF: discontinuous fiber; WH: whisker; HT: heat treated; —: unavailable.
[a]Polyacrylonitrile (PAN) precursor.
[b]Kevlar™ (DuPont).

Sources: [1]: *Composites* (Engineered Materials Handbook, Vol. 1) (ASM Int., Metals Park, 1987).
  [2]: *Handbook of Composites*, ed. G. Lubin (Van Nostrand, New York, 1982)
  [3]: *Composite Materials Handbook*, M.M. Schwartz (McGraw-Hill, New York, 1984).

Table I1.9 Composite Materials

Material	PMMA-2C	C60SiC	C60SiC (5%por.)	CFRC	CFRC (7%por.)	EP-12.5C	CFPSU	PE (UHMW)-10C
Condition	RT					ET		
Source	[1]	[2]	[3]	[2]	[3]	[4]	[5]	[6]
Density (gr/cm³)	—	2.6	2.4	1.7	1.78	—	—	0.98
E (tensile) (GPa)	5.52	100	80-90	140	40-58	14	110	1.94**
σ_{UCS}(MPa)	—	1000	250-370	800	230-320	—	—	14.2
σ_{UTS}(MPa)	38	220*	220-360*	800*	350-600	200*	1600*	22
Elong. (min.%)	0.7	<1	—	>4	—	—	1.3	150

Key: PMMA: polymethyl methacrylate; RT: room temperature cured; ET: 70°C cured; CFRC: carbon fiber reinforced carbon; CFPSU: continuous fiber carbon reinforced polysulfone; EP: epoxy; por.: open porosity; *: flexural strength; **: estimated; —: unavailable.

Sources: [1]: Pilliar et al. (1976).
[2]: Brückmann et al. (1980).
[3]: Christel et al. (1987).
[4]: Hastings (1978).
[5]: Claes et al. (1986).
[6]: Anonymous pamphlet, Zimmer, 1978.

REFERENCES

Boretos, J. W. and Pierce, W. S. (1968): J. Biomed. Mater. Res. 2:121.

Borowy, K.-H. and Kramer, K.-H. (1985): In: *Titanium Science and Technology*, Vol. 2. G. Luterjering, U. Zwicker and W. Bunk (Eds.). Deutsche Gesell. f. Metallkunde e. V., Obureresel, pp. 1301ff.

Boutin, P., Christel, P., Dorlot, J.-M., Meunier, A., de Roquancourt, A., Blanquaert, D., Herman, S., Sedel, L. and Witvoet, J. (1988): J. Biomed. Mater. Res. 22:1203.

Braley, S. (1970): J. Macromol. Sci.-Chem. A4(3):529.

Brückmann, H. and Hüttinger, K. J. (1980): Biomaterials 1:67.

Christel, P., Meunier, A., Heller, M., Torre, J. P. and Peille, C. N. (1989): J. Biomed. Mater. Res. 23:45.

Christel, P., Meunier, A., Leclercq, S., Bouquet, P. and Buttazzoni, B. (1987): Applied Biomater. 21(A2):191.

Christel, P., Vert, M., Chabot, F., Garreau, H. and Audion, M. (1985): *Proceedings of the 1st Internat. Conf. on Composites in Biomedical Engineering*, November 19–20, London, UK. Imprint of Luton, Luton, pp. 11/1ff.

Claes, L., Hüttner, W. and Weiss, R. (1986): In: *Biological and Biomechanical Performance of Biomaterials*. P. Christel, A. Meunier and A. J. C. Lee (Eds.). Elsevier, Amsterdam, pp. 81ff.

deGroot, K. (1983): *Bioceramics of Calcium Phosphate*. CRC Press, Boca Raton, Florida.

Dunkle, S. R. (1988): In: *Engineered Materials Handbook*, Vol. 2: *Engineering Plastics*. C. A. Dostal (Ed.). ASM Int., Metals Park, Ohio, pp. 200ff.

Frisch, E. E. (1984): In: *Polymeric Materials and Artificial Organs*. ACS Symp. 256. C. G. Gebelein (Ed.). ACS, Washington, DC, pp. 63ff.

Hastings, G. W. (1978): Composites July:193.

Hench, L. L. and Wilson, J. (1984): Science 226:630.

Lautenschlager, E. P., Stupp, S. I. and Keller, J. C. (1984): In: *Functional Behavior of Orthopedic Biomaterials*, Vol. II. P. Ducheyne and G. W. Hastings (Eds.). CRC Press, Boca Raton, Florida, pp. 87ff.

Pilliar, R. M., Blackwell, R., MacNab, I. and Cameron, H. U. (1976): J. Biomed. Mater. Res. 10:893.

Roe, R.-J., Grood, E. S., Shastri, R., Gosselin, C. A. and Noyes, F. R. (1981): J. Biomed. Mater. Res. 15: 209.

Semlitsch, M., Staub, F. and Weber, H. (1985): Biomed. Technik 30:334.

BIBLIOGRAPHY

ASTM (1990): *1990 Annual Book of ASTM Standards, Vol. 13.01, Medical Devices.* ASTM, Philadelphia.

Boretos, J. W. (1973): *Concise Guide to Biomedical Polymers.* C. C. Thomas, Springfield, Illinois.

Donachie, M. J., Jr. (Ed.) (1984): *Superalloys Source Book.* ASM Int., Metals Park, Ohio.

Donachie, M. J., Jr. (Ed.) (1988): *Titanium: A Technical Guide.* ASM International, Metals Park, Ohio.

Kingery, W. D. (1960): *Introduction to Ceramics.* John Wiley, New York.

Szycher, M. (Ed.) (1983): *Biocompatible Polymers, Metals, and Composites.* Technomic, Lancaster, Pennsylvania.

III

HOST RESPONSE: BIOLOGICAL EFFECTS OF IMPLANTS

8

The Inflammatory Process

8.1   INTRODUCTION

Inflammation is a nonspecific physiological response to tissue damage in animal systems. It arises as a response to trauma, infection, intrusion of foreign materials, local cell death, or as an adjunct to immune or neoplastic responses. If the initiating agent causes damage to, or frank rupture of, vascular tissue, blood coagulation, which is related, may be superimposed on the inflammatory response.

In this chapter, the general inflammatory response will be considered. Coagulation is dealt with in Chapter 9, immune, or specific, responses are dealt with in Chapter 11 and neoplastic transformation, whether chemically or mechanically mediated, is the subject of Chapter 12.

8.2   THE INFLAMMATORY RESPONSE

8.2.1   Clinical Signs

The four classical clinical signs of inflammation, in both animals and humans, are redness (rubor), swelling (tumor), pain (dolor), and heat (calor). The magnitude of these signs is related to the intensity and extent of the inflammatory process. The presence of an implant does not produce additional symptoms but may alter their severity and duration.

8.2.2   Initial Events

The first events that occur in response to an inflammatory stimulus are rapid dilation of the local capillaries and an increase in the permeability of their endothelial cell linings. The dilation (vasodilation) arises from the activation of a coagulation factor, factor XII (the Hageman factor), most probably by contact with collagen or a foreign protein or material. Through the intermediate activation of a polypeptide, kallikrein, this leads to conversion of a group of additional molecules to kinins. The kinins are strong mediators of vasodilation and endothelial permeation.

The vasodilation leads to an increase in blood entry into the capillary beds. Local aspect ratio changes in the capillaries, combined with loss of plasma through the capillary walls and a tendency for the platelets and erythrocytes to become "sticky," lead to slower flow and sludging. This results in the first clinical sign, redness, simply reflecting a higher local concentration of erythrocytes.

The increased permeability of the capillary endothelium allows fluid to move into the surrounding tissue bed. Under normal conditions there is a 10–15 mm Hg positive pressure differential between the arteriole end of a capillary and the external tissue bed. However, the endothelium is tight and permits only a very slow flow of water and small molecules into the surrounding tissue. This fluid is normally drained away by local lymphatic vessels, maintaining a constant tissue volume. As permeability increases, water and larger molecules, including normal plasma proteins and the locally activated kinins, move into the tissue. The increased fluid influx, if not promptly balanced by increased lymphatic drainage, leads to swelling, the second clinical sign. The local lymphatics may be constricted or blocked due to the original trauma or by occlusion by cell fragments. Additionally, the presence of plasma fractions raises the local osmotic pressure and tends to hold fluid in place. Thus, swelling (edema) is a usual early concomitant of inflammation.

Pain, the third clinical sign, results from at least two causes. First, the local edema may activate local deep pain receptors. In patients, this is felt as a throbbing pain, as it peaks repetitively with peak systolic pressure. Second, the kinins act directly on nerve ends to produce pain sensation. The familiar acute pain of a bee sting is associated with the activation of a kinin called bradykinin in bee venom.

The origin of the local heating effect, the fourth sign, is unclear. In the general case, it may be associated with local disturbances of fluid flow in the presence of increased cellular metabolic activity (and resulting heat production). Although not yet shown conclusively, a group of contaminants termed *pyrogens*, which are known to cause systemic fever, may be generated locally either by tissue necrosis or, in the presence of infection, as a result of activation by bacterial or viral toxins, especially endotoxin. Fine particles or bacterial fragments left on implants which are sterilized after inadequate cleaning may also be pyrogenic.

These early events of inflammation are largely chemical in origin and effect. Shortly after initiation, however, a series of cellular invasions take place. These cells are responsible for the removal of dead tissue and the repair of the resulting defect.

8.2.3 Cellular Invasion

The first new cells to appear at the site of injury are *neutrophils*. They are the most common of the granulocytes, also termed polymorphonuclear leukocytes (PMNs) due to their horseshoe-shaped multilobed nuclei, and can be distinguished from related cell types, basophils and acidophils, since their granules are stainable by neither basic or acidic dyes.

At the earliest stages of the inflammatory response, the neutrophils become sticky. At first they stick momentarily to the capillary endothelium and then are released. More stick and remain for longer periods.

Then, neutrophils begin to penetrate between the endothelial cells and move into the surrounding damaged tissue. Neutrophil emigration (*diapedisis*) begins minutes to hours after insult and may continue for as long as 24 hours. The time course varies with the nature and severity of the insult. Although following an initial but transient episode of increased vascular permeability, this emigration is accompanied by a longer sustained endothelial permeability. This second phase of permeability parallels the maturation of edema and erythema and subsides as neutrophil emigration draws to a halt.

The role of the neutrophil appears to be primarily that of phagocytosis. This process, the engulfing and degradation or digestion of fragments of tissue or material, is common to this and several other cell types. What distinguishes each cell type in regard to its phagocytic function is the nature and amount of the degradative enzymes that each has available. These are stored as cytoplasmic granules, perhaps 200 per neutrophil, that are readily seen in the electron microscope. Before encountering foreign particles, phagocytic cells are inactive; after such contact, they become *activated*. Activation is characterized by a change in metabolic activity, especially an increase in oxygen consumption (the "respiratory burst"), change in cell shape and an internal degranulation as lysozymes are released into phagosomes.

Neutrophilic phagocytosis proceeds by the contact of the cell membrane and a foreign material (inorganic or organic). Neutrophils are specialized to find and phagocytize bacteria; how they find the foreign material is largely unknown, but there appear to be several mechanisms. The neutrophil may be a attracted by a particular chemical composition (*chemotaxis*), by local pH differences, or by electrochemical factors associated with the foreign particle and its surroundings. In addition, a group of chemical species, opsonins, are thought to coat foreign surfaces (*opsonization*), thus rendering them chemotactic. These opsonins are small molecules excreted by a variety of cells and include the activated form of complement factor five (C5a). When in contact with and recognizing a foreign material through antigen (opsinin)-membrane receptor binding, a dimple develops in the cell wall (invagination). The particle is drawn in and the dimple closes, often breaking off to form a vacuole, termed a secondary lysosome, that leaves the particle inside the cell and surrounded by an everted unit membrane. This vacuole merges with primary lysosomes, whose granules are released into the vacuole. The enzymes in these granules are released into the interior of the vacuole in an active form and attack the vacuole contents. This process of degranulation is accompanied by a respiratory burst: a sharp metabolic rate increase and production of hydrogen peroxide and superoxide anions. The vacuole may clear and be absorbed, or its contents, presumably altered to be less objectionable, may be released outside the cell.

The complement system, of which factor C5 is an element, is a complex series of glycoproteins and protein inhibitors present in the circulating and interstitial fluids of mammals. It constitutes a powerful, noncellular part of the biological defense system against foreign materials. It can be activated by contact with antigen-antibody complexes (see Chapter 11), bacterial proteins and polysaccharides, endotoxins and many polymers, of natural or synthetic origin. Once activation is initiated, it proceeds along two pathways through a series of enzymatic cascade amplifying steps, much like the blood coagulation cascade (see Chapter 9). The end products are activated complement factors, which can mediate membrane damage directly, or molecular fragments which can influence inflammatory and immune processes (Anderson and Miller, 1984; Frank, 1979).

Accompanying the neutrophils is a far smaller number of another type of leukocyte, the *eosinophil*. Eosinophils are similar in structure to the neutrophils and, although they perform a similar phagocytic function, they can also phagocytize antigen-antibody complexes. Their name stems from the observation that, unlike the neutrophils, their granules can be stained by eosin, an acid dye. They contain a different distribution of enzymes and are usually only present in significant numbers in association with immune responses.

The neutrophils and eosinophils serve as the first active line of defense against foreign material in tissue. Leukocytes have a life span of only hours in blood and a few days in tissue. They are end-state cells and cannot divide. Although capable of oxidative phosphorylation, they obtain their energy primarily from the anaerobic metabolism of stored glycogen, and so they can persist in areas with highly disturbed metabolic states. After fulfilling their function, they die rapidly. Normally they constitute an "emergency squad" whose duties are later supplanted by another cell, the *monocyte*. Thus, the presence of extensive numbers of live neutrophils in tissue may be interpreted as evidence of continued inflammatory challenge.

The monocyte is the largest of the freely circulating leukocytes, extending to perhaps 15 μm in diameter. It is distinguished from the neutrophil by its larger size and its single, centrally located large nucleus. Once in tissue, it becomes a macrophage or mononuclear phagocyte (MNP) (Figure 8.1). In addition to circulating monocytes, there is a resident population of monocytes in tissue which can also rapidly become macrophages and migrate to the site of injury. From whatever source, macrophages arrive at the site of inflammation after the neutrophilic invasion

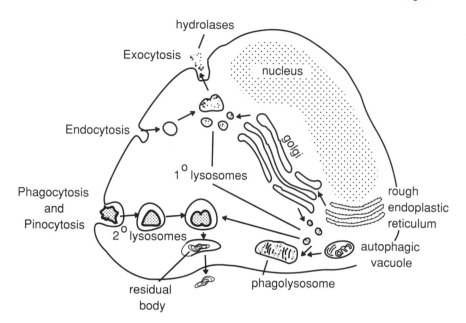

Figure 8.1 Phagocytic cell functions.

has begun to subside and concentrate in appreciably smaller numbers. Their role is similar to that of the neutrophil in that they can actively phagocytize materials and digest them. They also possess specialized membrane sites that mediate specific reactions. When activated by a suitable challenge, the macrophage possesses a number of activities, as seen in Figure 8.1.

The activated macrophage is not an end state cell; it has an aerobic glycolysis cycle and can undergo a number of transformations. In addition to releasing degradative enzymes (lysozymes) related to attempts to digest foreign materials, activated macrophages synthesize and release a wide range of biochemical factors which can mediate the activity of many other cells including lymphocytes, fibroblasts, osteoblasts, osteoclasts and foreign body giant cells (Ziats et al., 1988).

Macrophages may multiply by direct mitosis (Figure 8.2). By fusion, the macrophage is the progenitor of the next major cell type seen, the multinuclear *foreign body giant cell* (FBGC). This cell is up to 80 μm in diameter, is primarily found in foreign body or implant sites, and can be more aggressive and active than the neutrophil, eosinophil, or macrophage. Because of its larger size, it can phagocytize still larger particles. It has a relatively short life span that is on the order of days. Thus, the

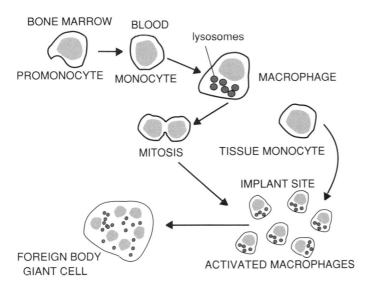

Figure 8.2 Development of the foreign body giant cell.

presence of multinuclear giant cells long after implantation would suggest the presence of a chronic FB reaction. It is possible that macrophage fusion to form multinuclear giant cells is specifically stimulated by foreign bodies (Mariano and Spector, 1974; Chambers, 1977).

The presence and activity of phagocytic cells is particularly related to the presence of "small" particles. The relationship between size and stimulus is not understood, except in a general way. Maximum stimulus seems to occur when the average particle size is in the 0.1–1.0 μm range. Larger particles, bigger than 50 μm, do not excite a reaction greater than bulk materials, unless they possess a dimension that is in the size range indicated (for instance, long, slender fibers). Large particles or even bulk implants may activate phagocytic cells resulting in external release of lysozymes and oxidative products. This process is termed *frustrated phagocytosis* and may prevent the phagocytes from performing their normal functions.

Although the chemical activity of phagocytozable particles does not seem to be responsible for their cellular stimulatory effect, the cellular response depends indirectly upon ease of degradation. If the particle is toxic to the ingesting cell or is expelled unchanged, the result may be massive accumulations of dying and dead neutrophils and macrophages. The resulting pus or caseation resembles that which accompanies massive bacterial infection. Phagocytosis, unsuccessful or successful, is a mechanism for moving foreign particulate material away from the implant sites. Cells and cell fragments pass into the local lymphatic drainage system and may accumulate in various lymph nodes in declining concentration as the distance from the implant site increases. Small particles may be cleared into the lungs via the lymphatics and either exhaled or swallowed (Styles and Wilson, 1976).

The progression described here, both in numbers and displacement of cell types, reflects the increased threat that the intrusive agent represents. In parallel with this progression, an invasion of blood-borne *lymphocytes* occurs. These cells, which normally pass directly through the endothelial lining and are taken up by the lymphatic processes, will pass through the lymphatic nodes and eventually return to arterial circulation. Their function in inflammation is not well known, and they are associated primarily with immunologic response (see Chapter 11). Their numbers do increase in areas of inflammation, and they are found associated with neutrophils and foreign particles in regional lymph nodes. It is possible that they respond to foreign proteins or to proteins denatured by foreign contact and have some ability to phagocytize particles. They may participate in fibroplasia, the final stage of inflammation.

8.2.4 Remodeling

Successful response to an inflammatory challenge will result in a locally decreased tissue mass. Dead cells have been phagocytized and removed by neutrophils and macrophages. Cells will be produced in situ by mitosis of the cell types present or by maturation of more primitive precursor cells.

This newly forming tissue is termed granulation tissue because of the pebbly specular appearance it has when seen growing at a free surface. The "pebbles" are vascular "buds," capillary and arteriole loops that grow rapidly out from stable tissue into the disturbed area. They "burrow" through, bringing the benefits of improved circulation and stimulating cellular activity.

In addition, large amounts of muccopolysaccharides and collagen will be synthesized. These materials, forming the familiar scar, form a scaffold for cellular reconstruction and remodeling of the damaged area. The mediators of this process (*fibroplasia*) are primarily fibroblasts, although studies in tissue culture suggest that a wide variety of cells can be stimulated to produce collagen and, to a lesser degree, muccopolysaccharides.

The remodeling of granulation tissue proceeds differently in different tissues. In skin, the reformation of tissue may be nearly complete, except for a possible absence of hair follicles and a small residual collagenous scar representing the collapsed remnant of the capsule which formed around the injury site. Bone has the ability to remodel completely to such a degree that it is said to regenerate. At the other extreme, articular cartilage never completely remodels but is repaired by formation of a loose tissue called *fibrocartilage*, which is inferior to normal cartilage and which deteriorates under repeated mechanical stress.

Remodeling is also the primary mechanism involved in tissue adaptation (see Chapter 10).

8.2.5 Capsule Formation

The maturation of the scar tissue (in soft tissue sites) marks the end of the process termed inflammation. The continuing presence of an implant prevents the collapse of the capsule which forms around the injured tissue and its maturation into a scar. The degree of scarring or capsule formation seen at later times depends upon the degree of original insult, the amount of subsequent cell death, and the location of the site.

Grading the host response at the implant site by a method such as that described in Appendix 2 in Chapter 17 actually evaluates the sum of two responses. These are the intrinsic inflammatory response to trauma and the host response to the implant. The response to trauma is relatively brief, passing through phases of insult, cellular proliferation, and reorganization in 1–2 weeks, as seen in the lower left of Figure 8.3. The host response to an implant is similar in its early phases but then enters a chronic phase (Figure 8.3). The time required to enter the chronic phase

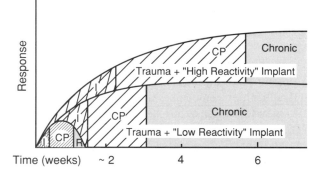

Key: I = insult, CP = cellular proliferation, R = reorganization.

Figure 8.3 Variations of local host response with implant "reactivity."

or in other words, achieve ''steady-state,'' varies with the degree of host response elicited. Thus, while it may be safe to evaluate a ''low-reactivity'' implant after 3–4 weeks, one may not gain a true picture of the host response to a ''high-reactivity'' implant unless studies are extended to six or even eight weeks after implantation.

Exploration of an implant site four or more weeks after implantation will usually reveal a relatively acellular fibrous capsule. This capsule is maintained by the continuing presence of the implant. In the absence of an implant (or after its removal or resorption) the capsule may collapse into a residual scar or be completely remodeled, as described above. Spindle-shaped fibroblasts are usually associated with the capsule, and small numbers of macrophages may be found in the vicinity. The presence of neutrophils suggests the possibility of a continuing inflammatory challenge, including infection. Detection of multinuclear (foreign body) giant cells (FBGCs) suggests the production of small particles by corrosion, depolymerization, dissolution or wear, and a continuing general tissue response. Observation of large numbers of lymphocytes suggests the possibility of a specific immune response (see Chapter 11).

The thickness of the fibrous capsule is related to several factors. Materials which are chemically active, such as metals which corrode freely or polymers with leachable constituents, will mediate formation of a capsule whose thickness is directly proportional to the rate (and thus, local concentration) of release of these small molecules. In addition to concentration, the chemical nature of the released material is important. It may be cytotoxic, inhibitory or merely neutral. The combination of these factors may lead to specific capsular morphology, at both a histological and ultrastructural level, which is characteristic of the particular material composition of the implant. These differences can only be fully appreciated when identical implants of different compositions are studied (McNamara and Williams, 1982).

Mechanical factors are also important in mediating capsule formation. Capsule thickness is presumed to increase with increased relative motion between implant and tissue. In extreme cases a fluid-filled bursa, mimicking a synovial capsule, may form; these bursae may be painful. The shape of the implant also affects the fibrous capsule thickness. The capsule will be thicker over edges and sharp changes in surface features. Thus, the capsule around a rectangular slab of reactive material will be dogbone or club-shaped. For this reason the phenomenon is called ''clubbing'' (Wood et al., 1970). (See Section 17.2.1 for a more complete discussion of this point.) Electrical currents, such as those emanating from an implanted stimulating electrode, also produce capsules whose thicknesses are related to current density. Since electrodes can also mediate changes in local pH and pO$_2$ as well as releasing corrosion products, effects due to direct electrical (*faradic*) and indirect electrochemical (*electrodic*) stimulation may easily be confused.

These general responses, in addition to depending upon tissue type, are species dependent, and in humans, may have an age dependence. Furthermore, identification of intrinsic tissue response to an implant may be complicated by the presence of an overlying acute or chronic infection. In Section 8.3 we will take up the relationship between implants and infection as a special case of the inflammatory response to implant materials.

8.2.6 Resolution

When injury occurs to tissue, the overall aim of the process initiated is to produce a return to the status quo ante, a reestablishment of homeostasis. The presence of an implant in an operative site must, of necessity, prevent attainment of the original condition but a steady state is possible in many cases. Reaching this condition is termed *resolution*.

Figure 8.4 shows an overall schematic of the progression of the local host response from insult through resolution in the presence of an im-

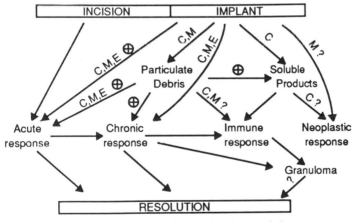

Key: C = chemical mediation, M = mechanical mediation, E = electrical mediation, ? = may be a mediating factor or may progress to resolution

⊕ = enhances effect present without implant

Figure 8.4   Schematic of development of local host response.

plant. The nature of the physical signals which mediate the local host response are still somewhat unclear; the symbols C, M and E are used here to reflect the most likely situation. It also seems probable that these signals interact in synergistic or antagonistic ways.

Four types of response are noted: acute, chronic, immune and neoplastic. The first two usually reach resolution, the third may lead to a *granuloma* (a benign tumorous condition due to continued tissue elaboration without progressive remodeling) while the fourth does not resolve and always represents a failure of the implant.

If resolution, that is, an achievement of a final state after which no more progressive biological changes occur, is a possibility, then a distinction may be made between four possible outcomes:

Extrusion: If the implant is in contact with epithelial tissue, the local host response will be the formation of a pocket or pouch continuous with the adjacent epithelial membrane. The process is termed *marsupialization* due to the similarity of the resulting structure to a kangaroo's pouch. In the case of the external epithelium, skin, this results in the effective externalization or extrusion of the implant from the host.

Resorption: If the implant is resorbable, then the implant site eventually resolves to a collapsed scar or, in the case of bone, may completely disappear.

Integration: In a very limited number of cases, such as the implantation of pure titanium in bone (Albrektsson et al., 1983), a close, possibly adhesive, approximation of nearly normal host tissue to the implant is possible, without an intervening capsule, although inflammatory cells may persist in small numbers.

Incapsulation: The most usual response as previously described. If the implant is placed in a location where bone may form, such as within a medullary space, and integration does not take place, the capsule may become mineralized, in which case the structure is called a *sequestrum*.

In addition, it is possible with the passage of time for granulomas to resolve, if the challenge presented by the implant changes or subsides.

Unlike the chronic granuloma and neoplastic transformation responses, whether each of these resolution outcomes represents success or failure of the implant depends on the circumstances, that is, upon the desired consequences of the insertion of the implant. This is the basic idea of biocompatibility, as discussed earlier: biological performance in a specific application that is judged suitable to that situation.

8.3 INFECTION

8.3.1 Types

A common phrase in both research and clinical situations is "infected implant." This is a misnomer since it is tissue surrounding the implant that is infected rather than the implant itself. *Infection* is specifically the invasion and multiplication of microorganisms in tissue. However, the misuse of the term serves to remind us of special problems that arise when bacterial infection is associated with an implant either in experimental animals or in humans. Implant site infections were a considerable clinical problem in the 1960s and 70s, with incidence rates approaching 5% in some series; now with a greater appreciation of the problem and better operating room techniques, rates are well below 1% for most permanent devices.

We distinguish three types of infection associated with implants. The first, the superficial immediate infection, is due to the growth of organisms on the skin (or near it) in association with an implant. Examples of this include suture infections and growth of microorganisms under burn dressings. These infections can usually be traced to bacteria residing normally on human skin, such as *Staphylococcus aureus* or *Staphylococcus epidermidis*, or to airborne bacteria that are trapped and cultured in the moist conditions at the superficial site.

The second and third types are, collectively, deep infections. The second, the deep immediate infection, is the usual type of low-frequency infection seen immediately after surgery. The bacteria responsible are usually skin dwelling types, generally *staphylococci*, carried into the implant site during the surgical procedure. Less commonly they are airborne types and, even less frequently, bacteria already present, but relatively inactive, in the operative site. This last source reflects the physical disruption of tissue during surgery, resulting in the release of material in cysts or sequestered sites into areas where bacteria experience better growth conditions. Infections of this type, while extremely infrequent, may arise in patients with a previous history of infection in the operative site, such as a septic joint, or a systemic chronic infection. The common external origin of deep immediate infections is underlined by the well-known positive correlation between rate of infection and length of the surgical procedure. Standard operating room environments contain 50–500 (cfu)/m^3 of bacteria (Antti-Poika et al., 1990). A cfu or *colony forming unit* is the minimum number of bacteria required to grow a cell cluster or colony on a suitable solid culture medium. Special precautions including air-tight gowns and hoods for surgeons, body exhaust systems for operating room personnel and control of air supply to cause laminar flow *away* from the patient can reduce the level of airborne bacteria to as low as 1–5 (cfu)/m^3, with a subsequent improvement in the incidence of clinical infection.

The last of the three types, the late infection, has the most cryptic origin. This infection, of the nonsuperficial or deep type, may occur months to years after surgery in sites with no prior history of infection. Its cause is generally believed to be the transport and "seeding" of blood-borne bacteria from an established infection at a remote site, such as a tooth-root abcess or a urinary tract infection. However, late infections which occur within three months of surgery are sometimes distinguished as *delayed* infections and ascribed to slow development of intra-operative bacterial contamination. Delayed and late infections represent a major problem in many procedures, such as total joint replacement or heart valve implantation. This is the case not because they are a frequent occurrence but because of the great difficulty encountered in treatment of the infection once it is established on and around the implant.

8.3.2   Use of Antibiotic-Impregnated Implants

The advent of modern bacteriostatic and bacteriocidal antibiotics has suggested the possibility of utilizing them to suppress or prevent infection associated with implants. An obvious approach would be to design a diffusion-controlled delivery system which could be made part of the clinical implant or implanted as an adjunctive device.

Since a common form of late infection is that associated with the use of polymethyl methacrylate cements in total joint replacement, it has been proposed by a number of investigators that appropriate antibiotics could be mixed with the cement at insertion. Release by dissolution or diffusion could then be depended upon to maintain an inhibitory antibiotic concentration in the vicinity of the joint replacement components and reduce the incidence of late infection.

This technique has been used clinically for some time but can be criticized on three grounds:

1.   It is difficult to be able to select the appropriate antibiotic since the late infection is produced by a range of bacteria with different sensitivities.
2.   The patient may develop an antibiotic sensitivity that would require the removal of the device to alleviate it.
3.   The antibiotic may not be able to penetrate effectively all of the infectible tissue around the implant.

Table 8.1 Influence of Antibiotic in PMMA Cement on Immediate and Late Infection in the Rat

Group[a]	Number[b] of legs	Antibiotic[c] (wt. %)	Results Infected (%)	Sterile (%)
A	60	—	42 (70)	18 (30)
	42	1.25	2 (5)	40 (95)
B	64	—	11 (17)	53 (83)
	12	1.25	1 (8)	11 (92)

[a]In group A the injection (2.5×10^7 staphylococcus aureus) was given $1/2$ hour after closure of leg wound and in Group B it was delayed for 6 weeks. Plugs of cement and adjacent bone were cultured after two weeks.
[b]Palacos^{TM} (Sulzer Bros., Bad Homburg, Germany).
[c]Gentamicin (0.5 g/40 g PMMA).
Source: Adapted from Elson et al. (1977).

Elson et al. (1977) performed an experiment that sheds some light on these concepts. Defects were created in rat tibiae, plugs of cement with and without antibiotic impregnation were inserted and allowed to cure in place before a bacterial innoculum was injected systemically, either within 1/2 hr of surgery (group A) or after a 6-week delay (group B) (Table 8.1).

The antibiotic-impregnated cement was highly effective ($X^2 = 48.55$, $p < 0.001$) in reducing immediate postoperative infection, but not to a statistically significantly extent by 6 weeks ($X^2 = 0.185$, n.s.). The latter comparison is complicated by the low incidence of infection in the animals receiving unimpregnated cement and the small size of the antibiotic-impregnated cement group at 6 weeks. Thus, the conclusion seems to be that the technique is effective if infection by a sensitive organism occurs immediately after the operation, but the effect on late infections is doubtful. This latter point is no surprise, as the diffusion release of antibiotic by cement must fall rapidly with time.

Bone cement, cured in place, seems especially to permit local infection but all implanted materials increase the risk of infection to some degree (Petty et al., 1985). The efficacy of antibiotic impregnation of bone cement in reducing or preventing immediate infection was verified in a more severe canine model using a direct infusion of *Staphylococcus aureus* into a bony site just before cement insertion (Petty et al., 1988). This principle has been extended to the use of pre-cured antibiotic impregnated PMMA ''beads'' as temporary implants to treat deep seated infections (Blaha et al., 1990). However, antibiotic impregnated cement has clearly been seen to be a failure in preventing late infection in more realistic models such as a clinically functional total knee replacement in the rabbit subjected to a systemic bacteremia 6–8 weeks after implantation (Blomgren, 1981).

It is this last point that attracts our attention to the question of associations between implants and infection. Bacterial growth in the presence of implants is tenacious and difficult to deal with. Some bacteria undergo a transformation of behavior and increase their activity and pathogenicity. Conventional treatments for bacterial infections that may be successful in nonsurgical cases often have reduced efficacy or fail in the presence of implants.

The difficulty of treatment often leads to the removal of the implant and the need to delay reimplantation until after the infection is successfully suppressed. This is possible in some cases, as total hip replacement, but not in others, as mitral (heart) valve replacement. Thus, "infection" constitutes a major source of implant "failure," despite the inability of an implant to become infected. A final concern is that the chronic use of antibiotics, in both treatment of patients and in decontamination of hospital facilities, has lead to the emergence of ever more resistant and virulent strains of micro-organisms. Thus, while the incidence of implant infection can be expected to continue to fall, the problems raised by each case may well become more difficult.

8.3.3 Geometric Factors

Geometric factors are extremely important to the consideration of interactions between implants and infecting micro-organisms. Initiation (*seeding* or *colonization*) and propagation of infection constitute a competitive process pitting the ability of the invading bacteria to replicate against the ability of the host tissues and support cells to annihilate. This competition defines the size of the cfu for each micro-organism. Therefore, the exposure of the bacteria to tissue and, especially, vascular processes is a critical factor in successful host defense against infection.

As we have seen, tissue about an implant is relatively acellular at maturity. Bacteria that can grow within the capsular membrane will encounter little active opposition until they break out into the surrounding tissues. In addition to immune responses, both general and specific, the immediate response to infection is the inflammatory response. The cells that mediate the response are blood borne. Thus, the reduction of the solid angle of surrounding tissue near an implant takes on extreme importance; it reduces the accessibility of the infected tissue near an implant to neutrophils and macrophages.

The existence of a "dead space," a volume filled with cell-free fluid rather than tissue, represents a special hazard of the geometric type. Design and implantation practices try to avoid such features since this fluid can act as an in vivo culture medium for bacteria.

Even in the absence of an acellular membrane and/or a dead space, some bacteria can produce their own geometric defense by forming a slimy coating, termed the *glycocalyx*, about themselves on the surface of the implant (Gristina and Costerton, 1984). This coating prevents host cells from approaching and phagocytizing the bacteria and reduces local effectiveness of antibiotics by affording a diffusion barrier.

While porous bodies represent no particular structural hazard when viewed from the aspect of carcinogenesis (see Chapter 12), they might be expected to present a geometric hazard in the initiation and/or support of infection. Kiechel et al. (1977) have studied the influence of porosity in polymethyl methacrylate implants in rabbits on the propagation of an aerobic infection by *Staphylococcus aureus* and its treatment by penicillin. Using a muscle site (corresponding to an immediate deep infection), they concluded that there was no effect on either the inoculum size needed to initiate infection or on the response of established infection to penicillin. Whether the outcome would have been different in the case of a late infection or an infection due to an anaerobic bacterium cannot be determined from this experiment.

A somewhat more sophisticated repetition of this study (Merritt et al., 1979) reached a somewhat different set of conclusions. Using a variety of porous and dense implant materials, but the same micro-organism, and comparing the effects of immediate (at implantation) injection of an inoculum to an inoculation after a 1-month delay, this later study concluded that porosity had an effect on infection rates. In the immediate injection (acute) group, the infection rate for porous materials was higher, while in the 1-month delay (chronic) group, the infection rate for dense materials was higher. The authors suggested that these findings support the hypothesis that bacteria can evade host defense mechanisms if they enter the pores of an implant before tissue invasion; but once the implant is invaded by host tissue, the bacteria are more exposed to cellular attack.

However, these conclusions are drawn from a comparison of porous polyethylene and dense alumina ceramic and devolve from relatively small experimental groups. Furthermore, the hypothesis, as the authors advance it, can explain the acute results but not the chronic results. Thus, the question is still open to debate. A recent study of the outcome of porous percutaneous carbon implants in rabbits for up to four years (Krouskop et al., 1988) further supports the view that implant porosity, if not immediately colonized by bacteria, represents no long term hazard for infection.

The clinical practice of removing the "infected implant" is based most directly on geometric arguments and on their physiological consequences. Two other types of arguments can be proposed: one based upon chemically mediated interactions between the implant and the bacterial infection and the other based upon chemical effects on the cell populations in the host that act against infection. Both of these are controversial today. I will discuss some of the more recent evidence of these effects.

It is thought that if implants can provide metals valuable to metabolism (see Section 13.1) through the corrosion of metallic parts, then they can also be shown to participate in bacterial metabolism. Thus, the view is that metallic implantation is similar in effect to putting fertilizer on a garden. However, one metal, iron, occupies a special position. Iron levels in the body are evidently subject to active control and change rapidly in response to infection (see Chapter 13). It has been suggested that the interference of endogenous iron sources with this control system plays a role in the propagation of certain infections (Weinberg, 1974).

8.4 EFFECTS OF IMPLANTS ON HOST DEFENSE SYSTEMS

8.4.1 Effects on Phagocytosis

We come then to a consideration of the effects of implants, primarily metals, on the cellular elements of the host defense system. There are many possible mechanisms for such effects but we will focus on one: alteration of phagocytic function. Phagocytic cells play key roles in inflammatory response, as previously discussed. We will consider several studies that explore various aspects of reduction of phagocytic function through a variety of animal models.

Graham et al. (1975) were interested in determining if the inhalation of metallic particles from industrial pollution would either reduce the number or the activity of lung macrophages. They used cultured rabbit alveolar macrophages, which were challenged by a 1-day incubation in the presence of metallic salts of various types, and examined the ability of the survivors to phagocytize 1.1-μm-diameter latex spheres. For the metals of interest in implant applications, they obtained the data shown in Table 8.2.

Each of the metal ions in Table 8.2 exhibited cytotoxic effects at the tested concentrations. All but vanadate also produced a reduction in phagocytotic behavior, generally independent of the cytotoxic effect. In

Table 8.2 Effects of Trace Metals (as Ions) on Alveolar Macrophages

	VO_3^-	Cr^{3+}	Mn^{3+}	Ni^{2+}	Ni^{2+}	Ni^{2+}
Ion concentration						
(M)	7×10^{-5}	3×10^{-3}	2×10^{-3}	5×10^{-4}	8×10^{-4}	1×10^{-3}
(ppm)	6.9	156	110	29	47	59
Phagocytic index (%)	100	70	75	50	20	13
Cell survival (%)	79	84	69	92	85	76
Phagocytic efficiency (%)	79	59	52	46	17	10

Note: The phagocytic index is the percentage of living cells that contain one or more latex spheres after a 1-hour exposure at a ratio of 120 spheres/cell. Phagocytic efficiency = phagocytic index × cell survival.
Source: Original data from Graham et al. (1975).

the case of nickel, where three concentrations were employed, the sign of both effects with the same, but the reduction of phagocytosis was more extreme. Note that the true in vivo effect would depend upon the product of the cell survival index and the phagocytic index, here called the *phagocytic efficiency*. Thus, for nickel, an increase of concentration by a factor of ~2 (from 29 to 59 ppm) produces a depression of macrophage effect of 78%.

The authors point out that the ion concentrations used were one to two orders of magnitude *below* those found in human lung tissue obtained in highly polluted industrial environments. However, this comment is based upon comparison to *average* tissue concentrations; in Chapter 14 it will be seen that concentrations immediately adjacent to corroding implants may be 2–3 orders of magnitude higher than average values. Thus the concentrations used by Graham et al. are relevant. They also report a study showing a positive correlation between metal aerosol inhalation in mice and susceptibility to infection, as well as studies in human populations showing positive correlations between metal dust inhalation and respiratory infection rate.

Rae (1975) has pursued this issue further. He incubated mouse macrophages with finely divided implant alloys and pure metals and examined the activity of two enzyme systems:

1. Lactic dehydrogenase (LDH). This is an enzyme normally contained wholly within cells. Thus, its detection is a measure of the leakiness of, or damage to, cell membranes.
2. Glucose-6-phosphate dehydrogenase (G6PD). This enzyme is required for synthetic processes and must be present for phagocytosis to take place. Thus, its reduction signals a reduction in the phagocytotic ability of cells.

Both cobalt and nickel particles were found to be cytotoxic. At the concentrations obtained in this study, no cytotoxic effect was observed for chromium or molybdenum. Nickel, cobalt, and cobalt-chromium alloy particles were found to significantly reduce the phagocytic ability of the surviving macrophages (as measured by G6PD activity).

The mechanisms of these two effects, cell death and reduction of phagocytotic ability, are unclear. In particular, they interacted in Rae's experiment, and an effect of particle size was displayed; smaller particles were at the same time more frequently phagocytosed and more toxic.

A study by Ward et al. (1975) suggests that a wide variety of effects exist that are not closely related. This view supports the results of the previous two studies. The authors studied a wide variety of metallic salts and their effects on neutrophil chemotaxis and on the incorporation of

Table 8.3 Chemotaxis Inhibition by Metallic Ions

Ion concentration for 50% inhibition of chemotaxis	Cr^{+3}	Mn^{+3}	$Fe^{+2,+3}$	W^{+4}	Mo^{+4}	Co^{+2}
M	3×10^{-4}	5×10^{-3}	10^{-3}	10^{-3}	10^{-3}	10^{-3}
ppm	16	27	56	18	96	59

Source: Data extracted from Ward et al. (1975).

amino acids into both neutrophils and fibroblasts. Rabbit neutrophils, HeLa (a human cancer cell line), and human gingival fibroblasts were used. An *Escherichia coli* filtrate was used as a neutrophil challenge to study chemotaxis by neutrophils. While incorporation inhibition (a reduction in phagocytic efficiency) was shown for a variety of metal salts, we are more interested here in the antichemotaxis results. An extract of the results of these studies is given in Table 8.3.

The authors discuss these chemical agents primarily as anti-inflammatory agents. It is interesting to note that these data do not support either of the two previous studies. This disagreement stems from several origins:

1. Chemotaxis is only one factor in phagocytosis.
2. Rae's study associated a small effect (reduction in phagocytic efficiency) of chromium and molybdenum with low solubility of metallic particles; this study bypassed the problem by using soluble salts.
3. Graham's study utilized a passive challenge to the macrophage rather than one of biological origin.

Remes and Williams (1990), using a bacterially-derived chemoattractant with human neutrophils, concluded that Cr^{+3} (2.5 ppm) (56% inhibition), Co^{+2} (≥ 20 ppm) (100%) and Ni^{+2} (≥ 30 ppm) (50%) all actively interfered with chemotaxis. The greater apparent sensitivity to chromium and cobalt than seen in Graham's study may reflect differences in experimental technique, challenge agent and/or cell source and underline the difficulties inherent in such studies.

8.4.2 A Final Comment

In the first edition of this work, I emphasized that these effects of apparent chemical interactions between implant and bacteria and implant and host cells were highly speculative and based almost completely on in vitro studies. Thus, they lack all humeral, metabolic, catabolic and excretory influences seen in whole animal experiments. I further pointed out that many critical experiments remained to be done before any of these effects could be either established or disproved. Unfortunately, the situation remains much the same today, if not more confused due to the growing awareness of the role of physical and electrical mediating factors which may act separately or in concert.

REFERENCES

Albrektsson, T., Brånnemark, P.-I., Hansson, H.-A., Kasemo, B., Larsson, K., Lundström, I., McQueen, D. H. and Skalak, R. (1983): Ann. Biomed. Eng. 11:1.

Anderson, J. M. and Miller, K. M. (1984): Biomats. 5:5.

Antti-Poika, I., Josefsson, G., Konttinen, Y., Lidgren, L., Santavirta, S. and Sanzén, L. (1990): Acta Orthop. Scand. 61:163.

Blaha, J. D., Nelson, C. L., Frevert, L. F., Henry, S. L., Seligson, D., Esterhai, J. L., Jr., Heppenstal, R. B., Calhoun, J., Cobos, J. and Mader, J. (1990): Instruct. Course Lect. 39:509.

Blomgren, G. (1981): Acta Orthop. Scand. 52 (Suppl. 187):1.

Chambers, T. J. (1977): J. Pathol. 122:71.

Elson, R. A., Jephcott, A. E., McGechie, D. B., and Verettas, D. (1977): J. Bone Joint Surg. 59B:452.

Frank, M. M. (1979): Rev. Infect. Dis. 1:483.

Graham, J. A., Gardner, D. E., Waters, M. D. and Coffin, D. L. (1975): Infect. Immun. 11:1278.

Gristina, A. G. and Costerton, J. W. (1984): Orth. Clin. N. Amer. 15:517.

Kiechel, S. F., Rodeheaver, G. T., Klawitter, J. J., Edgerton, M. T. and Edlich, R. F. (1977): Surg. Gynecol. Obstet. 144:58.

Krouskop, T. A., Brown, H. D., Gray, K., Shively, J., Romovacek, G. R., Spira, M. and Runyan, R. S. (1988): Biomats. 9:398.

Mariano, M. and Spector, W. G. (1974): J. Path. 113:1.

McNamara, A. and Williams, D. F. (1982): Biomats. 3:160.

Merritt, K., Shafer, J. W. and Brown, S. A. (1979): J. Biomed. Mater. Res. 13:101.

Petty, W., Spanier, S., Shuster, J. J. and Silverthorne, C. (1985): J. Bone Jt. Surg. 67A:1236.

Petty, W., Spanier, S. and Shuster, J. J. (1988): J. Bone Jt. Surg. 70A:536.

Rae, T. (1975): J. Bone Joint Surg. 57B:444.

Remes, A. and Williams, D. F. (1990): J. Mater. Sci.: Mater. in Med. 1:26.

Styles, J. A. and Wilson, J. (1976): Ann. Occup. Hyg. 19:63.

Ward, P. A., Goldschmidt, P. and Greene, N. D. (1975): J. Reticuloendothel. Soc. 18:313.

Weinberg, E. D. (1974): Science 184:952.

Wood, N. K., Kaminski, E. J. and Oglesby, R. J. (1970): J. Biomed. Mater. Res. 4:1.

Ziats, N. P., Miller, K. M. and Anderson, J. M. (1988): Biomats. 9:5.

BIBLIOGRAPHY

Bisno, A. L. and Waldvogel, F. A. (1989): *Infections Associated with Indwelling Medical Devices*. Amer. Soc. Microbiol., Washington, DC.

Coleman, D. L., King, R. N. and Andrade, J. D. (1974): J. Biomed. Mater. Res. 8:199.

Leibovich, S. J. and Ross, R. (1975): Am. J. Pathol. 78:71.

Rae, T. (1981): In: *Fundamental Aspects of Biocompatibility*, Vol. 1. D. F. Williams (Ed.). CRC Press, Boca Raton, FL, pp. 159ff.

Rae, T. (1981): In: *Fundamental Aspects of Biocompatibility*, Vol. 2. CRC Press, Boca Raton, FL, pp. 139ff.

Spector, W. G. and Wynne, K. M. (1976): Agents & Actions 6:123.

Trowbridge, H. O. and Emling, R. C. (1983): *Inflammation: A Review of the Process*, second edition. Comsource/Distribution Systems, Bristol, PA.

Van Furth, R. (Ed.) (1970): *Mononuclear Phagocytes*. Blackwell Scientific, Oxford.

Williams, D. F. (1987): J. Mater. Sci. 22:3421.

9

Coagulation and Hemolysis

9.1 INTRODUCTION

In the previous chapter, inflammation was presented as a nonspecific response to tissue damage. In this chapter, we will consider responses to two more specific events involving the circulatory system and its tributaries:

1. Damage to blood carrying vessels and/or contact with foreign materials leading to *coagulation*.
2. Damage to the tissue of blood, leading most generally to cellular destruction or *hemolysis*.

9.2 THE COAGULATION CASCADE

9.2.1 Intrinsic Pathway

The coagulation cascade may be throught of as a biological amplifier which permits an initiating event to be magnified into a process sufficiently widespread as to produce hemostasis and permit repair of the damage to the vascular system. The overall process is referred to as coagulation or *thrombosis* while the resulting hemostatic plug is termed a *thrombus*. If damage to the wall of a blood vessel (*endothelium*) is the initiation factor, then coagulation proceeds by an *intrinsic* pathway.

In this case, the first two events closely parallel those of inflammation. In fact, they are essentially those of the inflammatory process. Initially, the smooth muscle in the vessel wall dilates and then constricts, probably stimulated by activated Factor XII (also called Hageman factor) (see Figure 9.1). Although this is formally the first step in the intrinsic pathway, it is preceded by surface contact (and denaturation) of other molecules, including kininogen and prekallikrein. Endothelial permeation also increases but is masked by the release of serum and blood-borne cells if the vessel wall is ruptured. In addition, the endothelial lining becomes sticky.

The combination of the presence of activated Factor XII and the sticky quality of the endothelial lining triggers the first step unique to coagulation, the adhesion and aggregation of platelets. Adhered platelets rapidly lyse (undergo membrane rupture), through mechanical and biochemical paths, releasing adenosine diphosphate (ADP), serotonin, and epinephrine. ADP encourages further platelet adhesion, while the latter two agents cause vasoconstriction. The combination of these agents rapidly produces a platelet "plug" that serves to staunch further blood loss from the area.

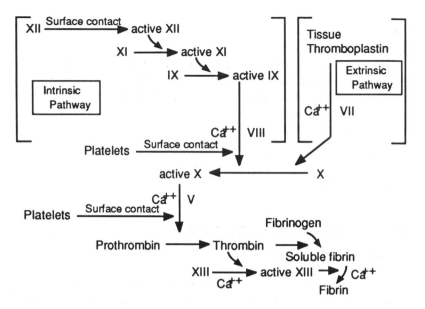

Figure 9.1 The coagulation cascade (adapted from Salzman, 1972).

Activation of Factor XII and platelet adhesion trigger a complex chain of events leading to transformation of inactive fibrinogen into an active molecule, fibrin. This transformation, accompanied by a shrinking or retraction of the platelet plug, produces a mature clot, a mesh of polymerized fibrin trapping leukocytes, erythrocytes, and platelet fragments. Within minutes to hours this clot is invaded by the same series of cells seen in pure inflammation and then later is invaded by numerous new capillaries. Eventually, the clot is removed and replaced by fibrous scar and remodeled tissue.

9.2.2 Extrinsic Pathway

The intrinsic pathway to coagulation depends upon interaction of normal blood components (macromolecules, cells, platelets, etc.) after alteration by an initiation event, primarily surface contact. There is an alternate initial coagulation pathway, involving release of materials from cells external to the vascular processes, the *extrinsic* pathway. The released material is termed *tissue thromboplastin*. This is a protein-phospholipid complex derived from normal cell contents (otherwise termed factor III) which, with factor VII and Ca^{++}, activates factor X (Figure 9.1). The

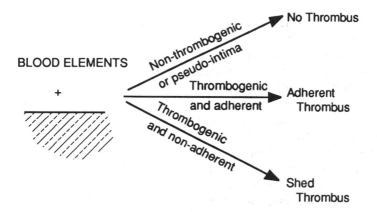

Figure 9.2 Interaction of implants with blood elements.

intrinsic and extrinsic coagulation pathways merge in this common event leading to the terminal event of fibrin production and clot formation.

9.2.3 Implant Induced Coagulation

The events of the intrinsic pathway are triggered, as previously stated, by contact between blood elements and a surface. The normal endothelial lining of blood vessels is not able, fortunately, to induce this response. Damage to the endothelium may expose collagen, its major structural molecule. Normally collagen is rendered nonthrombotic (not able to induce coagulation) by a covering of a family of macromolecules, glycopolysaccharides. This coating is electropositive, which attracts a layer of neutralizing negative ions and thus repels the negatively charged erythrocytes and platelets. Exposed collagen is electronegative, and thus highly thrombotic. This property is utilized in surgery when powdered crystalline collagen is used as a topical hemostatic agent on tissues where cauterization and/or ligation are not possible, for example, the liver.

The intrinsic pathway may also be triggered by contact with a foreign body* (an implant) and the resulting events are quite similar. In this case, however, the implant persists after the initial insult. This difference from the case of isolated vessel wall damage has several consequences (Figure 9.2):

---

*We speak here of contact with a ''foreign body;'' however, the surface of an implant so rapidly becomes coated with both serum proteins and coagulation factors that it is better to think of the foreign body contact effect as being mediated by surface adhered and denatured organic molecules.

1. Coating by blood-borne proteins retaining native structure may not permit activation of Factor XII or adhesion of platelets. Such a surface would have a high degree of blood compatibility.
2. Platelet adhesion may not be accompanied by sufficiently rapid lysis to promote further progression of the cascade. Such a surface would not actively form a thrombus. However, it would deplete the circulating blood of platelets and, thus, might reduce the coagulatibility of the host. Such an effect is common in chronic (repeated) hemodialysis, or intraoperatively when an ex-vivo blood oxygenator is used.
3. A thrombus may form but be rapidly removed by blood dynamic forces. Such an implant will "shed" emboli and cause damage by infarction at a remote site. This effect is utilized in the Kusserow test for blood compatibility of materials (see Section 17.2.3).
4. Remodeling will not remove the obstruction but will tend to encapsulate it, as in the case of the fibrous encapsulation found around implants in soft tissue. If the implant surface is structured (such as felt or velour) to encourage cellular trapping, the surface exposed to the blood may come to function as the endothelial wall of a normal vessel and is termed "pseudo-intima" (see Section 10.3.2). It differs, of course, in having no smooth muscle component and, thus, no ability to dilate or constrict.

These effects have been summarized by Baier (1972) (Figure 9.3). In this figure, Baier also indicates possible points and methods of intervention which may reduce the thrombogenic potential of implant surfaces.

In the conventional (implant-free) progression of the intrinsic pathway, the fibrin clot gradually isolates flowing blood from the site of possible surface contact insult. However, on a foreign (implant) surface, it is possible for fibrin by itself to initiate coagulation, due to its configuration after surface binding (Lindon et al., 1986). Implants may also initiate coagulation by binding (and possible activation) of molecules such as complement C3 which are not normally involved in the intrinsic pathway (Herzlinger et al., 1981).

9.3 APPROACHES TO THROMBORESISTANT MATERIALS DEVELOPMENT

9.3.1 General Considerations

In the same article from which Figure 9.3 is drawn (Baier, 1972), the various approaches that are taken to design new surfaces or to render surfaces of older materials less thrombogenic were summarized (Figure

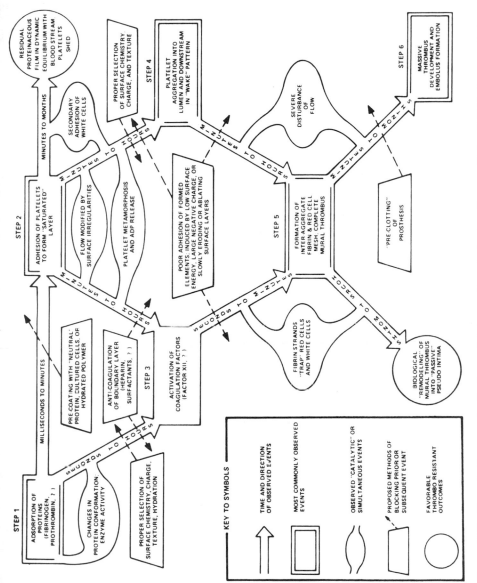

Figure 9.3 Coagulation events, timing and intervention strategies. Source: Baier (1972), by permission.

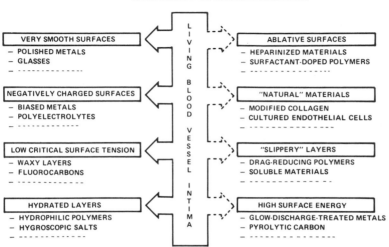

Figure 9.4 Approaches to producing thromboresistant surfaces. Source: Baier (1972), by permission.

9.4). The approaches shown by the solid-line arrows pointing to the left are those that mimic properties that the natural intact endothelium is known to have. Those shown by the dotted lines pointing to the right are other proposals not based so directly upon known natural properties but, rather, on theoretical approaches. All have met with mixed success and failure. With the limited exception of knitted polyester grafts, felted polyurethane, poly(tetrafluoro)ethylene and related materials that can sustain "pseudo-intimal" linings, certain bulk polymers, and graphites, most materials evoke unacceptable host responses for chronic exposure to blood.

It is of great interest that, although Figure 9.4 is now twenty years old, there are no really new approaches to solution of the so-called "blood contact" problem. The only addition which could be made to this schematic, other than subtopics under each of the eight main headings, would be the addition of viable biomaterials as a fifth mimicking approach (left side). This may be distinguished from the "natural" materials approach, which is specifically the modification of permanent implantable materials with natural molecules or a surface dwelling cell population. In the newer approach, a composite of resorbable materials and living cells (smooth muscle cells, endothelial cells, etc.) is made and implanted with the hope that it will eventually remodel into native (host) tissue (Weinberg and Bell, 1986).

The lack of new approaches to producing noncoagulating biomaterial surfaces reflects the complexity of the natural system and the difficulty in performing reproducible experiments with generality of results. This complexity has aroused considerable interest in finding simplifying or dominant factors in surface properties that affect thrombogenesis. Several of the directions of biomaterials efforts indicated in Figure 9.4 reflect such a search and are of both historical and practical interest. We shall consider three of these: negative surface charge, critical surface tension, and "natural" surfaces. Each has yielded important results and contributed to a better understanding of thrombogenesis, but each in turn has failed to produce the final answer in suppressing host response to materials in contact with blood.

9.3.2 Negative Surface Charge

The recognition that both erythrocytes and platelets have net negative surface charges has suggested to many investigators that electrostatic repulsion could be utilized to keep them away from implant surfaces, thus suppressing thrombogenesis. Sawyer and Srinivasan (1972) have been the most active proponents of this idea. A series of experiments suggests that a relative positive potential on a biomaterial surface exposed to blood promotes thrombogenesis, while negative potentials tend to suppress thrombogenesis, proportional to the potential depression below local neutral. These conclusions arise from a complex series of experiments. Perhaps the most convincing data are given in Figure 9.5 (Sawyer and Srinivasan, 1972). Here we see the results of implantation of Gott rings in the canine vena cava (see Section 17.2.3) and a related design (Edwards), made from various materials, in the canine aorta. Metals with a negative electromotive potential form a positive interfacial potential with blood (due to the attraction of counterions, as previously noted) and vice versa. Here we can see the rapid and complete occlusion (coagulation sufficient to prevent blood flow) for silver and platinum forming positive interface potentials, and the increasing patency (proportion of devices permitting flow) at increasing times for metals such as iron (in stainless steel), aluminum, and magnesium that form negative interface potentials.

The approach is, however, quite limited. The use of rigid metallic surfaces for vascular prostheses might prove nonthrombogenic by surface interaction but might produce a pseudo-extrinsic thrombogenesis through mechanical damage to blood borne cells. Furthermore, control of interfacial potential might require an active power source, thus making the implant far more complex and less reliable than the knitted prostheses discussed in Section 10.3.2.

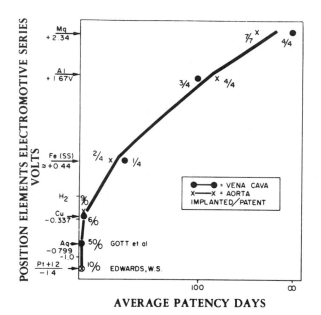

AVERAGE PATENCY DAYS

Figure 9.5 Relationship between interface potential of vascular implants and patency. Source: Sawyer et al. (1978).

9.3.3 Critical Surface Tension

The critical surface tension approach was proposed some time ago and has been strongly advanced by Baier (1972) as a significant but not necessarily dominant solution to the problem of surface thrombogenesis. The basic idea is to utilize the Young-Dupree equation [Section 5.4; Eq. (5.4)]:

$$\gamma_{SL} = \gamma_{PS} + \gamma_{PL} \cos \Theta \tag{9.1}$$

If there is some combination of surface tensions such that $\theta = 180°$, then no adhesion of the particle p may occur. Then, of necessity, cosine θ must equal -1. This condition can be achieved if:

$$\gamma_{SL} = \gamma_c = \gamma_{PS} - \gamma_{PL} \tag{9.2}$$

(γ_c = critical surface tension)

Since γ_c is determined by the biological system, the trick is to find a surface with a suitable surface tension such that Eq. (9.2) is satisfied; then no adhesion would occur for a particular molecular or cellular species and one or more of the surface contact steps in Figure 9.1 could be prevented. This can be done at two points in the coagulation cascade:

1. The surface may have the appropriate critical surface tension to prevent adhesion of Factor XII.
2. The surface may have the appropriate critical surface tension to prevent adhesion of platelets in either the intrinsic or the final (common) pathway.

This turns out to be far more difficult in practice than in principle. For instance, a surface might be selected that does not permit adherence of Factor XII. However, other molecular species may adhere, γ_{SL} may change, and adhesion of factor XII may then become possible. Measurements of surface tension of nonbiological surfaces, such as silicon, exposed in vitro or ex vivo to sera or whole blood suggest complex time-dependent changes in γ_{SL}.

Notwithstanding these practical problems, there is a theoretical range of material-blood surface tension that should suppress thrombogenesis (Figure 9.6). Studies of a wide variety of materials have provided evidence of relative thromboresistance in this region.

9.3.4 "Natural" Surfaces

The thromboresistance of the undamaged internal surfaces of blood vessels has attracted both the surgeon and the bioengineer for a long time. Direct transplants (*heterografts*) and implants of animal material (*xenografts*) are limited in utility by host rejection of the implant through an immune response (see Section 11.1) and by biological degradation of the foreign material. Various methods of processing have been tried to reduce the immune response and to improve resistance to degradation. There are a number of processes in use which involve cleaning the tissue, removal of cellular debris, and crosslinking the collagen component ("tanning") by a variety of agents such as glutaraldehyde, formaldehyde, etc. While human material has been used, the pig is a favorite donor. Kiraly and Nosé (1974) have summarized some of the applications of such materials.

These materials seem to have considerable degrees of thromboresistance. However, they are not incorporated into the body in the same way that knitted grafts are (Nosé et al., 1977). Thus, a pseudo-intima does not form and the surfaces exposed to blood are slow to mature. While cells are found on their surfaces, the same low surface tensions that suppress thrombogenesis seem to retard cellular adhesion. Additionally, these

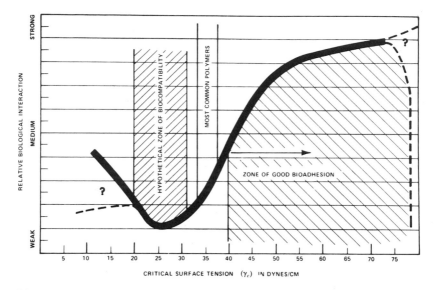

Figure 9.6 Proposed relationship of critical surface tension (γ_c) to biological response. Source: Baier (1972) by permission.

processed materials are relatively impervious to diffusion of fluids and tend to develop late calcification, similar to that which occurs naturally in arteriosclerosis. However, the approach remains interesting and bears considerable promise for the future, perhaps utilizing cultured (grown in vitro) tissue. The "natural" approach is an obvious precursor to the more recent ideas involving viable biomaterials for blood contacting surfaces (see above).

9.3.5   An Overview of Thromboresistant Materials Development

It is difficult to consider in further detail the efforts at bulk and surface modification that have been made to render implant surfaces more friendly to blood. In general, the experiments with bulk materials are straightforward, but their interpretation depends greatly on the validity of the blood exposure test(s) used to evaluate them. We shall come back to this point in Chapters 16 and 17.

I believe that the experiments in surface modification and coating are open to broad, general criticism. In the first place, there is rarely very good characterization of the bulk materials used. One may ask what are the actual (rather than calculated) structural and energetic properties of the surfaces of these materials, and how do these properties affect the resulting surface exposed to cells?

Furthermore, there is rarely good evidence that the surface treatments or coatings are either homogeneous or cover the surface completely. The "catalytic" nature of coagulation suggests that the presence of occasional high-energy defects may serve as foci for initiation of thrombogenesis and may be far more important than the anti- or nonthrombogenic properties of the balance of the surface.

As pointed out previously, it can be generally assumed that implant surfaces are rapidly protein coated after insertion and that the proteins are denatured in some degree. I have suggested (Section 5.4) that if a normally free protein adheres to a surface, it is by definition denatured. Whether this is mild and reversible (3° or 4°), moderate (2°), or severe and essentially irreversible (1°) depends upon the nature of the protein-surface interaction forces (see Chapter 5 for a more complete discussion). Thus, the central issue concerning adhesion of proteins to surfaces is whether they are uniformly distributed or associated with surface defects, and whether the type of denaturation which occurs is that which will evoke a specific cellular response, in this case, the initiation of coagulation.

Finally, most studies neglect to measure the important physiochemical properties of the actual material/surface/protein complex during and/or after blood contact and instead rely on secondary determinations or theoretical considerations.

It seems most likely that, in the absence of surface bound molecules for which target cells have specific surface receptors, cells involved in coagulation are affected by the following properties of biomaterials:

1.   Potential gradients and associated ion fluxes near surfaces.
2.   Features of surface geometry with dimensions between 0.1 and 5 γm.
3.   Actual surface/solution interfacial free energy.
4.   Stresses at impact (governed by 1, 2, and 3 above as well as surface hardness and conditions of the experiment or clinical exposure).
5.   Presence of specific sufficiently denatured proteins to evoke a biological response.

9.4   HEMOLYSIS

9.4.1   General Description

Foreign materials may trigger thrombosis by contact with blood in the absence of motion at the blood-surface interface. However, such motion may, in itself, trigger thrombus formation or may cause damage to blood cells in the absence of thrombosis. Such damage, resulting in cell death and release of cell contents, is termed *hemolysis*. While it can be detected by a reduction in white cell count or by the presence of cell "ghosts" (fragments of empty cell membranes), it is most usually followed by measuring the level of serum hemoglobin. Presence of hemoglobin in blood serum is a direct result of erythrocyte lysis. There is a normal level of serum hemoglobin (in humans) of 0–1.1 g/liter; this is on average about 0.4% of all hemoglobin in the blood. Serum hemoglobin is normally bound to a carrier molecule, haptoglobin; however, the maximum capacity of this fraction in normal serum is 1.4 g/liter. Hemolysis rates above the release due to cell death in normal turnover will raise the level of serum hemoglobin. Modest increases lead to increases in excretion and may cause anemia over long periods. Greater elevations (25 g/liter and higher) will cause systemic clinical symptoms including cyanosis and hematuria, while still higher levels may lead to kidney failure and toxemia.

Hemolysis may occur in freely flowing blood in the absence of foreign surfaces. Turbulent flow and shear stresses above 1500–3000 dyn/cm² will cause direct lysis. If stagnation points exist, as in some device designs, this lysis may lead directly to thrombus formation in the apparent absence of surface contact. Of course, in this case, the initiating surface is damaged cell membrane exposed by cell lysis.

9.4.2 Experimental Relation to Flow Velocity

However, flow effects can be seen in the contact between blood and foreign surfaces at much lower ranges of shear stress. A large number of studies have demonstrated this effect. One of the best is that of Lampert and Williams (1972). These investigations studied hemolysis rates with respect to a standard material, aluminum, at a variety of flow rates. Their apparatus consisted of a fixed disc with an opposed rotating disc at a fixed separation distance, both on a common axis (see Figure 9.7 for a schematic representation). The interdisc separation (H) and the speed of relative rotation (Ω) could be varied.

The Reynolds numbers (R_e) for this apparatus were calculated as shown below.

With respect to separation (H) at the edge of the disc (max r):

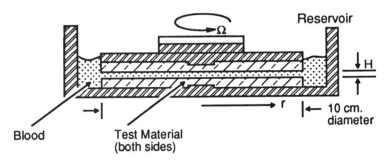

Figure 9.7 Schematic of apparatus for study of shear-induced hemolysis. Source: Adapted from Lampert and Williams (1972).

$$R_e(H) = \frac{\rho\Omega H^2}{\eta} \qquad (9.3)$$

where

ρ = density of blood
Ω = angular velocity
η = viscosity
$R_e(H) \leq 1.5$, laminar flow to $R_e(H) = 10^2$.

With respect to radius (r) at maximum separation (max H):

$$R_e(r) = \frac{\rho\Omega r^2}{\eta} \qquad (9.4)$$

where

ρ = density of blood
Ω = angular velocity
η = viscosity
$R_e(r) \leq 4.4 \times 10^4$, laminar flow to $R_e(r) = 10^5$

These conditions yield shear rates, $G \leq 11,2000$ sec^{-1}, ($r \leq 5$ cm.) and shear stresses that do not exceed 448 dyn/cm^2.

In this low shear regime, Lampert and Williams (1972) defined a relative plasma hemoglobin concentration increase ΔC with respect to aluminum:

$$\Delta C = \frac{[\Delta C_{Hb}]_{(m,t(\text{secs}))}}{[\Delta C_{Hb}]_{(Al,30\text{ sec})}}\Bigg|_{\text{same blood}} \tag{9.5}$$

For a number of materials tested, the following relationship was found:

$$\Delta C = At^\beta \tag{9.6}$$

For instance, for plexiglass:

$$\Delta C = 0.047t^{0.84} \tag{9.7}$$

In general, A was found to be apparatus dependent for this experiment and is given by the following relationship:

$$A = 0.020\beta^{-5.6} \tag{9.8}$$

On the other hand, β was found to be a constant characteristic of the test material and to have a single unique value for each material composition tested. For a group of five polymers (both test surfaces made from the same material), β was found to be related to the critical interfacial tension γ_c by a linear negative relationship of the form:

$$\beta = D - E\gamma_c \tag{9.9}$$

where D and E are experimentally derived constants.

This can be understood by the following argument. A high value of γ_c (with respect to a low value) favors rapid early protein absorption from the plasma. This increases the adhesion of platelets (increases the negative work of adhesion), leading to rapid maturation of a fibrin layer. This mature fibrin layer *discourages* further platelet and cell adhesion, thus *lowering* hemolysis rates with respect to those of the surface with lower γ_c.

Thus, while high γ_c leads to rapid thrombus formation, it may reduce free cell hemolysis. These experiments lead to a belief that the ideal blood compatible surface may not be the least reactive one.

These results were obtained with a constant shear stress. Performing experiments at varying values of Ω, Lampert and Williams (1972) formed the ratio R to examine shear stress effects:

$$R = \frac{\Delta C(m, W)}{\Delta C(m, 640)} \qquad t = 100\text{ sec} \tag{9.9}$$

The results of this analysis are equivocal due to uncertainties in the true fluid dynamics of the system. They are consistent with a linear rise of R at low stress (below 400 dyn/cm^2) and a nonlinear increase at higher apparent stresses. These results are interpreted as reflecting a constant boundary layer effect (material effect) and a superimposed bulk shear effect (turbulence effect) at higher rotational speeds. Thus, this experiment neatly shows the merging of the two flow regimes.

Although extremely enlightening, this series of studies may be criticized on two grounds:

1. The blood is exposed to an air-liquid interface and to materials other than the test materials. Thus there may be hemolysis associated with increased cell fragility and with contact with nontest surfaces. These effects are contained within the constant A; however, they prevent the derivation of an *absolute* hemolysis rate for a material.
2. Since the experiment is conducted in vitro and under conditions where significant hemolysis rates will occur in brief periods, it is very difficult to relate these results to those which might be obtained in vivo at more realistic low hemolysis rates.

9.5 COMMENT

It should be clear from the brief remarks in this chapter that understanding and controlling the host response of materials exposed to blood represents one of the great unsolved problems of biomaterials. The problem is complicated by several factors:

1. The phenomena involved are complex and multifactorial.
2. There seems to be a broad range of host response unlike the less specific response to inflammation.
3. Partly because of 1 and 2 above, as well as the rapidity of development of the coagulation cascade, there is an acute lack of adequate experimental models and techniques for research in this field.

We will return to this last point in Part IV of this book.

REFERENCES

Baier, R. E. (1972): Bull. N. Y. Acad. Med. 48:257.

Hayashi, K., Fukumura, H. and Yamamoto, N. (1990): J. Biomed. Mater. Res. 24:1385.

Herzlinger, G. A., Bing, D. H., Stein, R. and Cumming, R. D. (1981): Blood 57:764.

Kiraly, R. J. and Nosé, Y. (1974). Biomat. Med. Dev. Art. Org. 2(3):207.

Lampert, R. H. and Williams, M. C. (1972). J. Biomed. Mater. Res. 6:499.

Lindon, J. N., McManama, G., Kushner, L., Merrill, E. W. and Salzman, E. W. (1986): Blood 68:355.

Nosé, Y., Kiraly, R. J. and Picha, G. (1977). J. Biomed. Mater. Res. Symp. 8:85.

Salzman, E. W. (1972): In: *The Chemistry of Biosurfaces*, Vol. II. M. L. Hair (Ed.). Marcel Dekker, New York, pp. 489ff.

Sawyer, P. N., Stanczewski, B., Lucas, T. R., Srinivasan, S., Ramasamy, N. and Kirschenbaum, D. (1978): In: *Vascular Grafts*. P. N. Sawyer and M. J. Kaplitt (Eds.). Appleton-Crofts, New York, pp. 53ff.

Sawyer, P. N. and Srinivasan, S. (1972): Bull. N.Y. Acad. Med. 48:235.

Weinberg, C. B. and Bell, E. (1986): Science 231:397.

BIBLIOGRAPHY

Department of Health and Human Services (1985): *Guidelines for Blood-Material Interactions*. NIH Publication 85-2185, Public Health Service, National Institutes of Health. U. S. Government Printing Office, Washington, D. C.

Bruck, S. D. (1974): *Blood Compatible Synthetic Polymers*. C. C. Thomas, Springfield, Illinois.

Gott, V. L. and Furuse, A. (1971): Fed. Proc. 30:1679.

Hughes-Jones, N. C. (1973): *Lecture Notes on Haematology*, 2nd edition. Blackwell Scientific, Oxford.

Kambic, H. E., Kantrowitz, A. and Sung, P. (Eds.) (1986): *Vascular Graft Update: Safety and Performance*. ASTM STP 898. American Society for Testing and Materials, Philadelphia.

Lefrak, E. A. and Starr, A. (1979): *Cardiac Valve Prostheses*. Appleton-Century-Crofts, New York.

Merrill, E. W. (1977): Ann. N.Y. Acad. Sci. 283:6.

Sawyer, P. N. and Kaplitt, M. J. (Eds.) (1978): *Vascular Grafts*. Appleton-Crofts, New York.

Schoen, F. J. (1989): *Interventional and Surgical Cardiovascular Pathology*. W. B. Saunders, Philadelphia.

Van Kampen, C. L., Gibbons, D. F. and Jones, R. D. (1979): J. Biomed. Mater. Res. 13:517.

Vroman, L. (1968). *Blood*. American Museum Science Books, B26. Doubleday, New York.*

Blood by Leo Vroman is a marvelous, amusing, and witty account of blood biochemistry and surface interactions by one of the leading blood physiologists of this century. While one will enjoy it as recreational reading, one cannot avoid learning a great deal about blood.

10
Adaptation

10.1 INTRODUCTION

So far in Part III we have considered acute host responses to singular events. The insertion of an implant may evoke inflammation, with an acute course and a longer chronic phase. Interruption of a blood vessel by injury or insertion of an implant may trigger coagulation acutely and, perhaps, hemolysis chronically. In Chapter 12, we will take up the subject of neoplastic transformation: abnormal tissue development and elaboration as a result of either chemical or foreign body challenge.

Between these two types of events, acute response and abnormal development, there is another class of tissue response; that is, the presence of an implant, perhaps the implant's chemical, physical, or electrical properties, affects the organization and elaboration of tissue elements in the vicinity. We must consider these events because of the well-known ability of many of the tissues to remodel adaptively to reflect changes in demand and function. From the phrase "adaptively remodel" I shall purloin the term "adaptation" to describe such events as influenced by implants.

10.2 TISSUE GROWTH STRATEGIES

10.2.1 General Principles

Goss (1978, p. 2) discusses the strategy of growth of tissue in terms of three patterns previously recognized by Bizzozero (1894):

Expanding tissues: Those that grow by mitosis to increase cell number; for instance, the liver.

Static tissues: Those that retain essentially constant cell number but grow by individual hypertrophy; for instance, muscle.

Renewing tissues: Those that retain essentially constant cell number by replacing losses from differentiation of proliferating stem cells; for instance, skin.

In the rapidly developing immature individual, all tissues expand by mitosis. As maturation proceeds the cells of some tissues lose their mitotic ability and the tissues become either static or renewing. It is also possible for tissues to continue to display a combination of two of these patterns, varying their response to the challenge.

The question of control of these processes is still open. Since all of these processes tend towards limits, a variety of negative-feedback control systems have been proposed (see Section 10.3.5 for one example). However, a response is evoked in each cell type by either death of a portion of the tissue or by surgical removal. Less significant challenges, such as blunt trauma or change in mechanical functional requirements may also evoke responses. These observations suggest the existence of a wide variety of control processes regulating the quantity and, to some degree, the type of tissue in any location in a mammalian body. It is altogether reasonable to assume that the response of any such active biological control system facing a challenge may be affected by the presence of an implant.

10.2.2 Fracture Healing

The process of fracture healing, the restoration of the integrity of mineralized tissue after mechanical injury, includes an interesting example of natural adaptation in the absence of implants.

Successful fracture healing can be considered most generally to entail four phases:

1. Hematoma: from injury to formation of a "soft" callus.
2. Soft callus: from initiation of "soft" callus formation until its condensation into a "hard" callus.
3. Hard callus: from the initiation of "hard" callus until normal stiffness is restored.
4. Remodeling: from restoration of normal stiffness to full restoration of normal structure.

Stages 1 and 2 (see Figure 10.1) represent the acute or healing phase leading to the formation of a natural splint or *callus*. This is a weak

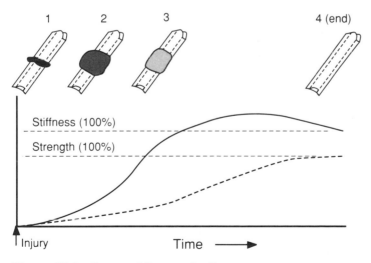

Figure 10.1 Stages of fracture healing.

provisional tissue, consisting of fibrocartilage and fibrous bone with little internal organization. However, when 50–75% of the normal stiffness is reached, the healing fracture undergoes a very dramatic and fairly rapid transformation. The soft callus rapidly condenses and shrinks to form a more organized hard callus, then is slowly resorbed through a progressive remodeling process leading to full restoration of normal structure (including re-opening of the medullary canal in long bones). These latter two stages appear to be mediated by mechanical requirements through Wolff's Law (see Section 10.3.1) and lead to an efficient structure with normal properties. Thus stages 3 and 4 of fracture healing can be thought of as a chronic phase and appear to represent adaptive remodeling of an essentially healed bone (end of stage 2).

Note that normal stiffness is obtained first (and in fact, exceeded, due to the combined presence of callus and healing cortical bone in the defect) while normal strength is obtained only towards the end of the remodeling phase.

Fracture healing, in any particular bone, with a certain degree of injury, can be expected to progress to completion in a mean time characteristic of the initial condition and, to some degree, the type of treatment (internal fixation, external splinting, etc.). If there is a significant delay but eventual complete healing and remodeling, the condition is termed *delayed union.* However, it is possible for the process to be interrupted at any stage. If this occurs in Phase 1 and 2, before any degree of structural integrity is obtained, and is permanent, the result is a *nonunion,* sometimes called a "pseudarthrosis."

10.3 EXAMPLES OF ADAPTATION IN IMPLANT APPLICATIONS

10.3.1 Introduction

There is a wide variety of possible examples of adaptation in implant applications. We shall restrict our consideration here to the following examples:

Growth of the "neointima" in arterial prostheses
Attachment of tendon prostheses to soft tissue
Hard tissue remodeling in the presence of implants
Bone response to electrified implants

There are many other examples of adaptation to the presence of implants. In fact, it may be useful to consider that accommodation of the local host tissues to implants, when resolution to a stable situation is possible (see Section 8.2.6), consists of two phases: an acute or *healing* phase and a chronic or *adaptive* phase.

In all of the adaptive processes, the control mechanism invoked is phenomenologically described by a unifying principle. In hard tissue, this principle is termed *Wolff's Law.* Julius Wolff, a 19th century German anatomist, suggested in 1892 that bone (and, by inference, other load-bearing tissues) remodels in an attempt to maintain a constant (optimal) pressure or stress. Although he did not make such a specific statement, it is usual to say that Wolff's Law is:

The form being given, tissue adapts to best fulfill its mechanical function.

Thus, as load increases over a period of time, bone mass would be expected to increase to maintain a constant level of stress. Conversely, a reduction in load should produce a loss of tissue. Observations in a variety of both static and renewing tissues suggests a considerable pragmatic generality for Wolff's Law.

10.3.2 Vascular Adaptation

The growth of a new tissue layer, or neointima, on the internal surfaces of knitted arterial prostheses is an interesting example of adaptation. Here the prosthesis is introduced to serve as a framework to support host tissue which will serve as a long-term nonthrombogenic blood contact surface. The body of the prosthesis, which is most commonly made of polyester fiber, provides the mechanical resistance to internal pressures that was once provided by the original vessel.

Upon insertion of the prosthesis, the features of the clotting cascade described in Section 9.2 take place. In fact, the surgeon usually preclots the prosthesis using the patient's blood in an effort to provide a smooth defect-free surface and to reduce blood loss. The fibrin surface, both inside and outside, matures and achieves stable dimensions within 24 hours.

At this point healing is clearly initiated. Within a few weeks the outside (tissue side) of the prosthesis is covered by granulation tissue, and a capsule of fibrous tissue matures with increasing organization. The resolution or healing response of the internal surface facing flowing blood is a different matter.

Schoen (1989, p. 36) distinguishes between two forms of resolution: *pseudointimal* formation, the mere coating of the implant's surface with proteins and cells other than endothelial cells; and *neointimal* formation, the formation of an endothelial lined surface, usually overlying a layer of smooth muscle cells. The choice of outcome is apparently governed by the nature of the biomaterial surface and, as we shall see, can be considered as an example of adaptation of the natural healing process to the properties of the prosthesis.

During ideal or optimal healing, blood vessels penetrate to the lumen of the prosthesis and "tufts" of tissue spread along the inner surface, merging to form a smooth neointimal surface that functions like the natural arterial lining. Some investigators (for instance, Annis et al., 1978) feel that this process of "through-growth" is not essential to the formation of the neointima, and longitudinal growth from the ends inward is possible in impermeable vascular prostheses. In a pig this process is completed within a month, but it takes up to a year in a human.

However, this healthy adaptive maturation has been shown (Wesolowski et al., 1968) to be critically dependent upon the prosthesis porosity. Figure 10.2a displays the relationship developed for a variety of porous prosthetic materials tested as arterial grafts in the pig. The porosity is given in units of liters/min/cm^2 of water expressed from the lumen through the wall of an unclotted non-blood-exposed graft with a pressure differential of 120 mmHg. The calcification index is the product of the average calcification (on a subjective scale of $1+$ to $4+$) of all animals implanted with a given material and the percentage of all specimens of the *same* material that display calcification. Here we see that high porosity, above a value of 1, favors maturation of the neointima, while lower porosity leads to failure of microvascular ingrowth, focal necrosis of the neointima, and possible calcification of the resulting pseudointima, in many respects mirroring the events of arteriosclerosis in natural tissue.

Attempts have been made to moderate this process by "seeding" the lumen wall with autologous cells (Kahn and Burkel, 1973), with cultured endothelial cells (Mansfield et al., 1975), and by various pretreatments; however, the relationship found by Wesolowski et al. (1968) appears to be well founded.

The generality of this relationship is suggested by Figure 10.2b. This displays the relationship of a "net acceptability index" and the previously discussed physical water porosity. It represents the results of the evaluation of a variety of knit and velour polyester arterial prostheses (Sawyer et al., 1979). The "net acceptability index" is a transformation of the "net evaluation index" used by the authors so that the low scores are satisfactory, as in Figure 10.2a. The "net evaluation index" is a combined score based upon in vivo performance and postimplantation evaluation. It is clear that the same result is obtained as that found by Wesolowski et al. (1968); high porosity favors good in vivo performance. This is clearly an example of a structural attribute of an implant strongly affecting tissue adaptation after surgery. Unlike the case of Wolff's Law

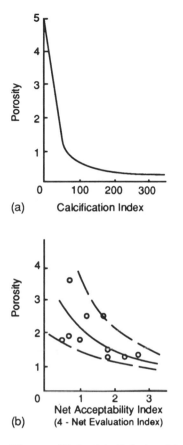

Figure 10.2 (a) Relationship between physical water porosity of arterial (woven) prostheses and calcification in the pig. (b) Relationship between physical water porosity and net acceptability index of arterial grafts (dashed lines indicate range). Sources: 10.2a adapted from Wesolowski et al. (1968); 10.2b adapted from Sawyer et al. (1979).

adaptation in bone, where the amount of tissue is affected by the implant attribute, here the type of tissue is affected by the (nonchemical) implant attribute.

10.3.3 Prosthetic Replacement of Tendons

Thes second area of adaptation that we shall consider is in the area of tendon prostheses. Here the natural system provides a junction between the soft tissue (tendon) and the hard tissue (bone) by a series of collage-

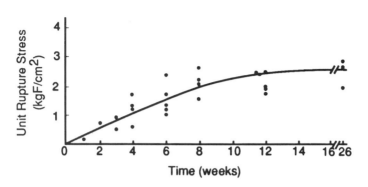

Figure 10.3 Prosthesis/tissue anastamosis strength. Source: Homsy et al. (1972).

nous fiber (Sharpey's fibers) that pass into the bone. There has been considerable interest in reproducing such a system either by inserting a porous material into the bone or by providing a porous bridge between the end of the natural tendon insertion and a tendon remnant or the muscle. In either case, a situation is desired in which the natural tissue will grow into the prosthesis and provide mechanical strength. We shall return to the topic of bony ingrowth later.

An interesting study of this subject (Homsy et al., 1972) was performed using a porous graphite-reinforced poly(tetrafluoro)ethylene (Proplast™, Vitek, Inc., Houston, Texas). Blocks of this material were inserted in the rabbit calcaneal tendon, replacing segments of the natural structure, and sutured in place to be load-bearing immediately. At periods of up to 26 weeks the rabbits were sacrificed and the strength of the bond between natural and prosthetic material was tested in tension. The results are shown in Figure 10.3. These results are reported to be the same for tendon/prosthesis and muscle (gluteal)/prosthesis interfaces and to reach a plateau value of rupture strength shortly after 8 weeks, about 5 weeks later than for collagenous ingrowth in bony sites.

Here the gradual development of strength, as contrasted with the rapid (within 3 weeks) complete penetration of unorganized collagen that the authors would have predicted from earlier studies suggests that an organized structure is forming under control of the axial mechanical load exerted by the muscle.

This picture is substantiated by studies of a different system in this application. Here the system is a complete tendon prosthesis that gradually disintegrates under use. Studies with a braided carbon ligament in

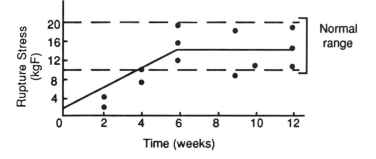

Figure 10.4 Breaking strength of pseudo-tendon. Source: Adapted from Jenkins et al. (1977).

both rabbits (Forster et al., 1978) and sheep (Jenkins et al., 1977) showed that, as the prosthesis frayed and fatigued, it served as a scaffold for the growth of fibrous tissue that took over the function of the prosthesis. Figure 10.4 shows the breaking strength of the rabbit calcaneal tendon prostheses from this study. The authors ascribe the rapid development of strength to a stimulatory effect of the carbon, but one might equally suggest that the gradual transfer of stress to the ingrowing tissue evoked this adaptive response. The development of maximum strength at 6 weeks, somewhat earlier than in the study of Homsy et al. (1972), suggests that the disintegration of the prosthesis served to transfer stress more rapidly than was the case with a more durable material.

This experiment has been carried to a natural conclusion by incorporating the carbon fibers in a matrix that is degradable in the internal environment (Alexander et al., 1979). The matrix used here was polylactic acid, and the implant was inserted as a replacement for the canine patellar tendon. A gradual dynamic replacement of prosthesis with organized fibrous tissue was reported as the prosthesis matrix disintegrated, with reasonable maturity and linear tissue arrangement after 2 months of implantation. However, in neither of these experiments did the resulting tissue come to closely resemble normal tendon. Thus it should be best termed *neotendon*, in parallel with Schoen's (1989) terms for healing endothelial tissue. In the more extreme cases, when the new tissue more resembles a fibrous scar or capsule, it is more appropriately termed *pseudotendon* and its long-term durability under mechanical loads must surely be questioned. Balduini at al. (1986) have summarized later clinical developments which have attempted to build on these observations.

10.3.4 Adaptive Remodeling of Bone Near Implants

10.3.4.1 "Stress-Shielding"

Wolff's Law suggests that the introduction of a load bearing implant coupled to a previously load bearing natural structure should result in some atrophy or tissue loss. Since the assumption of a portion of the tensile load or bending moment by the implant must necessarily reduce the local stress in the adjacent tissue, this phenomenon has come to be called "stress-shielding" (Huiskes, 1988).

The most commonly cited example is that of the progressive loss of bone material in the proximal medial femoral cortex, observed in both dogs and humans, after total replacement of the hip. However, observation of the remodeling response is complicated by a simultaneous presence of an osteolytic response mediated by the presence of wear debris released from the articulating interface.

A more easily examined example is an unfortunate concomitant of the use of metallic internal fracture fixation devices. These devices are usually far more rigid than the bone they are attached to, even if the bone is intact (as in experimental situations). While they provide excellent support and maintain reduction and fixation during healing, a considerable amount of *osteoporosis*, or loss of bone mineral mass, occurs in the bone under the plate or adjacent to the rod. A study of experimental fractures in rabbits (Brown and Mayor, 1978) seems to reinforce this point. Fractures were produced in the tibias of rabbits and were internally fixed with rods made of a variety of metals and polymers. The rods were a constant diameter to provide a range of stiffness relative to the intact bone from 10 to 0.03X. The animals were sacrificed and studied at 9 and 16 weeks after fracture and fixation. The results were confused by an anomalous response to one metal alloy (Ti6Al4V), but at 16 weeks those fractures fixed with rods that were *less* stiff than the intact bone were significantly stronger and tougher in torsion than those fixed with rods that were stiffer than the intact bone. Additionally, in the weaker, less tough bones (fixed with the stiffer rods), a greater degree of osteoporosis was seen histologically.

This experiment is difficult to interpret solely in terms of an adaptive response to implants since the changes in bone are presumably caused by two factors: a healing response and an adaptive response. Moyen et al. (1978) have done an experiment in dogs where no fracture was involved and obtained somewhat similar results. In this experiment, metallic plates of two different stiffnesses, varying by a factor of approximately 5, were attached to the midshaft of the femur in the dog. The bone mass under the plates was compared with the control (unplated) side after 6 and 9 months implantation. Interestingly, a small constant porosity of 1–3% was seen. However, after 6 months the bone mass under the rigid plate had decreased 26.4%, versus only a 16.4% decrease under the more flexible plate. The decreases in bone mass continued to develop more slowly up to 9 months, and showed modest reversals in another group implanted for 6 months and studied 3 months after plate removal. The authors ascribed the failure to observe the greater porosity seen in other studies (such as the one by Brown and Mayor, 1978) to the absence of the fracture healing process accompanying the remodeling (adaptive) process.

Perren et al. (1988) further criticized the possible adaptive role in the production of porosity near fracture fixation devices and suggested that the physical presence of the device interferes with revascularization after fracture. He showed an inverse relationship between the area of plate-bone contact and cortical porosity in a sheep tibial model. One may, in turn, criticize this work since reducing the plate-bone contact area probably reduces the coupling between plate and bone, thus rendering the plate effectively less stiff, that is, less able to remove load (and thus reduce local stress) from the bone.

Another study (Bradley et al., 1979) combined some of the aspects of the studies of Brown and Mayor (1978) and Moyen et al. (1978). In this case a variety of fracture fixation plates, with stiffnesses between 4 and 40% of the bone to be fixed, were used in a 16-week study of the healing of femoral midshaft osteotomies in dogs. A definite relationship between plate rigidity and strength of both the bony material and the femoral midshaft as a structure was seen. This is shown in Figure 10.5.

There are two comments that should be made about this study. In the first place, the parameter examined was *strength* rather than *rigidity*, as in the case of the previous two studies. While rigidity may parallel strength, the correlation is not exact, even in materials that are simpler in microstructure than bone. In the second place, porosity was not studied, as in the work of Moyen et al. (1978). Porosity, suggested by Moyen to be a concomitant of trauma rather than simply adaptation [my term] and by Perren et al. (1988) as secondary to interference with revascularization, may severely affect material and structural strength due to stress concentration effects, while having a modest volume-fraction effect upon modulus and, thus, upon material and structural bending rigidity (see Section 6.3.2). It is impossible to isolate this effect in this study.

One should not assume from these studies that large changes in stress are necessary to modify bone growth, that is, to produce adaptive changes in bone. Modest changes in stress, such as those that might be produced by simple, soft polymeric caps of bone shafts after segmental excision,

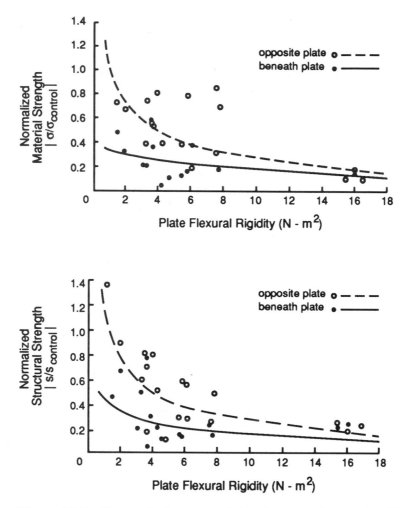

Figure 10.5 Changes in bone material and structural strength with fixation plate rigidity. Source: Adapted from Bradley et al. (1979).

have been shown to produce profound adaptive changes (Lusskin et al., 1972).

10.3.4.2 Ingrowth into Porous Biomaterials

It is an easy step from the earlier discussion of the ingrowth associated with the prosthetic replacement of tendons to consideration of the more general problem of ingrowth into porous bodies. Since one of the responses to implants is the formation of a fibrous capsule, it is no surprise that tissue will invade the internal spaces of an implant with an open, connected pore structure.

There have been extensive studies of this phenomenon. Ingrowth occurs into porous implants fabricated from a wide variety of metals, polymers, and ceramics. The nature of the ingrowing tissue, in the presence of sufficient interfacial mechanical stability,* is dependent upon the size of the pore, or more properly, on the minimum size of the interconnections between pores (Klawitter and Weinstein, 1974). Soft tissue elements will be found in interconnects as small as 1–5 μm; at some minimum interconnect diameter between 50 and 100 μm, mineralized tissue will be found and organized osteonal bone will grow into interconnects as small as 250 μm. Maximum interfacial shear strengths develop between 8 and 16 weeks after implantation, depending upon anatomical locations, species of animal, and type of tissue ingrowth. Velocity of ingrowth appears to increase with pore sizes above 50 μm and to reach a peak near pore sizes of 400 and 500 μm, as determined in a single pore model (Howe et al., 1974).

This latter phenomenon, the ability of tissue to mineralize and organize as interconnect size increases, is another clear example of adaptation. A study of Proplast™ (Vitek, Inc., Houston, Texas) by Spector et al. (1979) confirmed this finding and demonstrated that it is not a false conclusion based upon comparison of studies with different materials and/or test conditions. This material, if implanted directly, exhibits a pore size of 76 μm with an interconnect size of 50 μm. In a canine cortical bone site, only fibrous ingrowth was observed for periods of up to 20 weeks. However, if the material was "teased" before implantation to increase the size of interconnects, a variable degree of bony ingrowth occurred. There appears to be a difference of opinion over the interpretation of the findings in this report (Homsy, 1979; Spector, 1979). This finding, combined with earlier reports (Klawitter and Hulbert, 1971) suggests a practical lower interconnect limit of 100 μm for bone ingrowth. The mechanism of control of ingrowth and the manner in which pore interconnect size controls mineralization are unknown. While Wolff's Law arguments can be invoked to explain tissue remodeling near the bone-implant interface, it is presumed that tissue more than one pore diameter deep within the implant porosity will be essentially load-free, if the modulus of the implant ex-

---

*The issue of the role of interfacial shear, producing the so-called "micro-motion" on tissue ingrowth, is sufficiently confused at this time that an analytical discussion of this point is not possible. However, see Brunski (1988) for a contemporary discussion of the question.

ceed that of bone to any degree. However, tissue maturation internal to porous implants appears to have little dependence on implant material modulus.

10.3.4.3 Adhesion

Tissue is not inherently "sticky." Cells adhere to each other through the interaction of a variety of specific and non-specific adhesion molecules and specific cell surface receptors for portions of these molecules. However, there has been considerable interest in attempts to cause implants to adhere to tissue.

When such adhesion is produced by the mere close (molecular-scale) approximation of tissue and implant, without an intervening fibrous capsule or other elements of an inflammatory response, it is termed most generally *tissue integration*; in the case of bone, the more specific term is *osseointegration* (Albrektsson and Hanson, 1986). In this case, the apparent adhesion is produced by cellular binding to proteins adsorbed to the implant surface. Originally thought to be a property of pure titanium alone, such tissue integration has now been shown for a variety of metallic implant surfaces (Linder, 1989). In fact Linder (1989) has suggested that, in general, "osseointegration is a response of bone to a tolerable implant material inserted under tolerable conditions" without specifying the meaning of "tolerable" in either case.

When tissue adhesion to an implant is accompanied by a chemical alteration of the implant surface, a true bonding process with a continuous gradation of structure and composition across the tissue-implant interface may occur. While there is no generally accepted term for this condition, biomaterials which produce it have been termed *surface active* (Hench and Wilson, 1984) in recognition of the necessity of chemical reaction with the local host environment prior to bond formation.* A number of ceramic and glassy materials have been produced which develop such bonds to both bone and soft tissue (Figure 10.6).

In either the case of integration or bonding, the implant becomes mechanically coupled to the adjacent tissue. In the case of hard tissue, this results in a strain incompatibility, due to the differences in moduli between bone and the implant. Several adaptive changes are possible. The most common situations are:

---

*The term *bioactive* has also been used for such materials. However, this is an apparent misnomer as it appears that the necessary surface modification is a consequence of exposure to the physiological environment, rather than to life processes.

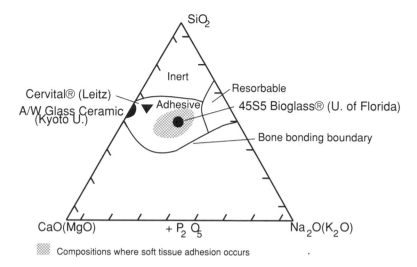

Figure 10.6 Ceramic and glassy tissue bonding materials (by permission: L. L. Hench).

1. The implant is placed in cancellous bone and is of very significantly higher modulus than the surrounding tissue. The frequently observed response in this case is the formation of a bony "plate," much resembling a subchondral plate in an articular joint, and a relative rarefaction of the cancellous trabeculae behind the plate. This results in introducing a relatively compliant tissue zone adjacent to the implant and shielding the bone further away from the mechanical consequences of the stiff implant.
2. The implant is placed in cortical bone and is of significantly higher modulus than surrounding tissue. In this case, bone near the implant becomes porous, much as in the stress shielding examples previously discussed (Section 10.3.4.1). However, this porosity may increase and proceed to a remodeled condition resembling cancellous bone. Thus, this process is termed *cancellization*.

In either case, the tissue structure changes are a result of the change in mechanical conditions near the newly formed interface and thus are true examples of adaptive remodeling.

10.3.5 Bone Response to Electrified Implants

It has been proposed that the mechanism of Wolff's Law in bone is electrically controlled. Bone and other tissues produce potentials when

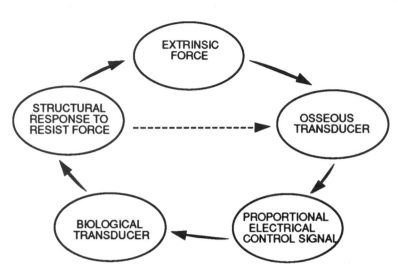

Figure 10.7 Negative feedback control system proposed as a basis for Wolff's Law. Source: adapted from Bassett (1971).

deformed; these potentials are called piezoelectric potentials, or more generally, strain-related potentials. A variety of other potential sources also exist.

Bassett (1971) proposed a generalized, closed loop control system as shown in Figure 10.7 to relate these signals to hard tissue remodeling. There is a conceptual error in his scheme, since, presumably, the structural response (to adjust stress) results in a change in the osseous transducer (see added dashed line), rather than in a modification of the extrinsic force as proposed. Nevertheless, this general idea, first proposed some years before this review was published, has been a motivating factor in the investigation of the effects that electrical phenomena have on the modification of bony growth and remodeling. Numerous studies have shown a correspondence between endogenous electrical phenomena and growth, repair and remodeling processes, but no critical experiments which show that these signals are both necessary and sufficient to serve as stimuli for the observed processes have yet been performed.

While it is not clear that electrical control of bone growth, remodeling, and repair is an example of adaptive response to implants, it is worth noting some general conclusions from research in this area for completeness of this discussion (Black, 1987).

A physical electrode may be considered as a special case of metallic implantation: one in which the electrode potential, instead of being allowed to find its appropriate mixed corrosion potential (see Section 4.6), is maintained under a controlled potential or current condition.

The responses to implantation of such a physical electrode in or near hard tissue are summarized below:

1. The presence of a metallic cathode, with a negative potential, stimulates the conduct of cells involved in bone formation.
2. In particular, bony repair in sites of trauma is accelerated, and cells in medullary sites may be induced to form bone in the absence of bony trauma.
3. In all cases there is apparently a narrow stimulatory "window" defined by limits on current and electrode potential.
4. Finally, monophasic negative pulses of frequencies from 10 to 750 Hz, with duty cycles between 50 and 5%, provide stimulation which approaches but does not exceed that of direct uninterrupted current.

While the relationship of this line of research to the more general problem of adaptive growth is unclear, it does represent one of the more systematic attempts to elucidate the phenomena involved.

10.4 A FINAL COMMENT ON ADAPTATION

It now seems clear that one of the original goals of implantation, that is, to produce minimal tissue (host) response, is an outmoded view that may limit further development of implant materials and devices. It is proper to ask that *adverse* host response be kept within bounds acceptable to the application. This is the necessary condition of biocompatibility. It is essential to rule out a priori certain classes of response, such as neoplastic transformation, as unacceptable. Now it appears equally reasonable to pursue active tissue response in the form of adaptation, so that natural tissue can take over the role of the implant as completely as possible.

The early experience with tendon replacement reported in this chapter suggests that the use of a degradable or "active" implant that can direct adaptation to form a new structure may be an early example of this most elegant and long-lasting type of implant application. It is an incomplete example, since the new structure, while able to serve in the place of the original one, is neither a true structural or functional replacement. Much work remains to be done until the principles of adaptation are fully understood and harnessed to the solution of patient problems.

REFERENCES

Albrektsson, T. and Hansson, H.-A. (1986): Biomats. 7:201.

Alexander, H., Weiss, A. B., Parsons, J. R., Strauchler, I. D. and Gona, O. (1979): Trans. ORS 4:27.

Annis, D., Bornat, A., Edwards, R. O., Higham, A., Loveday, B. and Wilson, J. (1978): Trans. Am. Soc. Artif. Intern. Organs XXIV:209.

Balduini, F. C., Clemow, A. J. T. and Lehman, R. C. (1986): *Synthetic Ligaments: Scaffolds, Stents, and Prostheses.* Slack, Thorofare, NJ.

Bassett, C. A. L. (1971): In: *Biochemistry and Physiology of Bone,* Vol. 2, 2nd ed. G. H. Bourne (Ed.). Academic Press, New York, pp. 1ff.

Bizzozero, G. (1894): Brit. Med. J. 1:728.

Black, J. (1987): *Electrical Stimulation: Its Role in Growth, Repair, and Remodeling of the Musculoskeletal System.* Praeger, New York.

Bradley, G. W., McKenna, G. B., Dunn, H. K., Daniels, A. U. and Statton, W. O. (1979): J. Bone Joint Surg. 61A:866.

Brown, S. A. and Mayor, M. B. (1978): J. Biomed. Mater. Res. 12:67.

Brunski, J. B. (1988): In: *Non-Cemented Total Hip Arthroplasty.* R. H. Fitzgerald, Jr. (Ed.). Raven, New York, p. 7ff.

Forster, I. W., Ráliš, Z. A., McKibbin, B. and Jenkins, D. H. R. (1978): Clin. Orthop. Rel. Res. 131:299.

Goss, R. J. (1978): *The Physiology of Growth.* Academic Press, New York.

Hench, L. L. and Wilson, J. (1984): Science 226:630.

Homsy, C. A. (1979): J. Biomed. Mater. Res. 13:987.

Homsy, C. A., Cain, T. E., Kessler, F. B., Anderson, M. S. and King, J. W. (1972): Clin. Orthop. Rel. Res. 89:220.

Howe, D. F., Svare, C. W. and Tock, R. W. (1974): J. Biomed. Mater. Res. 8:399.

Huiskes, R. (1988): In: *Non-Cemented Total Hip Arthroplasty.* R. H. Fitzgerald, Jr. (Ed.). Raven, New York, pp. 283ff.

Jenkins, D. H. R., Forster, I. W., McKibbin, B. and Ráliš, Z. A. (1977): J. Bone Joint Surg. 59B:53.

Kahn, R. H. and Burkel, W. E. (1973): In Vitro 8:451.

Klawitter, J. J. and Hulbert, S. F. (1971): J. Biomed. Mater. Res. Symp. 2:161.

Klawitter, J. J. and Weinstein, A. M. (1974): Acta Orthop. Belgica 40:755.

Linder, L. (1989): Acta Orthop. Scand. 60:129.

Lusskin, R., Thompson, W. A. L., Pena, A. and Sanwal, S. (1972): Clin. Orthop. Rel. Res. 83:300.

Mansfield, P. B., Wechezak, A. R. and Sauvage, L. R. (1975): Trans. Am. Soc. Artif. Intern. Organs XXI:264.

Moyen, B. J.-L., Lahey, P. J., Jr., Weinberg, E. H. and Harris, W. H. (1978): J. Bone Joint Surg. 60A:940.

Perren, S. M., Cordey, J., Rahn, B. A., Gautier, E. and Schneider, E. (1988): Clin. Orthop. Rel. Res. 232:139.

Sawyer, P. N., Stanczewski, B., Hoskin, G. P., Sophie, Z., Stillman, R. M., Turner, R. J. and Hoffman, H. L., Jr. (1979): J. Biomed. Mater. Res. 13:937.

Schoen, F. J. (1989): *Interventional and Surgical Cardiovascular Pathology: Clinical Correlations and Basic Principles.* W.B. Saunders, Philadelphia.

Spector, M. (1979): J. Biomed. Mater. Res. 13:991.

Spector, M., Harmon, S. L. and Kreutner, A. (1979): J. Biomed. Mater. Res. 13:677.

Wesolowski, S. A., Fries, C. C., Martinez, A. and McMahon, J. D. (1968): Ann. N.Y. Acad. Sci. 146(1):325.

Wolff, J. (1892): *Das Gesetz der Transformation der Knochen.* A. Hirschwald, Berlin.

BIBLIOGRAPHY

Brighton, C. T., Black, J. and Pollack, S. R. (Eds.) (1979): *Electrical Properties of Bone and Cartilage: Experimental Effects and Clinical Applications.* Grune & Stratton, New York.

Burke, J. F., Yannas, I. V., Quinby, W. C., Jr., Bondoc, C. C. and Jung, W. K. (1981): Ann. Surg. 194:413.

Fitzgerald, R. H., Jr. (Ed.) (1988): *Non-Cemented Total Hip Arthroplasty.* Raven, New York.

Homsy, C. A. (1973): Orthop. Clin. N. A. 4:295.

Lane, J. M. (Ed.) (1987): *Fracture Healing.* Churchill Livingstone, New York.

Rubin, C. T. and Hausman, M. R. (1988): Rheum. Dis. Clin. N.A. 14:503.

Thompson, D'A. W. (1917): *On Growth and Form.* Cambridge University Press, London.

Woo, S. L.-Y., Lothringer, K. S., Akeson, W. H., Coutts, R. D., Woo, Y. K., Simon, B. R. and Gomez, M. A. (1984): J. Orthop. Res. 1:431.

11
Allergic Foreign-Body Response

11.1 SPECIFIC VERSUS NONSPECIFIC RESPONSE

In our earlier discussion of the inflammatory response (Chapter 8), it was indicated that the actions of neutrophils and macrophages in response to a foreign material constitute a *nonspecific* defense mechanism. That is, as we have seen, their response is universal in nature and only slightly affected by the structure and chemical composition of the foreign material. In this chapter we shall discuss a second type of response to foreign materials, the *specific* or *immune* response.

The aspects of specific response to foreign or non-self materials are grouped together and collectively ascribed to a system of cells and mediating agents, collectively termed the *immune system. Immunity* is usually understood as the property of being secure or nonsusceptible to the adverse effects of a particular bacterium or foreign material. Conversely, allergy is the property of being especially sensitive (or *hypersensitive*) to such agents. Figure 11.1 shows the overall system and its general features.

The immune system is configured to distinguish between self (those things which are part of the natural, intact physiological system) and non-self (all other things). It normally ignores all aspects of self; this is termed *autoimmunity*. Introduction of foreign tissue produces an inflammatory response termed *rejection*. Resistance may be conferred by inheritance of a "memory" for certain non-self materials, such as proteins in bacterial cell walls, thus producing a *natural resistance* to certain infections. Perhaps the most important aspect of the system is its ability to acquire immunity by developing a specific memory for particular foreign bodies. This may be produced by deliberate exposure under nonpathogenic conditions (*vaccination*) or by prior exposure under sensitizing conditions (high dosage, physical stress, presence of an adjuvant material, etc.). The result of this specific memory may be desirable, as in affording acquired (vs. natural) resistance to infection or undesirable, in producing a form of adverse reaction to implants, termed *hypersensitivity*.

11.2 MECHANISMS OF IMMUNE RESPONSE

Specific or immune responses depend upon the exact details of the chemical composition and conformation of the foreign material. This class of response is directed primarily towards recognition of foreign proteins such as toxins, viruses, and bacterial cell wall components. The response to these foreign materials takes place through two mechanisms: *humoral* and *cell-mediated* response. In each case, the foreign body is referred to as an *antigen* and it assists the body in producing an *antibody*. Antibody is a generic name for a macromolecular complex formed by the association of large immunoglobulins present in serum into a Y-shaped molecule with a molecular weight exceeding 1×10^6. The stem of the Y is essentially the same for all antibodies, permitting it to bind to cell-surfaces; regions in the arms are variable in structure, producing the specificity, or ability to bind with a particular antigen, characteristic of an antibody.

The humoral mechanism is based upon the production of freely circulating antibodies. The high molecular weight antibodies are designed to unite with the foreign material and "denature" or neutralize it. That is, the antibody-antigen complex lacks the undesirable or destructive activity of the free antigen. The complex may accumulate in tissue, be carried to lymph nodes by phagocytes, or be more easily catabolized than the free

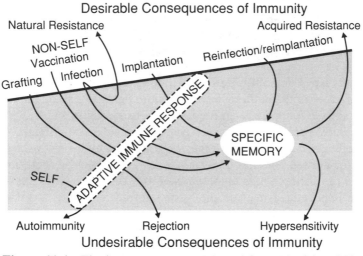

Figure 11.1   The immune system. Adapted from Playfair (1979).

antigen. Circulating antibody production is mediated by a class of lymphocytes called *B-cells* that arise from primitive mesenchymal cells in the bone marrow. Antibody production is initiated by introduction of a foreign material. It may itself be able to act as an antigen, it may combine with native proteins, especially complement fragments C3 and C5 (which may be activated), to form an antigen or it may undergo metabolic processing by a phagocytic cell, usually a macrophage, to form an antigen.

Once antibodies are produced and released into the blood stream, they may persist for long periods of time. This is the principle of immunization to bacterial and viral infection, giving a small provocative dose of either a specific antigen or one that produces antibodies specific to the infectious organism of interest. Then, when a later challenge is encountered, immediate antibody-antigen complexing occurs without the delay necessary for antibody production. The antibody production response is quite variable, depending upon the initiating agent, its concentration, the general state of health of the immune system, and the presence of sensitizing agents such as corticosteroids. Finally, a subpopulation of B-lymphocytes, memory B-cells, are capable of rapidly synthesizing additional amounts of the specific antibody (which is ''remembered'') upon stimulus by a later exposure to the same foreign body.

Cell-mediated response is dependent upon the action of another class of lymphocytes, the *T-cells*. These cells also arise in bone marrow but pass through the thymus gland where they undergo a conversion that improves their ability to differentiate. They then collect in lymph nodes and associated tissue. Their activity continues to be affected by a hormone secreted by epithelial cells in the thymus gland. The T-cells cannot be distinguished from B-cells morphologically until challenged by an antigen. As in the case of humoral response, a foreign material or one of its degradation products may itself be able to act as an antigen, it may combine with native proteins, especially complement fragment C5 (which may be activated to C5a [cf. Shepard et al., 1984]), to form an antigen or it may undergo metabolic processing by a phagocytic cell, usually a macrophage, to form an antigen. T-cells may then be distinguished by morphological changes which include the appearance of many polyribosomes and scant rough endoplastic reticulum. They produce and store a different class of antibodies, primarily bound to the cell membrane surface. These antibodies are highly specific and cannot survive with appreciable activity outside of or separate from the T-cell. T-cell antibodies act against intracellular infections, cancer, and foreign materials of nonbiological origin. However, T-cells must be present and aggregate in the region of antigen concentration to be effective.

In addition to direct neutralization of undesired effects of foreign materials, the formation of antibody-antigen complexes acts to enhance inflammatory response. This can happen directly through a nonspecific phagocytic response since the complex can grow to microscopic size by accretion of additional antibody and antigen molecules, or it can happen indirectly through complement activation (Chenoweth, 1986).

Since particulate foreign materials may be distributed widely in the host, and since lymphocytes and circulating antibodies are also widespread, local response to antibody-antigen complexing may occur anywhere. This results in the wide variety of physiological effects that we popularly term "allergic:" swelling of the membranes in the respiratory system (hayfever, asthma, etc.), rash or reddening (arthus, etc.), hives (swelling related to local kinin activation), and so forth.

11.3 CLASSES OF HYPERSENSITIVITY REACTIONS

Allergic responses are more properly categorized collectively as hypersensitivity reactions. These may be separated into *immediate hypersensitivity*, mediated by direct antibody-antigen combination, and *delayed hypersensitivity*, which is cell mediated. Immediate hypersensitivity is further subdivided into three types:

1. Anaphylactic or immediate shock response
2. Cytolytic/cytotoxic reactions
3. Toxic complex syndrome

Delayed hypersensitivity responses are usually called type 4 reactions. If a challenge evokes an immediate reaction (type 1, 2, or 3) and a delayed reaction, the net reaction is termed "mixed" and referred to as type 5. (Note: These responses are frequently designated by Roman numerals.)

The nature of supposed responses to implants (to be discussed later) suggests that they are, collectively, examples of delayed hypersensitivity, or type 4 reactions. The type 4 reaction is clinically similar to a chronic inflammatory response, except that both lymphocytes and neutrophils are seen in foci of antigen-antibody complex accumulation and additional concomitant symptoms, as discussed in Section 11.4, may occur.

11.4 HYPERSENSITIVITY REACTIONS ASSOCIATED WITH IMPLANTS

11.4.1 Polymers

While humoral and cell-mediated immune responses are usually directed toward materials of natural origin, it is of interest to us to inquire concerning the response to implants. As of now, little response to polymers is recognized, except for the special case of implants made from processed natural tissue. The use of such materials, including processed allografts (tissue from human donors) and xenografts (tissue from other species, such as porcine heart valves, etc.) is finally limited by the ability to denature or chemically mask the foreign materials so as to suppress their antigenic activity (Bajpai, 1983).

Attempts to produce immune responses to bulk polymers (Stern et al., 1972) have produced very mild effects. They are most suggestive of immune response to native proteins denatured by surface contact (see Chapter 5) rather than direct antibody production or T-cell activation by polymer molecules. Clinically, cell-mediated responses have been reported to poly(methyl methacrylate) "bone cement" (Clementi et al., 1980) and to silicone elastomers (Kossovsky et al., 1987). Despite these reports, concerns about immune or allergic responses to biomaterials have centered primarily on metallic materials, both on the skin and as implants.

11.4.2 Metals

General

The action of metals on the immune system is a bit of a puzzle. It is generally recognized that they must be dissolved to be active but the low molecular weight and simplicity of structure of the resulting ions argue

Table 11.1 Incidence of Metal Sensitivity

Allergen	Patients presenting with skin problems (%)			Normals (%)		
	Male	Female	Total	Male	Female	Total
Nickel	3.1	12.9	9.6	1.5	8.9	4.2
Chromium	12.7	7.1	9.3	2.0	1.5	1.7
Cobalt	4.7	5.3	6.0	nr	nr	nr

nr: not reported
Source: Adapted from Hildebrand et al. (1988).

against their being capable of directly activating either humoral or cell-mediated response. It is thought that they combine with organic molecules, such as albumin, to form complexes termed *haptens*, which possess antigenic qualities. It has also been suggested that increases in concentrations of naturally occurring metal-carrier protein complexes, such as Fe-transferrin or thioneins, can render them effective haptens. Details of metal-protein complexing affecting molecular shape changes are complex (Friedberg, 1974).

Metal sensitivity among the general population is not rare, with sensitivity to nickel being the most common, followed by cobalt and chromium. The incidence of sensitivity is estimated to be between 1 and 5%, with as much as 10% of the population being sensitive to at least one of these metals (Table 11.1). Incidence rates are different for men and women, reflecting differences in home and work place exposure, and are higher for individuals involved in certain industries, including mining, metal refining, electroplating, printing, etc.

In addition to acting as antigens through hapten formation, metals of interest for use in implants have been shown to affect directly the response of the host immune system to other antigens. Chromium and nickel have been shown to suppress antibody production, while the roles of cobalt and manganese are anomalous. Iron, chromium and nickel also have been shown in vitro to bind with T-cell surface antigens (Bravo et al., 1990); this binding may change the specificity of the previously formed antigens.

Dermatitis

We shall discuss only two aspects of immune response to metals. The first of these is the production of dermatitis as a direct response to metallic contact. This is of importance for a number of reasons. It may be a direct and unacceptable side effect of the use of external metallic support devices such as braces or dentures. It has come to be recognized as a general indicator of systemic challenge in the process of sensitivity to metals.

A modern account of metallic dermatitis is that of Fisher (1986). Fisher recognizes sensitization by nickel, chromium, cobalt, gold, and platinum among the metals of implant interest (the last two have extensive dental applications). With the exception of nickel, these do not provoke an initial sensitivity in the solid state due to their low solubility. However, in a previously sensitized individual, all can evoke skin inflammations, that is, dermatitis, of various types and degrees of severity. There appears to be little cross-sensitivity, that is, metal "A" causing a response in a patient sensitized to metal "B," although nickel sensitivity is often accompanied by cobalt sensitivity. However, this later finding may be due to the close association of these two metals in alloys, etc., and not to a true cross-sensitivity. Chromium appears to cause a response primarily when present as a chromate [Cr^{+6}], while the other metals are active in divalent and trivalent ionic forms.

Fisher's chapter is especially valuable for its catalogs of metal-bearing articles and environmental (exposure) settings. It also provides excellent clinical descriptions of the various dermatitides and of techniques for skin testing for sensitivity.

Wahlberg (1973) has tested a number of patients with identified clinical sensitivity to metals and has determined threshold concentrations for topical application of metal salts. Of interest is his finding that the carrier used in the patch test affects the threshold in many cases. That is, if the carrier aids penetration of the metallic ions, then a lower threshold is found. Table 11.2 presents these data for water-based solutions only.

At variance to Fisher's conclusion, we can see a subtle pattern of cross-sensitivity, at least with respect to the minimum dose required for response. These data should be regarded with some care since individual patients showed a response to concentrations one to two orders of magnitude below mean threshold values for these experimental groups.

Wahlberg also looked for relationships between clinical severity of responses and threshold values for particular patients and found poor correlations except in the case of cobalt, where the correlation coefficient $r = 0.61$. He concluded that:

> . . . renewed contact with cobalt plays a greater role in recurrence [of allergenic response] than [with] allergens such as chromium and nickel, where other factors, such as infections, heat, moisture, cold, stress, etc., . . . contribute to the recurrence.

Table 11.2 Sensitivity Thresholds for Metals

Clinical sensitivity	Challenge compound	Threshold concentration (mean, wt%)
Co	$CoCl_2$	0.27
Co, Ni	$CoCl_2$	0.25
Co, Ni, Cr	$CoCl_2$	0.51
Co, Cr	$CoCl_2$	0.30
Cr	K_2CrO_4	0.21
Cr, Co	K_2CrO_4	0.08
Cr, Co, Ni	K_2CrO_4	0.14
Cr	$K_2Cr_2O_7$	0.22
Cr, Co	$K_2Cr_2O_7$	0.22

Source: Adapted from Wahlberg (1973).

It should be noted here that clinical practice is to use 1 to 5 weight percent solutions for skin patch testing for metallic allergenic response. Comparing these concentrations with the threshold data in Table 11.1 lends real weight to the concern that patch testing, in the presence of a sensitizing agent or condition, may contribute to a later immune response in a previously insensitive individual.

There are many clinical reports of skin response, both local and remote, to metallic devices. One that is reported completely (Brendlinger and Tarsitano, 1970) will suffice as an example. The patient, a 25-year-old woman, appeared with generalized eruptions on her trunk, arms, and legs. Treatment with topical corticosteroids provided some relief. She sought treatment for a dermatitis on her ring finger 8 months later. Despite treatment, she continued to experience a spread and increase in intensity of symptoms. A rash appeared on her feet. She began to experience pain and soreness in her mouth 2 months later, that is, 11 months after her first symptoms appeared. On examination she was found to have a cobalt-chromium partial denture that she had acquired several months before her initial skin problems and had worn intermittently thereafter. Replacement of the metal denture with an acrylic denture resulted in prompt remission of symptoms. Reinsertion of the metal denture resulted in a return of symptoms within 24 hours. At that time, skin testing showed that she was sensitive to both chromium and nickel in pure, solid metallic form.

Sensitivity to chromium, cobalt and nickel is now well recognized in dental applications, where contact between the device and the oral mucosa occurs. Hildebrand et al. (1988) report a compilation of 149 cases which fulfill three strict criteria: (1) presence of one or more clinical features suggestive of immune response, such as eczema, redness, ulceration, etc.; (2) healing (resolution of the clinical features) after removal of the device; and (3) positive skin (epicutaneous) test for a metallic component of the device. Of particular interest in this report is the observation that only 28 patients ($\sim 20\%$) reported a prior history of symptoms referable to an established metal sensitivity. Thus, it is probable that the use of stainless steel and cobalt-base alloys in dental applications can result in sensitizing previously nonsensitive patients.

It should not be concluded that skin or mucosal contact by the implant is necessary to produce dermatitis. A report (Cramers and Lucht, 1977) documents the cases of three patients with 316L stainless steel and screw implants who developed dermatitis 3 to $3^1/2$ months after surgery. Two patients were found to be sensitive to chromium and cobalt by patch test and one was sensitive to nickel. On surgical exploration, no infection could be cultured in any case, and all immune response symptoms resolved promptly after the devices were removed.

Implant Site Inflammation

The second aspect of immune response to metals that we shall consider is the direct implant site inflammation. Hicks (1958) is probably the first to report this effect. Initially, it was thought to be a simple inflammatory response associated with the relatively high corrosion rates of alloys in use in the 1940s and 1950s. More recent interest in the problems of loosening of total joint replacements, especially of the hip, has reawakened interest in this problem, especially in relation to possible immune response to metals.

An initial study of the possible relationship between sensitivity to metal and problems with total joint replacement (Evans et al., 1974) aroused considerable interest.

Evans et al. reported the results shown in Table 11.3. They also found an apparently greater effect associated with metal-on-metal devices. Several of their conclusions are:

1. Evidence is presented which suggests that after replacement, bone necrosis and consequent loosening of the prosthesis may be due to the development of sensitivity to the metals used.
4. Examination of this material [tissue of joints from sensitive patients] showed necrosis of bone and soft tissue following obliterative changes in the vascular supply.

Table 11.3 Relationship Between THR Loosening and Metal Sensitivity[a]

Patients	Total (#)	Sensitive (#,%)	Insensitive (#,%)
Loose	14	9[b] (64)	5 (36)
Not loose	24	—	24 (100)

[a]By skin test.
[b]11 loose prostheses.
Source: Adapted from Evans et al. (1974).

7. We conclude that prostheses in which metal articulates with polyethylene should be preferred; that any patient in whom loosening or fragmentation occurs should be patch tested; and that if sensitivity is found the implant should be removed.

This study is suggestive of a linkage between loosening, perhaps secondary to inflammation, and sensitivity, especially for metal-on-metal devices. Such devices, which would be expected to shed larger amounts of metallic debris than metal-on-polymer ones, appeared to be involved more often in the supposed linkage than did metal-on-polymer devices. The study was poorly controlled, as was a smaller one later by Elves et al. (1975) who reported the following results:

> Sensitivity to chromium, cobalt, nickel, molybdenum, vanadium and titanium was studied by patch tests in 50 patients who had received total joint replacements. Nineteen (38%) were sensitive to one or more metals, primarily cobalt and nickel. In 23 patients, nontraumatic failure of the prosthesis had occurred and 15 of these failures were sensitive to metal. Out of 27 patients with no evidence of prosthesis loosening, four were sensitive to nickel and cobalt or nickel alone. Dermatological reactions occurred in 13 patients after surgery; however, only eight of these showed evidence of metal sensitivity.

However, several questions remain unanswered. One of particular interest is whether sensitization occurs from an implant or if it is a pre-existing condition. There are no accurate incidence rates for metal sensitivity in the general population. Studies have been done only on populations that appear at dermatological clinics, that is, patients with active skin problems. A report published with that of Elves et al. (Benson et al., 1975) favored the pre-sensitization position. Groups of patients awaiting total hip joint replacement were compared with those who had received the devices already and who were, in some cases, already symptomatic (device loosening). While "high" rates were found in both groups, no differential was seen.

Table 11.4   Metal Sensitivity[a] and Associated Complications

Patient classification	Total number	Sensitive[a]	
		preoperative	postoperative
I: No previous bone operation	173	10	14[b]
II: Previous metal implant	17	2	2
III: Loose THR (to be revised)	16	2	2
IV: Normal THR (contralateral)	6	—	—

THR = total hip replacement.
[a]Sensitive to at least one: Co^{2+}, Ni^{2+}, CrO_4^{2-} (by skin test).
[b]From retest, 6 months postoperatively of 66/168 patients with no preoperative sensitivity.
Source: Adapted from Deutman et al. (1977).

A later study of 212 patients awaiting total hip replacement (Deutman et al., 1977) produced more definite data. These patients could be divided into four groups, as shown in Table 11.4. This study shows a modest possibility of association between sensitivity and device loosening and a small possibility of either sensitization or activation of previous sensitization by metallic implantation.

Another clinical report (Brown et al., 1977) casts further doubt on an easy interpretation. This study reports a group of 20 American patients with 23 hip implants of the metal-on-metal type. At least one implant was loose in each patient. No patients were found to be sensitive to cobalt, nickel, or chromium. This difference in incidence rates of metal sensitivity has been ascribed to different environmental exposure between Brown's American patients and Elves' and Benson's British patients.

The questions of the relationship between immune sensitivity and device loosening and of possible sensitization by implanted device components have remained of interest to the clinical and research communities. Carlsson et al. (1980) examined a group of 112 patients before and 134 patients after metal-on-polymer hip replacement and concluded that there was little or no relationship between previous sensitivity and loosening. They also felt that it was doubtful that devices could induce sensitivity. Waterman and Schrik (1985) studied 85 patients before and after metal-on-polymer hip replacement and, although finding definite evidence of post-operative sensitization to Cr^{+6}, Co^{+2}, Ni^{+2} and to methyl methacrylate, also concluded that there was no relationship between these findings and device loosening.

These latter studies suggest that there is probably more than one mechanism involved in device loosening. The results of Brown et al. would be

Table 11.5 Relationship of Metal Sensitivity[a] to Loosening in Internal Fixation of Fractures

Postoperative complications	Number (#)	Sensitive			Sensitive to (#[b])		
		M(#)	F(#)	Total(%)	Ni	Cr	Co
None	208	—	8	8(3.9)	7	3	—
Delayed union	230	10	14	24(10.4)	21	4	3
Infected	267	14	13	27(10.1)	26	2	4

[a]By skin test.
[b]Some sensitive to more than one metal; no cobalt in alloy used.
Source: Hierholzer (1990).

explained by some, such as Willert and Semlitsch (1977), as being secondary to accumulations of wear debris. In small amounts, wear particles are either encapsulated or removed to regional lymph nodes. When this system is overwhelmed, a foreign body response with accumulation of giant cells may invade the tissues surrounding the joint, causing resorption of both soft and hard tissue. In such cases, device loosening might be secondary to cellular attack of the interface between implant and tissue, or as it were, it might be a biological analog of crevice corrosion.

There has been continuing interest in the question of a possible relationship between immune sensitivity to metal implants and bone damage in the absence of infection (aseptic loosening). Leynadier and Langlais (1988) reviewed the studies cited here, as well as others, and summarized the outcome of 300 patients. In these studies they found an overall 7.4% incidence of sensitivity (to Ni, Cr and/or Co in some valence state) in patients with a good clinical outcome (N = 163) versus a 46% incidence of sensitivity in patients with aseptic loosening (N = 137). They further noted an unexpectedly high rate of sensitivity to cobalt (30%) which was more than ten times that expected from their control population. On the basis of this review, they concluded that, except in exceptional cases, loosening was more likely to promote development of sensitivity rather than vice versa.

This view is not substantiated by Hierholzer (1990) in a study of patients with nickel-bearing (steel) fracture fixation implants. He found that while tissue nickel concentrations around fracture fixation hardware were as much as 100 times greater in infected than non-infected sites, there was a positive correlation between incidence of all prosthetic loosening (whether associated with delayed union or infection) and metal sensitivity. His results are summarized in Table 11.5.

Table 11.6   Metal Sensitivity[a] at Device Removal

Alloy	Number	Insensitive (%)	Sensitive (%)	Reacting (%)[b]
Stainless steel	187	43	37	20
Cobalt-base	55	26	36	38

[a]Sensitive to one or more Ni^{+2}, Co^{+2}, Cr^{+6} (by MIF test).
[b]"Non-migrators" (all chemotaxis suppressed); reverted to sensitive on retest 30–60 days after implant removal.
Source: Adapted from Merritt and Brown (1985).

A major criticism of this entire line of clinical investigation of possible immune responses to implants is the relative crudity of the skin test. Merritt and Brown (1980) have adapted a test termed the leukocyte Migration Inhibition Factor (MIF) Test to determine sensitivity to metallic ions more accurately. This test is an in vitro measure of the ability of metal ions (incorporated into haptens) to inhibit migration of human leukocytes towards a chemotactic attractant. Inhibition of migration is taken as a measure of production of a leukocyte migration inhibition factor by T lymphocytes presumably activated by the specific metal-bearing hapten involved. In a later report (Merritt and Brown, 1985) this group reviewed a study of 283 patients who underwent routine or cause-related device removal. Their data are summarized in Table 11.6. These data suggest a far higher incidence of metal sensitivity in patients with metallic devices than in any other study previously cited. In a parallel test of 629 patients coming to surgery (without prior history of metal implantation) they found, again by the MIF test, that 25% were sensitive to at least one metal among nickel, chromium and cobalt.

Summary

It appears possible to draw the following conclusions concerning sensitivity to metallic biomaterials:

1. There is a significant level of immune sensitivity to metals in the general population.
2. Groups of patients who have had nickel, chromium and cobalt bearing implants display higher than expected incidences of sensitivity.
3. There are specific documented instances of symptoms consistent with a type 4 delayed hypersensitivity reaction being specifically related to the presence of a metallic implant.

4. There is a general correlation between sensitivity to metal and other clinical symptoms, especially device loosening; however, which is cause and which is effect is unclear at this time.

The entire issue of the role of specific (i.e., immune) versus general response to implants remains an exciting one. Basic and clinical research results can be expected to shed more light on it in the future.

REFERENCES

Bajpai, P. K. (1983). In: *Biomaterials in Reconstructive Surgery*. L. R. Rubin (Ed.). C. V. Mosby, St. Louis, pp. 243ff.

Benson, M. K. D., Goodwin, P. G. and Brostoff, J. (1975): Br. Med. J. 4(Nov. 15):374.

Bravo, I., Carvalho, G. S., Barbosa, M. A. and de Sousa, M. (1990): J. Biomed. Mater. Res. 24:1059.

Brendlinger, D. L. and Tarsitano, J. J. (1970): J. Am. Dent. Assoc. 81:392.

Brown, G. C., Lockshin, M. D., Salvati, E. A. and Bullough, P. G. (1977): J. Bone Joint Surg. 59A:164.

Carlsson, Å. S., Magnusson, B. and Möller, H. (1980): Acta Orthop. Scand. 51:57.

Chenoweth, D. E. (1986): Trans. Am. Soc. Artif. Intern. Organs 32:226.

Clementi, D., Surace, A., Celestini, M. and Pietrogrande, V. (1980): Ital. J. Orthop. Traumatol. 6:97.

Cramers, M. and Lucht, U. (1977): Acta. Orthop. Scand. 48:245.

Deutman, R., Mulder, T. J., Brian, R. and Nater, J. P. (1977): J. Bone Joint Surg. 59A:862.

Elves, M. W., Wilson, J. N., Scales, J. T. and Kemp, H. B. S. (1975): Brit. Med. J. 4(Nov. 15):376.

Evans, E. M., Freeman, M. A. R., Miller, A. J. and Vernon-Roberts, B. (1974): J. Bone Joint Surg. 56B:626.

Fisher, A. A. (1986): *Contact Dermatitis*, 3rd ed. Lea & Febiger, Philadelphia, pp. 710ff.

Friedberg, F. (1974). Q. Rev. Biophys. 7(1):1.

Hicks, J. H. (1958). In: *Modern Trends in Surgical Materials*. L. Gillis (Ed.). Butterworth, London, pp. 29ff.

Hierholzer, S. (1990): Personal communication.

Hildebrand, H. F., Veron, C. and Martin, P. (1988): In: *Biocompatibility of Co-Cr-Ni Alloys.* H. F. Hildebrand and M. Champy (Eds.). Plenum Press, New York, pp. 201ff.

Kossovsky, N., Heggers, J. P. and Robson, M. C. (1987): CRC Crit. Rev. Biocompat. 3:53.

Leynadier, F. and Langlais, F. (1988): In: *Biocompatibility of Co-Cr-Ni Alloys.* H. F. Hildebrand and M. Champy (Eds.). Plenum Press, New York, pp. 193ff.

Merritt, K. and Brown. S. A. (1980): Acta Orthop. Scand. 51:403.

Merritt, K. and Brown, S. A. (1985): In: *Corrosion and Degradation of Implant Materials: Second Symposium.* ASTM STP 859. A. C. Fraker and C. D. Griffin (Eds.). American Society of Testing and Materials, Philadelphia, PA, pp. 195ff.

Playfair, J. H. L. (1979): *Immunology at a Glance.* Blackwell Scientific Publications, Oxford.

Shepard, A. D., Gelfand, J. A., Callow, A. D. and O'Donnell, T. F., Jr. (1984): J. Vasc. Surg. 1:829.

Stern, I. J., Kapsalis, A. A., DeLuca, B. L. and Pieczynski, W. (1972): Nature 238:151.

Wahlberg, J. E. (1973): Berufsdermatosen 21:22.

Waterman, A. H. and Schrik, J. J. (1985): Contact Derm. 13:294.

Willert, H.-G. and Semlitsch, M. (1977): J. Biomed. Mater. Res. 11:157.

BIBLIOGRAPHY

Abbas, A. K., Lichtman, A. H. and Pober, J. S. (1991): *Cellular and Molecular Immunology.* Saunders, Philadelphia.

DeLustro, F., Dasch, J., Keefe, J. and Ellingsworth, L. (1990): Clin. Orthop. Rel. Res. 260:263.

Guyton, A. C. (1991): Textbook of Medical Physiology, 8th ed. Saunders, Philadelphia, pp. 374ff.

Hildebrand, H. F., Veron, C. and Martin, P. (1989): Biomats. 10:545.

Hood, L. E., Weissman, I. L. and Wood, W. B. (1978): *Immunology.* Benjamin/Cummings, Menlo Park, CA.

Kumar, P., Bryan, C. E., Leech, S. H., Mathews, R., Bowler, J. and D'Ambrosia, R. D. (1983): Orthopedics 6:1455.

Merritt, K. and Brown, S. A. (1981): In: *Systemic Aspects of Biocompatibility,* Vol. II. D. F. Williams (Ed.). CRC Press, Boca Raton, FL, pp. 33ff.

Merritt, K. (1984): Biomats. 5:47

Rostoker, G., Robin, J., Binet, O., Blamoutier, J., Paupe, J., Lessana-Leibowitch, M., Bedouelle, J., Sonneck, J. M., Garrel, J. B. and Millet, P. (1987): J. Bone Jt. Surg. 69A:1408.

Weir, D. M. (1977): *Immunology: An Outline for Students of Medicine and Biology*, 4th ed. Churchill-Livingstone, Edinburgh.

12

Chemical and Foreign-Body Carcinogenesis

12.1 DEFINITIONS

In this chapter, we shall consider the role of implants in carcinogenesis. Some definitions are needed to make the arguments clear and unambiguous.

Carcinogen: An agent capable of causing cancer.

Cancer: Perhaps the best definition is that of Roe (1966): "Cancer is a disease of multicellular organisms which is characterized by the seemingly uncontrolled multiplication and spread within the organism of apparently abnormal forms of the organism's own cells." Roe further states that the three key characteristics of cancer are *cellular multiplication*, *autonomy*, and *invasiveness*.

Neoplasm: A tissue mass arising from an abnormal, uncoordinated proliferation of cells.

Primary neoplasm: A locally arising neoplasm.

Tumor: Literally a swelling; used to refer to neoplasms.

Benign: Possessing controlled self-limiting growth without invasiveness or ability to metastasize.

Malignant: Possessing uncontrolled growth, invasiveness, and ability to metastasize.

Metastasis: A neoplasm arising by "seeding" from a primary malignant neoplasm to a remote site. Also called a secondary neoplasm.

Leukemia or lymphoma: A malignant neoplastic transformation of cells of the circulatory system.

Sarcoma: A malignant neoplasm arising from cells of connective tissue.

Carcinoma: A malignant neoplasm arising from cells of epithelial origin.

Mutagenesis: The production of inheritable (genotypic) changes in cells.

Carcinogenic/Tumorigenic: Used interchangeably to connote agents capable of causing cancer.

Carcinogenesis: The production of cancer (more properly but less commonly termed cancerogenesis).

With these terms in mind, we shall first examine the chemical origins of carcinogenesis as it relates to implants. Initially, all neoplasms associated with implants in experimental animals and in patients were thought to be chemical in origin. It is now recognized that these tumors can arise from both chemical and nonchemical origins. The nonchemical or so-called "solid-state" origin will be taken up in the later parts of this chapter.

12.2 CHEMICAL CARCINOGENESIS

12.2.1 Introduction

Chemical carcinogens have many different forms, attack a variety of cells, and produce a variety of neoplasms. The nature of their action leads to effects being possible near implants by direct solution or diffusion, at a distance by transport and concentration, or in the absence of implants by ingestion or inhalation. It is not within the scope of this book to discuss the details of neoplastic transformation. However, we should recognize that there is more than one transformation effect. Neoplastic growth may be initiated by alteration of metabolic processes, by alteration of replication processes (either by growth stimulation or by reduction of contact inhibition), or by mutagenesis. Although all carcinogens are now thought to be mutagens, not all mutagenic agents are carcinogenic. A mutation may be lethal (to a cell), prevent cellular replication or simply not affect metabolic or growth processes sufficiently to produce malignant behavior.

12.2.2 What "Everybody Knows" About Cancer

Before proceeding to a discussion of classes and types of chemical carcinogens, it would be well to discuss some popular misunderstandings about carcinogenesis in general and chemical carcinogenesis in particular. There are at least three things about cancer that "everybody knows," but probably are not so:

1. "Cancer is increasing." Figure 12.1 presents a summary of cancer death rates in the United States for the period 1930–1986, adjusted for age (American Cancer Society, 1990). The adjustment for age is necessary since, as life expectancy and, thus, age at death increase, a *fixed annual incidence rate* of mortality from one source will produce an *increase* in actual deaths. Some death rates due to particular cancers are decreasing, especially stomach cancer in men and stomach and uterine cancer in women. Some rates are increasing, in particular the rate of death due to lung cancer in both sexes. However, excluding lung cancer, which is related to environmental effects and tobacco smoking, death rates due to cancer have either been constant or have decreased since 1945 for men and since 1930 for women. Furthermore, although nearly 30% of Americans alive today will develop cancer, the five year survival rate, despite an aging population, has increased to 40% compared with 33% in the 1960s. The increased survival rate is due in part to earlier detection, permitting

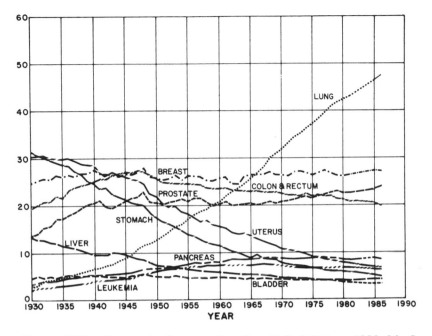

Figure 12.1 Cancer death rates, by site, United States, 1930–86. Source: American Cancer Society, 1990 (by permission).

both a longer normal course until death and earlier medical intervention, and to improved therapies, especially for some specific cancers.
2. "Everything causes cancer." In fact, the contrary seems to be true. Chemical carcinogenesis seems to be the exception rather than the rule. The statistics are as follows:
There are approximately 1,500,000 compounds and substances that are individually identifiable. An exhaustive literature search concerning the 6,000 most likely candidates uncovered evidence that only 17% or approximately 1,000 were *possible* carcinogens. A survey by the National Institute of Occupational Safety and Health (Christiansen and Fairchild, 1976) of 2,415 *suspected* carcinogens produced evidence of 1,905 having reported carcinogenic effects but only 1,000 thought to be carcinogenic in animals. In the most recent compilation available, 24 substances or groups of related substances and 5 occupational exposures are listed as *known* to cause cancer in humans; 119 additional ones can *reasonably be anticipated* to be carcinogens (U. S. Department of Health and Human Services, 1985).

3. "Toxic materials cause cancer." The classic study is that of Innes et al. (1969). He and his coworkers selected 120 pesticides and toxic industrial chemicals for evaluation. They were fed to two strains of mice in the maximum tolerable doses. The animals were sacrificed after a standard period and evaluated for tumor incidence. The results were as follows:

> Eleven compounds (including five insecticides) were significantly carcinogenic.
>
> Twenty compounds were equivocal, that is, did not show *significant* elevation of cancer incidence rates in this study.
>
> Eighty-nine compounds showed no elevation of cancer incidence rates.

Thus, in this study, of the 120 compounds selected for their toxicity and given in maximum possible doses, fewer than 10% proved to be carcinogenic. Furthermore, these findings have come under increasing criticism as being too pessimistic. It has been suggested that testing toxic potential carcinogens at high dosages may artificially accentuate their activity by inducing increased rates of cell division (Ames and Gold, 1990).

In classifying materials which are chemical carcinogens, we recognize three types of agents:

1. The *complete carcinogen* that produces neoplastic transformation by itself.
2. The *procarcinogen*, or carcinogen precursor, which is not in itself a carcinogen but is converted to one by metabolic processes in the body of the test animal or human.
3. The *cocarcinogen*, which is a weak carcinogen or has no inherent carcinogenic activity but increases the activity of complete carcinogens or procarcinogens when it appears in their company.

The exact roles and functions of both pro- and cocarcinogens remain unclear. It has been suggested that neoplastic transformation is a two-step process (Friedewald and Rous, 1944):

1. *Initiation*, which produces the primary cellular transformation. The cells enter a *latent* period and do not ordinarily develop into a tumor.
2. *Promotion*, which is characterized by the development of previously transformed cells into an active visible tumor.

More recently, this process has come to be viewed as having three steps, with specific conditions necessary during the latent period if subsequent expression (development of a tumor) is to occur.

Thus, the complete carcinogen can be viewed as one that is both an initiator and promoter, while the cocarcinogen may be either promoter or initiator but probably not both. Similarly, the procarcinogen may not be a complete carcinogen after metabolic conversion but its action may depend upon the presence of other initiators and promoters. These issues have been discussed more completely by Berenblum (1969). The details of the discussion make it clear that potential carcinogens must be considered in their roles as both complete or incomplete agents as well as possible promoters of previously initiated processes of neoplastic transformation.

12.2.3 Types of Chemical Carcinogens

An excellent review of these three types of agents among organic compounds is that of Weisburger and Williams (1975). Figures 12.2 and 12.3 and Table 12.1 are drawn from this study. Figure 12.2 lists some of the typical, stronger, pure organic carcinogens with their chemical structures. Table 12.1 lists some of the better known procarcinogens. The details of metabolic conversion are still unclear for many of these agents. Table 12.1 suggests the form of the converted carcinogen while Figure 12.3 provides examples of possible intermediates and structures. Of particular interest to us is the inclusion of vinyl halide or acetate in the procarcinogen list and a variety of epoxides in the activated list in Table 12.1. Polyvinyl chloride (PVC) and polyvinyl acetate (PVA) have some popularity in medical applications, while epoxide conversion is possible for many polymeric implant materials.

Testing for possible agents, especially of the pro- or co-type, is quite difficult due to the necessity of following the products through the various steps of the metabolic chain. Also, many of the small animals used for these tests have significant, and not inconsiderable, rates of spontaneous neoplastic transformation. While many agents that produce neoplastic transformation in test animals have not been definitely shown to be carcinogenic in man, it is essentially correct to assume that *all* human carcinogens also produce neoplastic transformation in animals. All but 2 of the 29 known (either specifically identified or associated with occupational exposure) chemical carcinogens in humans have been shown to have carcinogenic activity in at least one test animal species. However, the neoplasms may vary widely in location, malignancy, etc.

β-Propiolactone	$\begin{array}{ccc} O & \!\!\!-\!\!\! & CO \\ | & & | \\ CH_2 & \!\!\!-\!\!\! & CH_2 \end{array}$
1,2,3,4-Diepoxybutane	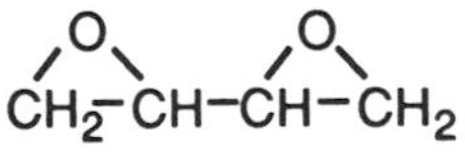 $CH_2\!-\!CH\!-\!CH\!-\!CH_2$ with epoxide O bridges
Ethyleneimine	$\begin{array}{c} NH \\ CH_2\!-\!CH_2 \end{array}$
Propane sulfone	$\begin{array}{ccc} O & \!\!\!-\!\!\!-\!\!\!- & SO_2 \\ | & & | \\ CH_2\!-\!CH_2 & \!\!\!-\!\!\! & CH_2 \end{array}$
Dimethyl sulfate	$CH_3OSO_2OCH_3$
Methyl methanesulfonate	$CH_3SO_2OCH_3$
Bis(2-chloroethyl) sulfide (mustard gas or yperite)	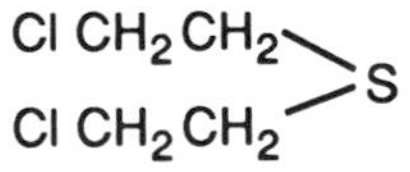 $\begin{array}{c} Cl\,CH_2\,CH_2 \\ Cl\,CH_2\,CH_2 \end{array}\!\!\!>\!\!S$
Nitrogen mustard (HN$_2$)	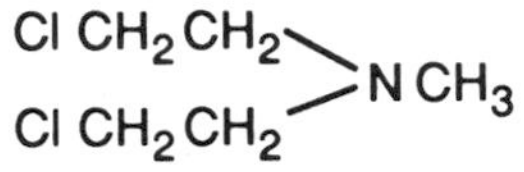 $\begin{array}{c} Cl\,CH_2\,CH_2 \\ Cl\,CH_2\,CH_2 \end{array}\!\!\!>\!\!N\,CH_3$
Bis(chloromethyl) ether	$ClCH_2OCH_2Cl$
Benzyl chloride	$C_6H_5CH_2Cl$
Dimethylcarbamyl chloride	$(CH_3)_2NCOCl$

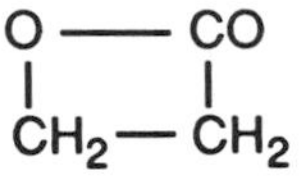

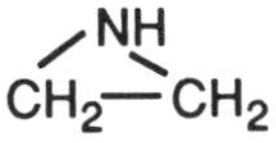

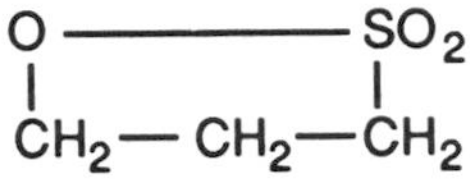

Figure 12.2 Typical direct-acting chemical carcinogens. Source: Weisburger and Williams, 1975 (by permission).

12.2.4   Metals as Chemical Carcinogens

The position of metals as carcinogens is less clear. One of the difficulties in determining this is the problem of distinguishing between *chemical* and *foreign-body* (FB) (see Section 12.3) action. Mechanical implantation or inhalation of metal dust may proceed to neoplastic transformation by a chemical route after corrosion, by an FB route by the presence of the residual metal, or perhaps due to aggregation of corrosion products at the implant or at a remote site. Furthermore, unlike most organic molecules, metals can display a wide range of electronic valences.

Procarcinogen --> (proximate carcinogen) --> Ultimate Carcinogen

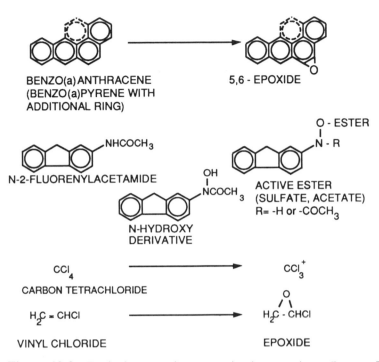

BENZO(a)ANTHRACENE
(BENZO(a)PYRENE WITH
ADDITIONAL RING)

5,6 - EPOXIDE

N-2-FLUORENYLACETAMIDE

NHCOCH$_3$

N-HYDROXY
DERIVATIVE

OH
|
NCOCH$_3$

O - ESTER
|
N - R

ACTIVE ESTER
(SULFATE, ACETATE)
R= -H or -COCH$_3$

CCl$_4$
CARBON TETRACHLORIDE

CCl$_3^+$

H$_2$C = CHCl
VINYL CHLORIDE

H$_2$C - CHCl
EPOXIDE

Figure 12.3 Typical procarcinogen activation reactions. Source: Weisburger and Williams, 1975 (by permission).

Metals may be placed in a classification system as given in the earlier discussion. They may be directly (or completely) carcinogenic (pure action) or they may potentiate other agents and their compounds. Reaction products or organometallic complexes may be carcinogenic, thus classing the original form as a procarcinogen. Potentiation, classing metals as cocarcinogens, is a very broad, nonspecific activity since many forms of neoplasms tend to concentrate metallic ions and complexes. It is difficult to distinguish cause and effect here: the concentration of metals may be causal or merely the consequence of the higher level of metabolic activity of the neoplastic cells.

Sunderman (1971), summarizing a broad range of animal studies, has made a strong case for carcinogenic roles for chromium, cobalt, iron, nickel, titanium, and for some metals not found in implant alloys. Environmental and industrial work place studies support the presumed carcinogenicity of chromium, cobalt, nickel, and, perhaps, iron.

Table 12.1 Principal Procarcinogens and Key Derived Active Metabolites

Procarcinogen	Actual or proposed proximate or ultimate carcinogen
Polycyclic aromatic hydrocarbons	Epoxide Radical ion?
Aflatoxin	Epoxide
Arylamine or amide; azo dyes	N-Hydroxylamino-O-esters; radical ion (?); epoxide (special case: cutaneous cancers)
Nitro aryl or heterocyclic compounds	N-Hydroxylamino-O-esters
3-Hydroxyxanthine, related purines	O-Esters
Safrole	1'-Hydroxy-O-Esters
Urethane, alkylcarbamates	Active esters
Pyrrolizidine alkaloids	Pyrrolic esters
Alkylnitrosamines or -amides, alkylhydrazines or -triazenes	Alkyl carbonium ion
Halogenated hydrocarbons	Haloalkyl carbonium ions
Vinyl halide or acetate	Epoxide?

Source: Adapted from Weisburger and Williams (1975).

Furst (1978) has extensively reviewed the status of metals as carcinogenic agents. This review is noteworthy because the author had previously (1969) proposed strict criteria that a material should meet before it could be considered carcinogenic:

> Tumors must appear both at the site and at a distance from the point of application; more than one route [of application] must be effective; more than one species must respond; the growth should be transplantable; and, if malignant, invasion and/or metastasis must be noted. Most important, all histological slides must be evaluated by a pathologist knowledgeable in animal tumors.

These criteria, although more than 20 years old, are still applicable and very relevant today. Of importance to us are the following conclusions drawn from studies subject to the above criteria:

Metals for which both pure metal and compounds are carcinogenic: Ni.
Metals for which pure metal is carcinogenic but no carcinogenic compound is known: Co.
Metals for which pure metal is not a carcinogen but which have carcinogenic compounds (given in parentheses): Cr (CrO_4^{2-}), Fe (dextran, dextrin), Ti (titanocene ?), Mn ($MnCl_2$?) (? = there is some doubt).

Only metals of interest in implant applications are included above. Furst (1978) lists several others, including cadmium, lead, and beryllium, that fall into one of these categories. However, it is fair to state that Furst regards metallic carcinogenesis as a well-established, real effect.

The complexity of the problem presented by potentially carcinogenic metallic implants is shown in a study by Gaechter et al. (1977). These investigators implanted polished rods of seven alloys, including common stainless steel, cobalt- and titanium-base implant alloys, in rats and followed them for two years. Each of these alloys contains at least one element recognised by Furst as carcinogenic. Neoplasms of a wide variety of types were found, but no statistical elevation above the control (nonimplanted) group incidence rates was seen. This study was possibly suggested by an earlier one by Heath et al. (1971) in which, 4 to 15 months after implantation, wear produced particles from a Co-Cr metal-on-metal total joint replacement were shown to be carcinogenic in rat muscle.

There are two possible arguments to explain these conflicting findings. In the first place, the rods used by Gaechter et al. (1977) may have released metal at a slower rate than seen in the works referenced by Sunderman (1971) or in the study of Heath et al. (1971). Thus, dilution may have prevented either direct chemical carcinogenesis or indirect (pro-) carcinogenesis by maintaining pool concentrations below critical levels. This possibility is supported by a later, much larger, somewhat longer rat implant study (Memoli et al., 1986) using both rods and powder, which demonstrated a small but significant increase in incidences of sarcomas and lymphomas in animals with implants containing cobalt, chromium or nickel.

The possibility that dilution may reduce the risk of neoplastic transformation leads directly to the question of whether a "threshold" of effect exists. That is, is there a concentration of a carcinogenic agent below which it loses its effectiveness? This is a matter of considerable importance in the implant field because corrosion rates of successful alloys are relatively quite low. A high-corrosion-rate alloy would probably be rejected for implant applications because of an acute tissue response. These low corrosion rates result in modest serum and tissue concentration increases, except in instances of local concentration as will be discussed in Chapter 14.

A great deal of attention has been paid to this possibility of threshold levels by legislators and administrators concerned with food purity and work place safety. The common view is that no threshold exists; that is, the transforming effect is like a molecular "trigger" and reduced concentration simply reduces the likelihood that the critical event will take place.

Therefore, given random chance enhanced by continued exposure, *any* concentration of a carcinogen can eventually evoke a neoplastic response. This view was the precipitating factor in the adoption in 1958 of the now famous Delaney Amendment to the Pure Food, Drug and Cosmetic Act. This statement imposed a *zero* (!) permissible level of carcinogenic agents as deliberate food additives. It states:

> . . . no additive shall be deemed to be safe if it is found to induce cancer when ingested by man or animal, or if it is found, after tests which are appropriate for the evaluation of the safety of food additives, to induce cancer in man or animals. . . .*

Contrast this with the discussion in Chapter 1 on value judgments inherent in definitions and the carefully enunciated position of Furst (1978) on the carcinogenic status of metals and their compounds.

It is clear that, applying the Delaney criteria, none of the metals listed by Furst (1978) could be judged satisfactory, even for short-term implantation. However, the wide utility of probably carcinogenic food substances, such as certain dyes and saccharin, has resulted in a case by case relaxation of the Delaney criteria for food additives. These decisions have been made by balancing risk against benefit with, admittedly, a portion of political judgment added in some cases.

We shall probably have to make the same careful judgments concerning metallic implants. To do so we need to accurately know what the dose-response relationship for carcinogenesis is in the various animal models, how to project this to low dose/long response time conditions (where animal experiments become prohibitively expensive), and, most importantly, how to translate the animal projections to rate predictions in humans. Very little of the required information is now available. Moreover, one disappointing factor is emerging. It appears that *linear* projections of response rates to low dose rates, even when the dose-response curve is itself linear, provide *underestimates* of the effect. One reason for this is discussed briefly in the following section.

An excellent recent review of metal-associated chemical neoplastic transformation has been provided by Sky-Peck (1986).

12.2.5 The Latent Period

A second argument which may shed light on the results of Gaechter et al. (1977) raises the issue of *latency*. It is common in both animal and human neoplastic transformation for a period of time to pass between exposure (initiation) and manifestation of neoplastic transformation. This waiting

---

*Cited in Federal Register 42(192), Tuesday Oct. 4, 1977, Part VI, page 54166.

or latent period differs from species to species and is different for each agent. In humans, latency periods are typically 15–20 years and may be as long as 40 years (Schottenfeld and Haas, 1979). Furthermore, there is no simple way to "scale" the effect, that is, to predict the latent period for an agent in one species from that observed in another species. Thus, one may argue that the latent period in the Gaechter et al. experiment exceeded the test period, despite the fact that 2 years is more than half the lifespan of the laboratory rat.

The latency argument is particularly important in the the application of conclusions such as those of Sunderman (1971) to expectations of implant site tumor incidences in patients. The vast majority of implants have been in patients for only 15 or fewer years because of the advanced age of the average implant patient and the relatively recent advances in total joint replacement. For instance, Table 12.2 illustrates that patients 65 years or older have a less than 50% chance of outliving a twenty year latency period. Thus, the appearance of metal carcinogenesis in humans may be awaiting the passage of an unelapsed latency period in the younger patients who have received implants in large numbers in the last decade (however, see Section 12.4). The differences in life expectancy at any age between men and women depend upon a number of factors, including occupational exposure, recreational pursuits, etc. and may not be related merely to gender difference.

Ceramic-body induction of carcinogenesis through a chemical route has not been reliably identified in animals at this time. This is probably due both to the low solubility of ceramics used in implants and the paucity of testing.

12.3 FOREIGN-BODY CARCINOGENESIS

12.3.1 Early Observations of Foreign-Body Carcinogenesis

So far we have considered the various classes of chemical carcinogens and have summarized the available information concerning their metabolism and ability to produce neoplastic transformation. Across all the classes of pro-, co-, and complete chemical carcinogens, it may be stated that the risk of neoplastic transformation increases *at least* linearly with the concentration and period of exposure.

Studies of chemical carcinogenesis show interesting differences in action depending upon the manner and form of administration of the agent. Such differences early on led investigators to study the influence of the physical form of the carcinogenic agent on its ability to induce transformation. A startling finding was that many agents not previously

Table 12.2 Life Expectancy* by Age in the United States

Age (years)	Male	Female
at birth	71.3	78.3
5	67.3	74.1
10	62.4	69.2
18	54.6	61.3
21	51.9	58.4
30	43.7	49.7
40	34.5	40.2
50	25.8	31.0
60	18.0	22.5
65	14.7	18.6
70	11.7	15.0
75	9.1	11.7
80	6.9	8.8
85	5.2	6.4

*Mean; all races; alive in 1986.
Source: National Center for Health Statistics (1988).

thought to be carcinogens produced dramatic neoplasm incidence rates in rodents when implanted in a solid form rather than injected or fed in soluble or dispersed form. This effect was called foreign-body (FB) carcinogenesis and is known more recently as solid state† carcinogenesis.

Among the early investigators of FB carcinogenesis were E. Oppenheimer and B. S. Oppenheimer who, in conjunction with a number of co-investigators, published a long series of papers in the 1940s and 1950s (cf. Oppenheimer et al., 1955). Their studies and those of other investigators of the period established the following points:

1. Solid materials without chemical carcinogenic activity can induce a variety of neoplasms in several small rodent species.
2. The induction activity generally increases with the size of the implant.
3. The induction activity varies inversely as the inflammatory response; that is, *well*-tolerated materials are, in the long run, better FB carcinogens.
4. Porosity with an average diameter above 0.22 μm (the smallest size studied) reduces the risk of transformation.

†I dislike this phrase due to its confusion with semi-conducting materials and, thus, the implication of electronic causality.

12.3.2 Mechanisms of Foreign-Body Carcinogenesis

An excellent contemporary summary of these early investigations, primarily using plastic films as challenge agents, is that of Alexander and Horning (1959) who proposed the following:

> The most likely process [of neoplasm induction] would appear to be that the film alters the normal environment of the neighboring cells in such a way as to favor the induction (or selection) of discontinuous variations leading to malignancy.

That is, the survival of viable products of normally occurring cell damage or mutation are somehow favored by the presence of the solid body and protected from physiological processes until they are ready to enter the rapid growth phase characteristic of malignancy.

We can recognize two considerations that appear to favor this argument. The first is geometric and was discussed briefly in Chapter 9 when we examined the role of implants in infection. A cell is normally surrounded by a volume of tissue subtending a 4π solid angle. As we approach an implant, the solid angle decreases to 2π. Furthermore, if the implant is invaginated with a surface roughness that has a characteristic dimension on the order of cell sizes (2–20 μm), the solid angle might even be less than 2π for some selected cells. The results would be less access to microvasculature, poorer diffusional supply, and reduced cell contact inhibition. The flaw in this general line of argument is, of course, the observation that materials with distributed porosity of cellular dimensions are *less* carcinogenic than smooth nonporous materials. Perhaps one of these geometric factors is dominant or perhaps the improved diffusion and cellular activity associated with microporous surfaces offsets the other aspects of the near surface geometry.

The second consideration that favors the argument proposed by Alexander and Horning (1959) is that chemical and electrical conditions near an implant-tissue interface are different from those at a distance; this has been discussed extensively in previous chapters. The question remains, however, of which field and concentration effects might favor protection of deviant cells. Russian investigators (cited by Bischoff and Bryson, 1964) felt that piezoelectric materials with sharp points and asperities were more tumorigenic than the same materials in smooth or colloidal form. This suggested a role for very high gradient electric fields in FB carcinogenesis. A study by Andrews et al. (1979) attempted to investigate this question by subcutaneous implantation of plates of polystyrene resin in mice. The resin was implanted in either neutral condition or as poled electrets of various strengths. Tumors associated with the control plates were evenly distributed on both sides, while those associated with the electrets were found predominantly on the electronegative side (although the difference, as well as all other conclusions reported for this study, was not statistically significant). The investigators felt that there was a trend to higher incidence rates and shorter latency periods for electrets with higher fields.

A more complete review (Bischoff and Bryson, 1964) updated and expanded this critique. The authors posed and answered three questions:

Question 1: "Is the concept of nonspecific (rather than by a specific chemical agent) solid state carcinogenesis justified?"

Answer: After considerable criticism of experimental method and test subject, they conclude that "on the basis of the responses to rather stable, unrelated substances, there is a type of nonspecific carcinogenesis in rodents that is dependent upon a minimum surface requirement."

Question 2: "Does solid state carcinogenesis occur in humans?"

Answer: For humans, the authors note the low reported incidence of neoplasms associated with implants, natural deposits (cholesterol plaques, gall stones, etc.), and chronic low-level inflammatory processes (leading to acellular fibrotic tissue which has FB attributes). They admit an exception to this basic pattern in the observation of carcinoma associated with silicosis and asbestosis. [Note however that in 1964 the very high correlation between a specific type of asbestos (chrysotile) and mesothelioma was not yet known.] They also recognize the problem of the latency period and of the relatively smaller implants (with respect to body weight) used clinically as compared with those in the experiments of Oppenheimer and others. Thus, they concluded that "the incidence of sarcoma (in humans) arising from subcutaneous FB granuloma is minimal."

Question 3: "Is the subcutaneous site in rodents valid for testing for carcinogenic hazards?"

Answer: This question reflects their observation of a "widespread disenchantment" with subcutaneous studies in rodents. Their arguments are rather vague, but, essentially, they arrive at the point of view that, through nonspecific irritation, FB carcinogenesis and chemical carcinogenesis may occur together and, without adequate controls, cannot be easily distinguished in the subcutaneous site. They suggest, however, that the determining factor is the difference between response to irritation in the chemical case, and transformation after *noninflammatory* isolation of the foreign body in the FB case. In this latter comment they presage the more modern studies of FB carcinogenesis.

12.3.3 Additional Studies of Foreign-Body Carcinogenesis

Further studies by Ott and by Brand and their students have focused upon the mechanisms of neoplastic transformation in rodents in order to distinguish FB from chemical carcinogenesis. Brand (1975) has more fully summarized his research and theoretical ideas. He has made extensive use of related species of mice that will accept tissue transplants without immune response, but in which cells and their daughters can be identified, by species, through an examination of chromosomes *(karyotyping)*. I will paraphrase two sections of his paper dealing with the mechanism of the tumorigenic process and hypotheses concerning initiation and promulgation of the process.

Brand reaches the following conclusions, which were documented by thorough and ingenious research:

1. The most probable target cell in FB carcinogenesis is the pericyte, a small cell type associated with microvasculature.
2. After implantation, the transformation needed to produce a preneoplastic parent cell, with all the genetic information for later expression in the active neoplasm, occurs quite rapidly. In his mice populations this occurs within 4–8 weeks of implantation.
3. While the transformation occurs near the FB-tissue interface, actual close contact with the FB is not required.
4. Transformation is quite uncommon; thus, neoplasms appear to develop from single parent cells representing one in the several million affected by the presence of the implant.
5. While neoplasm production will occur in a capsule after FB removal, a significant period of implantation after the initial transformation event is required.
6. A latent period always occurs between transformation and neoplastic expression. However, this period is characterized more by inactivity of microphages than of the transformed parent cell which may be cloning (undergoing mitosis without inheritable change) at a slow rate. In the presence of active macrophages, as in moderate to severe chronic inflammation, later neoplastic expression is suppressed.
7. When the latent period is over, and rapid malignant growth begins, all daughter cells, even if transplanted, appear to act in synchrony.

Brand's in vivo studies have been partially verified by a set of in vitro tissue culture experiments (Boone et al., 1979). Boone and coworkers studied the effects of attachment of mouse fibroblasts to polycarbonate plates in an in vitro tissue culture system.

Cells implanted after in vitro exposure produced transplantable, undifferentiated sarcomas. Notwithstanding a decrease in latent period with increased time in tissue culture, the authors concluded, as had Brand before them, that the smooth surfaces of the plates acted as an FB carcinogen for at least initiation, independent of chemical composition.

Brand (1975) cited six proposed mechanistic origins of FB carcinogenesis (with accompanying criticisms):

1. Chemical activity of components of the FB. While he suggested a moderating or modifying role for chemical agents, Brand felt that the nonchemical mechanism for FB carcinogenesis was well established.
2. Physiochemical surface properties of the FB. Again Brand recognized a possible role for interface physical effects but suggested that they are overpowered in importance by other physical factors, such as porosity.
3. Interruption of cellular contact or communication. Brand felt that this is an open question but indicates, however, that it would be expected to play a more important role in neoplasm expression and maturation than in induction.
4. Tissue anoxia and insufficient exchange of metabolites. Brand rejected this hypothesis based upon comparison of cell distances from vascular processes in normal and neoplastic tissue and upon induction studies with vascularized microporous surfaces.
5. Virus (as an unseen contaminant of FB's). The evidence is still weak for this mechanism; Brand evidently did not favor it.
6. Disturbance of cellular growth regulation. Brand clearly favored this mechanism, based on the heritability of neoplastic behavior in the growing cell population. He suggested that there is a wide variety of possible aberrations in growth control and communication processes in cells. Thus, his view is that the nonspecific surface effect, whatever its origins, acts in a mutagenic fashion on cell populations.

This discussion of FB carcinogenesis has focused on nonchemical neoplastic transformation effects produced by materials *external* to cells. However, solid materials in a form which can penetrate cells can also produce FB transformation. The best known example is crysotile asbestos which was recognised as a carcinogen only because it produced a relatively rare lung tumor (mesothelioma) (U. S. Department of Health and Human Services, 1985). Studies of asbestos and other fibers in animal models, has led to the Stanton Hypothesis: Mesothelioma can be induced by fibers less than 1 μm in diameter and more than 8 μm in diameter, regardless of fiber composition (Lipkin, 1980). Lipkin has also shown that in vitro fiber cytotoxicity correlates well with these dimensions rather than with fiber composition. Thus, slender stiff fibers, such as mineral whiskers, which are apparently able to penetrate cells and produce direct mechanical damage, appear to be undesirable components of biomaterials.

Perhaps the best place to end this discussion is with a quotation from Brand (1975):

> Despite the rarity of FB tumors in man, it would be irresponsible to look at the situation with complacency. Several measures are at our disposal which would minimize the probability of FB tumors in man. . . . These include (1) a more restrictive approach to artificial implantations, especially the exclusion of medically unnecessary cosmetic procedures, unless they are indicated for psychiatric reasons; (2) smallest possible size of implants; (3) reexamination of implant carriers at frequent intervals; (4) a centralized registry for gathering information on general complications as well as instances of neoplasia; (5) continued research (a) on implant materials regarding their suitability for specific surgical purposes and (b) on etiological questions concerning this type of neoplasia [p. 487].

There is certainly food for thought for the bioengineer in this passage.

12.4 NONSPECIFIC CARCINOGENESIS

A final form of neoplastic stimulation is also recognized. Neoplasms can arise in response to chronic irritation (leading to chronic inflammation). Chemicals (as well as foreign bodies), infection, and mechanical trauma have all been recognized as leading to this type of neoplastic transformation. It is characterized by an infidelity of replication (producing a daughter cell not identical to its parent). The formation of keloids (hyperplastic, expansive scars) is a nonmalignant example of this effect. The occasional, apparently spontaneous, malignant transformation of benign lesions, such as fibrous histiocytomas, is a somewhat more ominous example (Heselson et al., 1983).

12.5 EVIDENCE FOR IMPLANT CARCINOGENESIS IN HUMANS

In light of the long use of metallic implants in clinical orthopedics and other surgical specialties, it is fair to ask if there is any evidence of either chemical or FB carcinogenesis associated with implants.

There is a mounting number of reports of tumors at implant sites in animals. Sinibaldi et al. (1976) reported sarcomas in animals (7 dogs and

1 cat) occurring 6 months to 4 years after implantation of stainless steel devices. Of interest is the fact that five of the eight cases accompanied the use of the Jonas telescoping splint. This device, with its integral deep "crevice" between parts, would be expected to display an abnormally high rate of corrosion (see Chapter 4). Harrison et al. (1976) had earlier reported two cases in dogs, one 6 years and one 12 years after stainless steel implantation during clinical treatment of fractures. An additional case of sarcoma occurring 12 years after Jonas splint implantation in a dog has been reported more recently (Madewell et al., 1977). A survey by Stevenson et al. (1982), including these cases, reported 35 fracture-associated sarcomas in animals with an average of 5.8 years between injury (and internal fixation) and diagnosis. Finally, there is an apparent association between these tumors in animals and implant site infection. Such infection may be expected to produce elevated rates of corrosion and thus elevated concentrations of metal bearing species near implants, due to the resulting local acidosis (see also Chapter 14).

The early orthopedic literature reports only three cases of implant site tumors: Delgado (1958), sarcoma after fracture of tibia, internally fixed; Dube and Fisher (1972), hemangioendothelioma after fracture of tibia, internally fixed; and McDougall (1956), sarcoma (Ewing's type) in fractured humerus, after internal fixation. The last case is the best known and occurred more than 30 years after the initial injury and metallic implantation. Additional cases continue to be reported, typically occurring more than five years after implantation. Since there is now a general recognition that fracture fixation hardware should be removed within two years, such cases are expected to continue to be rare.

There are now seventeen known cases of tumors in humans associated with partial or total joint replacements, with an average postoperative period before diagnosis of seven years. The early reports have been discussed previously (Black, 1988) and some more recent reports are listed in Table 12.3. These tumors fall into two general groups:

1. Tumors of various etiologies, occurring in fairly short periods after implantation.
2. Primarily malignant fibrous histiocytomas, occurring 10-15 years after implantation.

To date all of these tumors have been associated with either stainless steel or cobalt-base alloy devices. The origin of the former group is somewhat obscure; however, the latter group may reflect either direct chemical carcinogenesis (associated with elevated tissue concentrations of metals near the implant; see Chapter 14) or possibly malignant transformation of the previously benign implant capsule.

Table 12.3 Recent Reports of Tumors Associated with Human Joint Replacements

Series	Type of implant	Site	Interval from insertion of implant to diagnosis of tumor (months)	Tumor
Haag and Adler	Weber-Huggler THR	Femur	10 years	MFH
Kolstad and Högstrop	Freeman-Samuelson TKR	Femoral region	3 months	Adenocarcinoma (2° to gastric carcinoma)
Tait et al.	Charnley-Muller THR	Gluteal region	11 years	MFH
Troop et al.	Charnley-Muller THR	Distal femur	15 years	MFH
van der List et al.	Charnley-Muller THR	Femur and acetabulum	11 years	Malignant epithelioid hemangio-endothelioma

Orthopaedic devices are placed in soft and hard connective tissues which are not especially sensitive to primary neoplastic transformation in humans (Black, 1985). Until recently, there was no evidence of remote site tumors possibly resulting from concentrations of suspected carcinogens, such as chromates, since no large epidemiological study had been done to detect their presence or absence. Such a study has now been performed (Gillespie et al., 1988).

Gillespie et al. identified 1358 patients who had total hip replacements in New Zealand between 1967 and 1978. They investigated the health status of over one thousand of these patients who could be followed for more than ten years post-implantation and found a highly significant 70% elevated incidence of tumors of lymphatic and hemopoietic origin. In addition they observed a significant *suppression* of soft tissue (colon, bowel and breast) tumor incidence up to ten years post-implantation followed by an apparent but nonsignificant increase in incidence. Even ten year follow-up may be insufficient for expression of primary (chemical) tumors at the low doses encountered at remote sites in implant-bearing human patients. These results may simply reflect an effect of corrosion products producing a chronic immune system stress; that is, playing the part of indirect promoters of soft tissue neoplasias produced by other causes. Longer term follow-ups and larger, better defined study groups will be required to explore these preliminary results.

The question of the occurrence of FB tumors in patients also remains a mystery, due in part to the absence of large scale clinical epidemiology. Brand (1983) reviewed 43 tumors occurring in humans at implant sites, at up to 53 years after implantation. Twenty-five percent occurred within 15 years and 50% within 25 years of implantation. In light of the large upsurge in medical and cosmetic implant use in the 1950s and 60s, he would have expected "several ten-thousand cases, if there were a significant etiological association as . . . in experimental rats or mice." Thus, although continuing to express concern about the possibilities for FB carcinogenesis in humans, he concluded that there was little clinical evidence for its occurrence.

The traditional problems of projecting animal experience to human occupational and clinical situations must surely apply in the consideration of carcinogenesis as a possible consequence of biomaterial implantation. In particular it should be noted that, despite Brand's comments in his 1983 study, it is only now that significant human populations with implants in place for more than 10–15 years are coming into existence. During the next decade, the survivors of these large, modern cohorts will begin to pass the average latency period for chemical carcinogensis. There are no animal experiments to guide us accurately in what our expectations should be. Perhaps Brand's previous remarks might be paraphrased "despite the rarity of *implant* associated tumors in man . . . ," and more careful attention should be given to the challenging and difficult issues of possible chemical and FB carcinogenesis by clinical implants.

REFERENCES

Alexander, P. and Horning, E. S. (1959): In: *Carcinogenesis: Mechanisms of Action*, Ciba Foundation Symposium. G. E. W. Wolstenholme and M. O'Connor (Eds.). Little, Brown, Boston, pp. 12ff.

American Cancer Society (1990): *1990 Cancer Facts and Figures*. American Cancer Society, Atlanta.

Ames, B. N. and Gold, L. S. (1990): Science 249:970.

Andrews, E. J., Todd, P. W. and Kukulinsky, N. E. (1979): J. Biomed. Mater. Res. 13:173.

Berenblum, I. (1969): Prog. Exp. Tumor Res. 11:21.

Bischoff, F. and Bryson, G. (1964): Prog. Exp. Tumor Res. 5:85.

Black, J. (1985): *The Hip*, Vol. 13. C. V. Mosby, St. Louis, pp. 199ff.

Black, J. (1988): *Orthopaedic Biomaterials in Research and Practice*. Churchill Livingstone, New York, pp. 292ff.

Boone, C. W., Takeichi, N., Eaton, S. del A. and Paranjpe, M. (1979): Science 204:177.

Brand, K. G. (1975): In: *Cancer: A Comprehensive Treatise*, Vol. 1. F. F. Becker (Ed.). Plenum, New York, pp. 485ff.

Brand, K. G. (1983): In: *Biomaterials in Reconstructive Surgery*. L. R. Rubin (Ed.). C. V. Mosby, St. Louis, pp. 36ff.

Christensen, H. E. and Fairchild, E. J. (Eds.) (1976): *Suspected Carcinogens*, 2nd ed. CDC, National Institute for Occupational Safety and Health, Cincinnati, Ohio.

Delgado, E. R. (1958): Clin. Orthop. 12:315.

Dube, V. E. and Fisher, D. E. (1972): Cancer 30:1260.

Friedewald, W. F. and Rous, P. (1944): J. Exp. Med. 80:101.

Furst, A. and Haro, R. T. (1969): Prog. Exp. Tumor Res. 12:102.

Furst, A. (1978): Adv. Exp. Med. Biol. 91:1.

Gaechter, A., Alroy, J., Andersson, G. B. J., Galante, J., Rostoker, W. and Schajowicz, F. (1977): J. Bone Joint Surg. 59A:622.

Gillespie, W. J., Frampton, C. M. A., Henderson, R. J. and Ryan, P. M. (1988): J. Bone Joint Surg. 70B:539.

Haag, M. and Adler, C. P. (1989): J. Bone Joint Surg. 71B:701.

Harrison, J. W., McLain, D. L., Hohn, R. B., Wilson, G. P., III, Chalman, J. A. and MacGowan, K. N. (1976): Clin. Orthop. 116:253.

Heath, J. C., Freeman, M. A. R. and Swanson, S. A. V. (1971): Lancet (March 20):564.

Heselson, N. G., Price, S. K., Mills, E. E. D., Conway, S. S. M. and Marks, R. K. (1983): J. Bone Joint Surg. 65A:1166.

Innes, J. R. M., Ulland, B. M., Valerio, M. G., Petrucelli, L., Fishbein, L., Hart, E. R., Pallotta, A. J., Bates, R. R., Falk, H. L., Gart, J. J., Klein, M., Mitchell, I., and Peters, J. (1969): J. Natl. Cancer Inst. 42:1101.

Kolstad, K. and Högstorp, H. (1990): Acta Orthop. Scand. 61:369.

Lipkin, L. E. (1980): Environ. Health Persp. 34:91.

Madewell, B. R., Pool, R. R. and Leighton, R. L. (1977): J. Am. Vet. Med. Assoc. 171:187.

McDougall, A. (1956): J. Bone Joint Surg. 38B:709.

Memoli, V. A., Urban, R. M., Alroy, J. and Galante, J. O. (1986): J. Orthop. Res. 4:346.

National Center for Health Statistics (1988): *Vital Statistics of the United States, 1986*, Vol. II, Sec. 6: Life Tables. DHHS Pub. No. (PHS) 88-1147. U. S. Government Printing Office, Washington, D.C.

Oppenheimer, B. S., Oppenheimer, E. T., Danishefsky, I., Stout, A. P. and Eirich, F. R. (1955): Can. Res. 15:333.

Roe, F. J. C. (1966): In: *The Biology of Cancer*. E. J. Ambrose and F. J. C. Roe (Eds.). D. Van Nostrand, New York, p. 28.

Schottenfeld, D. and Haas, J. F. (1979): CA 29:144.

Sinibaldi, K., Rosen, H., Liu, S.-K. and DeAngelis, M. (1976): Clin. Orthop. 118:257.

Sky-Peck, H. H. (1986): Clin. Physiol. Biochem. 4:99.

Stevenson, S., Hohn, R. B., Pohler, O. E. M., Fetter, A. W., Olmstead, M. L. and Wind, A. P. (1982): J. Amer. Vet. Med. Assoc. 180:1189.

Sunderman, F. W., Jr. (1971): Food Cosmet. Toxicol. 9:105.

Tait, N. P., Hacking, P. M. and Malcolm, A. J. (1988): Brit. J. Radiol. 61:73.

Troop, J. K., Mallory, T. H., Fisher, D. A. and Vaughn, B. K. (1990): Clin. Orthop. 253:297.

U. S. Department of Health and Human Services (1985): *Fourth Annual Report on Carcinogens*. DHHS Pub. No. (PHS) PB85-134633. NTIS, Springfield, VA.

van der List, J. J. J., van Horn, J. R., Slooff, T. J. J. H. and ten Cate, L. N. (1988): Acta Orthop. Scand. 59:328.

Weisburger, J. H. and Williams, G. M. (1975): In: *Cancer: A Comprehensive Treatise*, Vol. 1. F. F. Becker (Ed.). Plenum Press, New York, pp. 185ff.

BIBLIOGRAPHY

Ambrose, E. J. and Roe, F. J. C. (1966): *The Biology of Cancer*. D. Van Nostrand, New York.

Berenblum, I. (1974): *Carcinogenesis as a Biological Problem*. North-Holland Pub. Co., Amsterdam.

Becker, F. F. (1975): *Cancer: A Comprehensive Treatise*, Vol. 1: *Etiology: Chemical and Physical Carcinogenesis*, Vol. 4: *Biology of Tumors: Surfaces, Immunology, and Comparative Pathology*. Plenum, New York.

Brodeur, P. (1980): *The Asbestos Hazard*. New York Academy of Sciences, New York.

U.S. Department of Health and Human Services (1981). Proceedings of a Workshop/Conference on the Role of Metals in Carcinogenesis, Atlanta, GA, March 24–28, 1980. *Environmental Health Perspectives*, Vol. 40.

13
Mineral Metabolism

13.1 INTRODUCTION

The next chapter will deal with the distribution of metallic ions and some simple models for their dispersion. In this chapter we will consider one

*Prepared in part by Dr. G. K. Smith.

well-known metal, iron, in detail and a lesser known metal, chromium. This consideration is both important in its own right, as well as being an indicator of the complexity of the metabolism of metals. As in the case of most other metals, the details of metabolic pathways and kinetics of these two metals are not fully known. However, we shall attempt to contrast iron metabolism with chromium metabolism.

As discussed in Sections 2.2 and 2.4, the human body is primarily composed of four nonmetallic elements (in declining order of abundance): oxygen, carbon, hydrogen and nitrogen, which make up 96.9% of body tissues by weight. Six additional elements, of which only three are metals (calcium, potassium and sodium) play major physiological roles and contribute an additional ∼3.2% of body weight. All other constituents, contributing together no more than 30 grams, are termed *trace elements*. These include at least thirteen metals, of which ten are used routinely as non-trace constituents in human implants: iron, copper,* aluminum, vanadium, manganese, nickel, molybdenum, titanium, chromium and cobalt.

Most of these trace elements (with the possible exception of titanium) play vital physiological roles, and thus are termed *essential* trace elements. The role of each is characterized by three important attributes (Mertz, 1981):

1. *Amplification:* All known essential trace elements exert their biological actions through a succession of regulatory and/or synthetic steps which produce a manyfold amplification function and lead to effects on the whole body.
2. *Specificity:* Each essential trace element has a specific role, as a moiety in a molecule or as an enzyme cofactor. This specificity depends both on ionic size and valence. Other ions may interfere with the specific role of a trace element but not replace its function.
3. *Homeostatic regulation:* Without exception, for each essential trace element, there is a panoply of absorption, transport, storage and excretion mechanisms which regulate concentration at the site of action within an optimum range.

Our interest in trace elements is related to this third attribute, to the possibility that the introduction of an endogenous source, i. e., release of material from an implant, may interfere with homeostatic regulation and produce adverse effects, either at the site of normal action or at other

---

*Copper is not used in human implants, due to its cytotoxicity, but is a component of some designs of semi-permanent intrauterine contraceptive devices (IUDs).

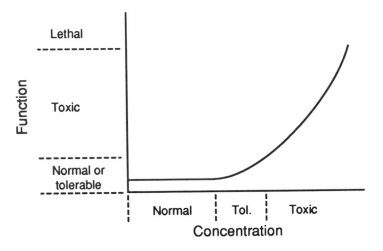

Figure 13.1   Relation between metal concentration and its functional consequences.

sites, either directly or by interference with other trace element mediated
processes.

Schwarz (1977), in discussing the toxic effects of mercury and its
organometallic compounds, makes the following point:

> Below a certain threshold (of concentration), the organism can maintain an
> equilibrium. However, once the threshold level is reached, small increases in
> doses lead to great increases of toxic effects. . . . Indeed, this relationship
> pertains not only to all metals but anything. It is universal, with the possible
> exception of mutagenicity and carcinogenicity, but even there repair mecha-
> nisms are at work which may give a small area of tolerance. [p. 3]

The general relationship between metal concentration level and functional
effect is shown in Figure 13.1. It should be emphasized that this is only
schematic in nature: the details and relative width of each range depends
upon the nature of the metal involved and may also depend upon individual differences between patients.

Thus, although calculations may be performed that predict various
levels of metallic ions in blood, tissues, etc. (Section 14.4), the real need
is to know the details of the metabolic and excretory pathways. The next
two major sections outline iron metabolism, whose details are well
known, and chromium metabolism, which is less well known and understood. Until the comparable systems are as well known for the other
major metallic components of common implant alloys, such as aluminum,

chromium, cobalt, nickel, titanium, etc., as they are for iron, great care will have to be taken in the interpretation of animal and clinical determinations of metallic content in vivo.

13.2 IRON METABOLISM

13.2.1 Introduction

Iron, a biologically ubiquitous metal, is essential to all higher forms of life owing to its central role in the heme molecule, facilitating oxygen and electron transport. The capacity of porphyrin-Fe-protein complexes to bind large quantities of oxygen reversibly makes hemoglobin and myoglobin well suited to the transport and storage of oxygen in higher organisms. Iron-containing enzymes include the cytochromes, catalase, cytochrome c reductase, succinic dehydrogenase, and fumaric dehydrogenase. Total body iron in an average adult ranges from 2 to 6 g, depending on body weight, hemoglobin concentration, age, sex, and size of the storage compartment. As a trace element, its presence in the body is exceeded only by magnesium (see Table 2.3).

On the basis of function, two iron metabolic compartments are recognized:

1. An *essential* compartment, containing 70% of the total body iron, is composed of hemoglobin, myoglobin, heme enzymes, cofactor and transport iron.
2. A *nonessential* storage compartment accounts for 30% of the total body iron in a normal individual and consists of iron storage in the form of ferritin and hemosiderin, primarily in the liver, spleen, and bone marrow.

The essential compartment can be further subdivided into the following distribution: 85% in hemoglobin, 5% in myoglobin, 10% in intracellular heme enzymes and iron cofactors in other enzyme systems, and 0.1% as transport iron bound to transferrin.

13.2.2 Absorption

Absorption of iron (see Figure 13.2) represents the single most important factor maintaining the normal balance of iron in the body. It is influenced by age, state of health, current body iron status, and conditions within the gastrointestinal tract, as well as by the amount and chemical form of iron ingested and by the relative iron chelating nature of other dietary constituents, e.g., phosphates, phytates, ascorbic acid, and amino acids. Dietary intake varies between 10 and 30 mg/day, with a common range of 12–15

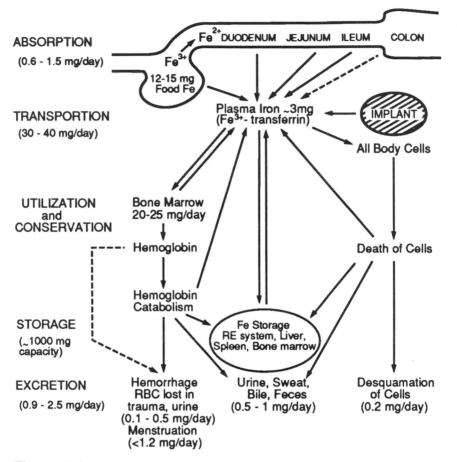

Figure 13.2 Iron metabolism in adults. Data from Fairbanks and Beutler (1988).

mg/day. Normally, only 0.6–1.5 mg/day are absorbed by the gastrointestinal mucosa representing only 5–10% of total dietary iron intake.

The actual mechanisms of absorption and transport of iron or iron chelates at the mucosal cell level are unclear; however, it appears that iron enters the mucosal brush border by a passive diffusion process and exits on the serosal surface to the plasma transferrin by an energy-requiring step. Intracellular absorbed iron in excess of immediate physiological needs is combined with a protein, *apoferritin*, to form *ferritin*, a water-soluble iron storage complex. As the mucosal cells become laden with ferritin, further absorption is impeded consistent with diffusion kinetics until ferritin iron is released to the plasma in response to body needs. Most absorbed iron, in the form of intracellular ferritin, is lost into the intestinal lumen when the crypt cells complete their 2–3 day maturation and migration to the tips of the villi and are sloughed. Intraluminal factors which decrease iron absorption include: rapid gastrointestinal transit time; achylia; malabsorption syndrome; precipitation by alkalinization by phosphates and phytates; and ingested alkaline clays and antacid preparations. Despite intensive research, the systemic factors regulating iron absorption have not been identified. In general, iron absorption increases whenever erythropoiesis (red blood cell production) is stimulated, during pregnancy, and in patients with hemochromatosis, while decreased absorption is associated with depressed erythropoiesis and iron overload.

13.2.3 Transport

After an iron atom enters the physiological system it is virtually trapped, cycling almost endlessly from plasma to developing erythroblasts. Iron is then released into the circulation for 100 to 160 days, and then is moved to phagocytic cells where it is cleaved from hemoglobin and finally released into the plasma to repeat the cycle. From the standpoint of distribution of total body iron, the transport compartment is the smallest ($\sim 0.008\%$). However, kinetically it is by far the most active, turning over as often as 10 times every 24 hours. The vehicle of this rapid transport and turnover of iron is *transferrin*, a β_1-globulin of approximately 86000 molecular weight, having a half-life of 8 to 10.5 days. Transferrin is synthesized in the liver, and the total body transferrin content of 7 to 15 g is nearly equally distributed between the intra- and extravascular spaces. It functions both to *accept* iron from gut absorption, storage sites, and phagocytic cells and to *deliver* iron to erythroid marrow for hemoglobin synthesis, to cellular reticuloendothelium for storage, to the developing fetus, and to all cells for incorporation into iron metalloenzymes. Normally, approximately one-third of the total body transferrin (termed *total iron-binding capacity*, TIBC = 300–360 µg/100 ml) is saturated with iron. The remaining transferrin represents a latent or unbound reserve [unbound iron-binding capacity (UIBC)]. The degree of saturation (%) and the TIBC are important parameters in the study of iron metabolism and related disease syndromes.

For example, *increased* TIBC is characteristically found in iron deficiency, in the third trimester of pregnancy, and in response to hypoxic states, whereas *decreased* TIBC is evident in infection, protein malnutrition, iron overload conditions, malignancy, cirrhosis of the liver, nephrosis, and protein-losing exteropathies. Figure 13.3 illustrates the relation-

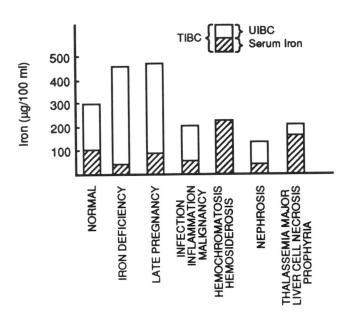

Figure 13.3 Serum iron concentration and the specific size of the transport compartment in a variety of clinical conditions. Source: Adapted from Fairbanks and Beutler (1988).

ships between plasma iron and transferrin in a variety of clinical conditions. In general, the level of plasma iron and saturation of TIBC are determined by the sum of factors extracting iron from the blood for storage and utilization balanced against those factors releasing iron into the blood, e.g., absorption, hemolysis, and storage site release.

13.2.4 Utilization

It can be estimated from blood volume (typically ~5 liters) and erythrocyte lifetime (~120 days) that, although more than 2.5 grams of iron exists in hemoglobin in the form of blood cells, only 20–25 mg/day is used in hemoglobin synthesis in bone marrow. This is supplied, through the transferrin transport pathway, from either absorption or release from storage. Defects in absorption, release and/or transport may produce suppression of hemoglobin synthesis and, over time, lead to an erythrocyte deficiency or *anemia*.

13.2.5 Storage

Iron in excess of metabolic needs is stored intracellularly as either ferritin or hemosiderin in various tissues of the body. Ferritin is normally found in many tissues of the body; however, the reticuloendothelial system of the liver and cells in the intestinal mucosa is the most significant metabolic storage site.

Hemosiderin, a granular water-soluble compound, is thought to be an aggregation of ferritin molecules. It can be seen microscopically in unstained tissue sections of bone marrow as clumps or granules of golden refractile pigment. Similar material is occasionally found in vivo in association with iron-bearing implants, such as stainless steel device components (Winter, 1976) and is thought by some investigators to be locally produced hemosiderin.* Although macrophages adjacent to implants can accumulate hemosiderin from endogenous sources, the source of this particular material is thought to be a combination of iron released from the original hematoma (formed during surgery) and iron oxides and hydroxides formed during corrosion. Primary storage sites are the hepatic parenchymal cells and reticuloendothelial cells of the bone marrow, liver, and spleen.

The relationship governing deposition of iron as ferritin or as hemosiderin is unclear. However, it is postulated that the relative content of iron in either storage form is a function of the total storage iron concentration. Both ferritin and hemosiderin, by chemically binding or shielding iron from the surrounding intracellular environment, serve to reduce its inherent toxicity. Iron in its ionic form, in excess of the TIBC of the blood, is extremely and instantaneously toxic as demonstrated by Gitlow and Beyers (1952). Slow intravenous injection of only 10 mg of ferric-ammonium citrate totally saturated the blood's iron-binding capacity of all patients tested, the excess randomly diffusing into all body tissues. The toxic response was manifested by coughing, sneezing, nausea, and, occasionally, vomiting.

13.2.6 Excretion

The body tenaciously conserves its content of iron, losing less than 0.01% of the total amount daily through excretory and other mechanisms (See Figure 13.2). Iron loss from the body consists of routine excretion (in urine, bile, feces, sweat), periodic loss (desquammation of cells and menstrual flow [in women]) and occasional loss (hemorrhage secondary

*Titanium-containing hemosiderin-like complexes are seen adjacent to titanium alloy implants, further suggesting that hemosiderin is a nonspecific precipitate.

to trauma or disease). Iron excretion is essentially passive and increases only very slowly with increases in body stores of iron.

13.2.7 Iron Overload

Considering the precisely regulated intestinal absorption of iron coupled with limited physiological excretory capabilities, one can easily envision the development of an iron overload state should iron, by whatever route, pathological or iatrogenic, gain access to the endogenous system. For example, excessive absorption of iron occurs in idiopathic hemochromatosis, a situation in which iron is continually deposited in the parenchymal cells of various organs, often resulting in arthritis, liver disease, cardiac failure, and diabetes.

Excessive intake of iron may lead to hepatic and reticuloendothelial involvement manifesting as portal cirrhosis. This is demonstrated by the Bantu tribespeople of Africa, who consume large quantities of food cooked and stored in iron pots. Additionally, a few reported instances of parenteral iron administration to treat misdiagnosed anemias have resulted in iatrogenic iron overload states with consequent toxic signs indistinguishable from those of hemochromatosis. Therefore, evidence indicates that excessive amounts of intracellular iron stemming from an iron overload condition may cause progressive destruction of parenchymal cells and subsequent fibrotic replacement. The positive relationship of cardiomyopathies and diabetes mellitus in iron overload to the deposition of hemosiderin in the myocardium and the pancreas further substantiate the toxicity of iron even in its bound storage form. In an iron overload state, the serum iron and transferrin saturation are usually increased, while the TIBC is somewhat depressed (Figure 13.3).

13.2.8 Iron and Susceptibility to Infectious Disease

For nearly half a century it has been recognized that an element of host response to bacterial invasion is a reduction in the iron content of the blood serum (reduction in SI; see Figure 13.3 [Fairbanks and Beutler, 1988]). The mechanism of this reduction has been identified as a *suppression* of intestinal absorption of iron concurrent with an *increase* in storage of iron in the liver and a concomitant reduction in transferrin saturation. The net effect is that growth-essential iron is made less available to microbial invaders, thus producing a so-called "nutritional immunity" (Kochan, 1973) for the host. To illustrate the strength of this argument, patients with infection and inflammation are unable to mobilize iron from reticuloendothelial cell depots irrespective of a normal total body iron content (Shils and Young, 1988) and thus experience a transient anemia.

This suggests that the physiological system would rather endure a short period of iron deficiency than risk a microbial invasion. In a survey of pathogen-host metal interrelationships Weinberg (1971) concluded that

> in the contest between the establishment of a bacterial or mycotic disease and the successful suppression of the disease by animal hosts, iron is the metal whose concentration in host fluids appears to be most important.

To acquire iron, microbial cells must often synthesize *siderophores*, phenolates or hydroxamates, whose function is to solubilize ferric iron at neutral pH and assimilate the metal. In addition, many microorganisms have the potential to produce their own powerful iron-binding ligands which compete with the host compounds, e.g., transferrin, for the available iron. Organisms which have the ability to solubilize, assimilate, and bind iron independently are termed *autosequesteric*.

Since most bacteria and fungi require only 0.3 to 4.0 μM concentrations of iron for growth, human blood plasma, having a concentration of 10 to 65 μM, would appear suitable to support bacterial growth and multiplication, leading to bacteremia. That bacteremias are the exception rather than the rule illustrates the profound role plasma transferrin plays in resistance to disease by production of "nutritional immunity." Transferrin has an association constant for iron of approximately 10^{30}. This indicates that, in the normal physiological situation where transferrin is 30% saturated with iron, the equilibrium free, ionic iron concentration is approximately 6×10^{-9} μM or 10^8-fold less than that required for microbial growth. Thus, for microbial invaders to have any chance of survival in a foreign host, they must have evolved the capability to synthesize siderophores with association constants for iron of 10^{30} or greater in order to compete for the available iron.

Indeed, many bacterial invaders have iron-binding ligands capable of extracting iron from host transferrin that is saturated 30% or more. Kochan (1973) has demonstrated the microbiostatic action of various mammalian serums in culture mediums with respect to bacterial growth of tubercle bacilli (Table 13.1). For example, human serum containing 30% saturated transferrin inhibited bacterial multiplication of the BGG strain of *M. tuberculosis*. Addition of 36 μM iron neutralized this bacteriostatic activity. Similar results were observed for bovine, mouse, and rabbit sera. However, guinea pigs, owing to their normally high transferrin saturation, had sera equally susceptible to tuberculosis before *and* after iron addition. Other microorganisms demonstrating similar behavior include species of *Candida*, *Clostridium*, *Escherichia*, *Pasteurella*, *Shigella*, and *Staphylococcus*. In vivo studies using strains of *Pseudomonas*

Table 13.1 Correlation of Transferrin Saturation (TR) with Bacterial Growth[a] in Mammalian Sera in Vitro

Source of sera	Number of samples	Fe concentration in serum (μm)	Saturation TR (%)	Bacterial growth[a] No added Fe	Bacterial growth[a] 36 μm added Fe
Human	10	17	30.0	0–1	10–14
Cow	4	34	39.0	0–1	10–14
Mouse	10	41	60.2	1–5	9–15
Rabbit	8	36	64.3	1–5	10–15
Guinea pig	20	49	84.2	9–14	9–14

[a]Bacterial growth expressed as the number of generations of *M. tuberculosis* in 14 days.
Source: Adapted from Kochan (1973).

aeruginosa, Salmonella typhimurium, Listeria monocytogenes, and *E. coli* have corroborated the in vitro results.

Weinberg (1974) summarized the role that iron plays in nutritional immunity:

> A very consistent finding is that the intricate checks and balances between the iron chelators of the microbes and of hosts are readily and markedly upset by changes in the environmental concentration of iron. If the metal is added, microbial growth is enhanced; if the metal is deleted, host defense is strengthened. This situation obtains not only in experimental systems in vitro and in vivo, but also in clinical disease situations.

"Might cryptic disturbances in iron metabolism during the lifetime of individual hosts permit resurgence of latent infections such as tubercular lesions?" (Weinberg, 1974). In light of a report concerning tuberculosis in dialysis patients (Pradhan et al., 1974), perhaps this question has been unknowingly addressed. Pradhan et al. have attributed the 15 times higher incidence of tuberculosis in dialysis patients to an increased susceptibility stemming from the uremic state and a consequent decreased immunological responsiveness. These conditions notwithstanding, it may be possible that sufficient iron (owing to the inherent iron concentration of the dialysate) enters the blood stream during repeated dialysis and constitutes a "cryptic disturbance," thus rendering the patient more susceptible to tubercular infection and/or relapse.

13.2.9   Role of Implants in TIBC Saturation

As mentioned previously, Gitlow and Beyers (1952) have shown that 10 mg of intravenous ferric-ammonium citrate was sufficient to saturate the TIBC of the blood and produce immediate signs of toxicity. Weinberg (1974) has expressed the belief that even small additions of iron may increase the transferrin saturation of the blood, rendering an individual more susceptible to infection. The use of stainless steel in joint replacement and fracture fixation applications, coupled with the evidence of Lux and Zeisler (1974) demonstrating the nature and relative proportions of corrosion products in metallotic tissue, would seem at this junction to warrant investigation of the possibility that the well-recognized (and accepted) local corrosion may eventually elicit subtle long-term systemic consequences.

To illustrate the minute quantities of iron involved in these considerations, and similarly the amount of corrosion that these quantities represent, a simple calculation can be done. For a standard 316L stainless steel orthopedic total hip replacement prosthesis having a total surface area of approximately 200 cm^2, corrosion equivalent to 10 mg of iron release would constitute a general surface dissolution to a depth of 625 Å, an amount beyond light microscopic resolution. In terms of local corrosion, a pit of 1 mm depth would release 10 mg of iron. Naturally, for bilateral hip implantation or for devices having a porous sintered stainless steel surface, the increased surface area would mean that proportionately less iron would have to corrode per unit area for equivalent effects. As suggested in Figure 13.2, the most likely mode of release is into the transferrin transport system.

It is clear, however, that such corrosion would not be equivalent to the clinical experiment of Gitlow and Beyers (1952). Corrosion would release iron over a period of time rather than in a single "dose." Furthermore, other implant-derived metals, such as chromium and aluminum, can also bind to transferrin, further reducing the UIBC and contributing to a higher apparent transferrin saturation (TR).

13.3   CHROMIUM METABOLISM
(LANGÅRD AND NORSETH [1986])

Chromium differs from iron by only two atomic numbers (24 vs. 26) and by less than 7% in average atomic weight. Both elements can form divalent and trivalent ions, although chromium can additionally form a hexavalent ion. However there are profound differences in their biological

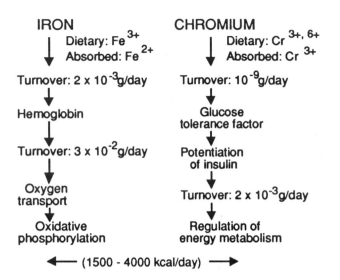

Figure 13.4 Comparison of iron and chromium amplification. Source: Cr data adapted from Mertz (1981).

roles. Iron is well known for its primary role in hemoglobin, the primary oxygen transport molecule in mammals, while chromium plays a no less vital role in regulating metabolism. These differences in biological roles and metabolic pathways are an excellent example of the principle of specificity in essential trace element function. Figure 13.4 compares and contrasts these roles and also illustrates the principle of amplification of function: extremely small daily intakes of each element have led to vital physiological roles in the whole organism.

It would be desirable to be able to depict a complete and systemic overview of chromium uptake, transportation, utilization and conservation, storage and excretion, as shown in Figure 13.2 for iron, but this is not possible. Table 13.2, based in part on estimates, summarizes the overall picture of the situation.

Absorption of Cr^{+3} takes place in the gastrointestinal tract. Although the diet may contain Cr^{+2} and Cr^{+6} as well as Cr^{+3}, the latter valence is the predominate form in the acidic conditions of the stomach and upper digestive tract (Mertz, 1983). Since Cr^{+3} is essentially excluded from cellular contents due to an inability to cross cellular membranes (Sanderson, 1976), it is very poorly absorbed. However, chromium can be found within cells, suggesting that mechanisms exist in vivo to oxidize Cr^{+3} to Cr^{+6} (Rogers, 1984).

Table 13.2 Chromium Metabolism

Daily intake:	
Dietary supply[a]:	200 µg
Absorbed (@ 1.5% absorbance)[b]	3 µg
Transport	
Serum content (3.2 L @ 0.16 ppb[c]):	0.5 µg
Daily utilization	
Unknown, leads to 2 mg insulin synthesis/day[d]	
Storage	
Daily addition	1 µg
Total body burden (70 kg individual)	7 mg
Daily excretion	
Urinary (1.5 L @ 0.2 ppb[e])	0.3 µg
Fecal, desquammation, etc. (balance)	~ 1.7 µg

[a]Table 13.3, maximum recommended.
[b]Table 13.3, mean value.
[c]Versieck and Cornelis (1980).
[d]Mertz (1981).
[e]Cornelis and Wallaeys (1984).

While daily absorption of chromium is very small compared to that of iron (3 µg vs 1–2 mg), the serum content is even lower (5 µg [0.16 ppb] vs 3 mg [0.94 ppm]). As in the case of iron, serum chromium is primarily bound to transferrin (Hertel, 1986) although it can also be nonspecifically bound to albumin.

The turnover of biosynthetic chromium, as an element in a glucose tolerance factor (GTF), leads to the synthesis of insulin and to the ability to metabolize carbohydrates (Schroeder, 1966).

About one third of the daily absorption, perhaps that portion reduced to Cr^{+6}, is stored in the reticuloendothelial structures in cells and non-specifically in other cells, such as erythrocytes. The balance is excreted, primarily fecally and through desquammation of cells.

The overall picture, in comparing chromium metabolism to iron metabolism, is one of radical differences, despite the similarity of the elements, even in the absence of implants. This should alert us to the difficulties in understanding release, storage and excretion of metals from implants where there is still less knowledge of the metabolic pathways involved.

Table 13.3 Daily Recommended and Safe Mineral Intake

Mineral	Recommended dietary allowances[a,b] Dietary content	Absorbed (%)	Internal[c]
Phosphorus	800 mg	50–70	400–560 mg
Calcium	800 mg	20–40	160–320 mg
Magnesium	315 mg	40–60	125–190 mg
Zinc	13.5 mg	2–38	0.3–5.1 mg
Iron	10 mg	heme[d] 20	
		non-heme 6–18	0.5–1.0 mg
Iodine	150 µg	30–50	45–75 µg
Selenium	62.5 µg	80	50 µg

Mineral	Safe and Adequate Intake[a] Dietary content	Absorbed (%)	Internal[c]
Fluorine	1.5–4.0 mg	35–100	0.5–4.0 mg
Manganese	2.0–5.0 mg	?	<5.0 mg
Copper	1.5–3.0 mg	36	0.5–1.1 mg
Molybdenum	75–250 µg	?	<250 µg
Chromium	50–200 µg	0.5–2[e]	0.25–1.0 µg

[a]Source: National Research Council (1989).
[b]Adjusted for average of male (79 kg) and female (63 kg), 25–50 years old.
[c]Calculated; presumably equal to loss by all routes.
[d]Decreases with increasing non-heme iron in meal.
[e]Decreases with increasing daily chromium intake.

13.4 HUMAN DIETARY METAL INTAKE

The only significant source of essential trace elements, including metals, for human metabolic processes, is through the gastrointestinal tract. Transdermal absorption and inhalation, although capable of causing local host responses, rarely can be shown to contribute to internal metal concentrations or remote storage. Primarily, humans depend upon dietary sources for essential trace elements. Table 13.3 lists the recommended dietary allowances (RDAs) and suggested safe and adequate intakes (SAIs) for two physiological and ten essential trace elements, as determined by the US National Research Council. Note that fluorine is included, although its natural concentration in the body is very low and despite the absence of a known normal biological role in mammals, because of its strong action in preventing tooth decay. This table also lists ranges of percent absorption and resulting internal availability. It is against this latter amount that release by an implant should be judged.

Two final points deserve to be made to complete this discussion of normal mineral metabolism. In the first place, even in the absence of the routine common consumption of the "one-a-day" type of vitamin and mineral supplements (which usually contain 50–150% of the RDA and SAI of all essential trace elements), modern diets in the developed nations generally provide all essential minerals required for normal physiological functions in healthy individuals. The widespread "enrichment" of food-stuffs, such as milk, bread and breakfast cereals, the increasing amounts of fresh foods consumed as well as modern food preservation and preparation techniques virtually guarantee this result.* Furthermore, the homeostatic systems which control metal absorption, transport, storage and excretion render meaningless the occasional practice of consuming "mega-doses" of essential trace elements: in the general case the excess will not be absorbed and will simply be excreted directly. For some easily absorbed minerals, excess consumption may lead to transient toxicity. If defects in regulation exist, then excessive intakes may lead to one or another of various metal storage diseases (Underwood, 1977). Fortunately these are rare and in many cases have familial, presumably genetic, predispositions.

In the second place, it must be emphasized that concern about in vivo metal released from implants centers primarily on two situations: (1) the possibility of very elevated (10 to 100-fold normal) concentrations occurring either near implants or in remote storage sites; and (2) the possibility of release of metal in a different valence state than normally occurs in the body with subsequent formation of biologically active organometallic species.

As will be discussed at further length in Chapter 14, measurements of serum concentrations and urinary excretion of metals, while useful, do not provide a true picture of either of these effects. Unfortunately, with the advent of modern, highly sensitive techniques for determining metal content of biological materials (atomic absorption spectroscopy, neutron activation analysis, inductively coupled plasma-mass spectroscopy, etc.), a number of commercial concerns have become involved in diagnostic studies of trace element profiles in patients and in normal individuals. Homeostasis may permit a broad range of concentrations about the optimal concentration for any essential trace element (c.f. Figure 13.1); thus, in the absence of a metallic implant which could produce elevated concentrations and/or different valence states, small changes in serum metal

---

*One exception to this assertion may be chromium: increasing dietary use of refined (white) sugar which contains less chromium than raw sugars but requires insulin for metabolic conversion (Mertz, 1983) may lead to progressive chromium deficiency.

concentrations within a normal range may only reflect daily variations in intake, utilization, etc. It is highly unlikely that such studies can lead to primary diagnosis of actual metal deficiencies or overloads, with a concomitant need for corrective therapy, that have not already been detected by clinical indications (Kruse-Jarres, 1987). Thus the validity of such studies in healthy individuals or in non-implant patients, other than for verification of prior diagnoses, must be viewed with great skepticism.

Not withstanding these caveats, in the next chapter we shall take up in detail the issues of release, distribution, and excretion of ions from implants.

REFERENCES

Cornelis, R. and Wallaeys, B. (1984): In: *Trace Element—Analytical Chemistry in Medicine and Biology*, Vol. 3. Walter de Gruyter & Co., Berlin, pp. 219ff.

Fairbanks, V. F. and Beutler, E. (1988): In: *Modern Nutrition in Health and Disease*, 7th ed. M. E. Shils and V. R. Young (Eds.). Lea & Febiger, Philadelphia, pp. 193ff.

Gitlow, S. E. and Beyers, M. R. (1952): J. Lab. Clin. Med. 39:337.

Hertel, R. F. (1986): In: *Environmental Carcinogens: Selected Methods of Analysis*, Vol. 8. I. K. O'Neill, P. Schuller and L. Fishbein (Eds.). IARC Scientific Publication 71. International Agency for Research on Cancer, Lyon, pp. 63ff.

Kochan, I. (1973): Curr. Top. Microbiol. Immunol. 60:1.

Kruse-Jarres, J. D. (1987): J. Trace Elem. Electrolytes Health Dis. 1:5.

Langård, S. and Norseth, T. (1986): In: *Handbook on the Toxicology of Metals*, 2nd ed. Vol. II: *Specific Metals*. L. Friberg, G. F. Nordberg and V. B. Vouk (Eds.). Elsevier, Amsterdam, pp. 185ff.

Lux, F. and Zeisler, R. (1974): J. Radioanal. Chem. 19:289.

Mertz, W. (1981): Science 213:1332.

Mertz, W. (1983): Chemica Scripta 21:145.

National Research Council (1989): *Recommended Dietary Allowances*, 10th ed. National Academy Press, Washington, DC.

Pradhan, R. P., Katz, L. A., Nidus, B. D., Matalon, R. and Eisinger, R. P. (1974): J. Am. Med. Assoc. 229:798.

Rogers, G. T. (1984): Biomats. 5:244.

Sanderson, C. J. (1976): Transplantation 21:526.

Schroeder, H. A. (1966): J. Nutrition 88:439.

Schwarz, K. (1977): In: *Clinical Chemistry and Chemical Toxicology of Metals.* S. S. Brown (Ed.). Elsevier/North-Holland, Amsterdam.

Shils, M. E. and Young, V. R. (Eds.) (1988): *Modern Nutrition in Health and Disease*, 7th ed. Lea & Febiger, Philadelphia.

Underwood, E. J. (1977): *Trace Elements in Human and Animal Nutrition*, 4th ed. Academic Press, New York.

Versieck, J. and Cornelis, R. (1980): Anal. Chim. Acta 116:217.

Weinberg, E. D. (1971): J. Infect. Dis. 124:401.

Weinberg, E. D. (1974): Science 184:952.

Winter, G. D. (1976): In: *Biocompatibility of Implant Materials.* D. Williams (Ed.). Sector Pub., London.

BIBLIOGRAPHY

Brown, S. S. (Ed.) (1977): *Clinical Chemistry and Chemical Toxicology of Metals.* Elsevier/North-Holland, Amsterdam.

Burrows, D. (1983): *Chromium: Metabolism and Toxicity.* CRC Press, Boca Raton, FL.

Davies, I. J. T. (1972): *The Clinical Significance of the Essential Biological Metals.* C. C. Thomas, Great Britain.

Friberg, L., Nordberg, G. F. and Vouk, V. B. (Eds.) (1986): *Handbook on the Toxicology of Metals*, 2nd ed. Vol. I: *General Aspects.* Vol. II: *Specific Metals.* Elsevier, Amsterdam.

Gibson, R. S. (1990): J. Can. Diet. Assoc. 51:292.

Langård, S. and Hensten-Petterson, A. (1981): In: *Systemic Aspects of Biocompatibility*, Vol. I. D. F. Williams (Ed.). CRC Press, Boca Raton, FL, pp. 143ff.

Mertz, W. (Ed.) (1986–1987): *Trace Elements in Human and Animal Nutrition*, 5th ed., Vol. 1 & 2. Academic Press, San Diego & Orlando.

Schroeder, H. A. (1973): *The Trace Elements and Man: Some Positive and Negative Aspects.* Devin-Adair, Old Greenwich, CT.

Williams, D. R. (1971): *The Metals of Life: The Solution Chemistry of Metal Ions in Biological Systems.* Van Nostrand Reinhold, London.

14
Systemic Distribution and Excretion

14.1   INTRODUCTION

The traditional approach to consideration of biological performance is to focus on the implant-host interface. Thus, material response studies have dealt with degradation in implant properties and host response studies with formation of a capsule and other events within the adjacent tissue.

More modern considerations recognize that a mammal, such as a test animal or a human patient, is an interconnected structure with various mechanisms permitting exchange between its various tissues and organs. The systemic and remote site results of such exchanges involving implants and implant degradation products will be dealt with in Chapter 15.

Previously in Chapters 3–5 and 7, we have considered mechanisms which can modify native proteins or release materials from implants. In this chapter, we shall examine some aspects of the distribution and excretion of these implant related products. Their distribution through the various systems of the body can take place in a number of different ways:

1. Movement of solid bodies.
2. Movement of particulate materials, either passively or actively (cell-mediated).
3. Movement of dissolution or corrosion products by passive diffusion or by active circulatory transport.

14.2 MOVEMENT OF SOLID BODIES

14.2.1 Large Particles

Large particles or portions of implants can move through soft tissue if they possess a certain degree of asymmetry. A sphere, such as a buckshot pellet, will stay in position for an indefinite time. An asymmetric "needle," such as a sewing needle or a porcupine quill, will move point first and may travel for long distances due to the action of muscle forces on it.

It is also possible for large material particles to become involved in blood circulation. Wear particles from vascular prostheses will move "downstream" until they are trapped in reduced vessel diameters on the arterial side of capillary beds or in the lungs on the venous side of the circulatory path. Much larger particles can also be transported. A report of four cases of shell fragments transported into the cerebral circulation is a dramatic illustration of this possibility (Kapp et al., 1973).

Pins, wires and other implants used for internal fixation of fractures and for adjunctive tissue immobilization during placement of permanent implants may also become dislodged and migrate. Lyons and Rockwood (1990) reviewed reports of 47 such occurrences after surgery in the vicinity of the shoulder. Smooth pins and wires were more likely to be reported as migrating than threaded ones; screws and staples, on the other hand, were not reported as migrating. Eight of the patients died, 6 suddenly, due to damage to heart and blood vessels near the heart by the migrating implants. In a majority of the reports (35 of 39) in which a time postoperative could be determined, migration apparently occurred within

8 months of implantation, although the mean time to diagnosis was 22 months.

Migrating device fragments are no respecter of organs, having been reported to enter the lungs (Aalders et al., 1985), the heart (Lyons and Rockwood, 1990) and to move as far as from the shoulder to the spleen (Potter et al., 1988). Therefore thoughtful design of materials and implants using them should be such as to minimize or eliminate the possibility of release of macroscopic fragments.

14.2.2 Phagocytic Transport

As we have already seen in Section 8.2.3, particles that are sufficiently small are phagocytosed by a variety of cells. There are four possible results of cellular phagocytosis:

1. The phagocytic cell (PC) can successfully digest the particle. Partial digestion and externalization (*exocytosis*) of the particle may also occur, but rarely so.
2. The PC attempts to digest the particle but the degradation products prove to be cytotoxic. Then the PC dies, its cell membrane and phagosomes lyse and another PC may attempt to phagocytose the particle and digest it. If this progression continues through many repetitions, the dead PCs accumulate, resulting in *caseation*; the resulting mass of dead cells resembles cheese.
3. The PC may transport the particle, by passing either into the blood or lymphatic circulation, but most usually to regional lymph nodes where particle-loaded cells accumulate and may produce granulomas, such as the "teflonomas" reported by Charnley (1961) after the use of a poly(tetra fluoro)ethylene as a bearing surface in total hip replacement.
4. The PC may be able to transport the particle to the lungs. There it is possible for the particle to be extruded through the lung wall and exhaled through the airway (Styles and Wilson, 1976).

It should additionally be noted that all four of these outcomes will result in activation of the PC, with concomitant release of biologically active agents which may alter local host response to the implant (see Section 8.2.3) (Anderson and Miller, 1984; Schnyder and Baggiolini, 1978).

In the context of this chapter, it would be very desirable to make some statements about the rates of transport of particles by phagocytes in the third and fourth situation above. It is probably not possible to generalize, but some extension of the comments in Section 8.2.3 is desirable.

The transport of particles by phagocytes (primarily macrophages since neutrophils are short-lived and FBGCs tend to remain near the implant site) consists of at least two major steps: uptake and transport. Very little is known about transport rates in the lymphatics, primarily since most uptake studies have been done either in vitro or by systemic injection of a colloid of particles in vivo and sequential sampling of the phagocyte population in the circulatory system.

There are some details known about the first step, uptake. Two general approaches have been taken to describe the uptake process. The first is to fit the kinetics of phagocytic removal of particles to the Michaelis-Menten model used in studies of enzyme activity (Normann, 1974). In this approach, uptake is considered to have two phases: *attachment* to the PC and *engulfment*.

Attachment is modeled as consisting of a reversible attachment step and an irreversible engulfment step:

$$P + PC \underset{k_2}{\overset{k_1}{\longleftrightarrow}} [PC - P] \overset{k_3}{\longrightarrow} PCp \qquad (14.1)$$

$$\text{attachment} \qquad\qquad \text{engulfment}$$

where P is an extracellular particle and p is an intracellular one. Previous studies have shown, for a wide variety of animal species, the velocity of clearance of circulating particles (uptake), V, follows a first order proportional absorption law dependent upon particle concentration, C (Biozzi et al. 1953, 1957; Normann 1974):

$$V = \frac{dC}{dt} = -kC \qquad (14.2)$$

The following equation is typical of the results of this approach:

$$\frac{1}{V} = \frac{1}{k_3 E_0} + \left[\frac{K_c}{k_3 E_0}\right] \frac{1}{[C]} \qquad (14.3)$$

where

V = clearance velocity (uptake rate of particles by PCs)

C = concentration of extracellular particles

E_0 = total available attachment sites

K_C = overall kinetic constant = $(k_2 + k_3/k_1)$

A saturation effect (presence of a fixed maximum rate) is attributed to the fact that the number of sites on the phagocyte membrane that can initiate invagination and engulfment (E_0) is limited.

There are two major criticisms of this result. The first is that the uptake velocity is not a first order process for all particle concentrations and the second is that it seems unlikely that specific membrane loci exist for such a nonspecific process.

The second approach (Stiffel et al., 1970) is to describe the observed kinetics as exponential in the form given by Eq. (14.4):

$$C = C_0 10^{-Kt} \tag{14.4}$$

where

  K = total body phagocytic index (essentially, the particle clearance velocity)

  C_0 = initial particle concentration

  t = time

The major difficulty with this result is that, once again, it is not a good description of the kinetics of phagocytic behavior. What is actually observed is a complex uptake velocity behavior with three domains, provided that the particles are within a given size range and that the initial concentration is high enough so that the kinetics are not dictated by flow processes (Vernon-Roberts, 1972). The initial phase seems to be first order and dictated by either the adsorption of serum opsonins to the particles or the adsorption of the coated particles to the phagocyte. The second phase is exponential, dictated by the dose (concentration of particles) as in any other dose-response situation. The third phase is a slowly disappearing component seen when a heterogeneous distribution of particles is injected. What this "tail" actually represents is the removal of smaller particles. A possible explanation for this strange kinetic pattern is that there are two opposing effects. One is a saturation effect attributed to a limited number of binding sites by some and, more appropriately, to a limited amount of serum opsonins by others (Jenkin and Rowley, 1961). The second effect is the increased efficacy of clearance as the blood (or presumably tissue) concentration of particles goes down. The limit on the particle size range mentioned in Chapter 8 restricts these investigations to particles on the order of the size of leukocytes, approximately 4–7 μm.

A somewhat more pragmatic approach to the kinetics of uptake (Korn and Weisman, 1967; Weisman and Korn, 1967) has led to the conclusion that the kinetics of uptake are determined by a constant (absorption) vesicle volume. Careful studies with well-controlled particle size ranges led to the conclusion that, in an amoeba model, while larger particles are taken up singly, small particles are accumulated external to the cell until a critical volume is reached, whereupon the "cemented" mass is absorbed simultaneously. Typical results, obtained by earlier investigators in a mammalian model which also led to this conclusion, are given in Table 14.1.

In Table 14.1 it is also useful to note that oxygen consumption is required during phagocytosis by PMNs (since the process is energy requiring and PMNs are aerobic cells) and that it increases with particle size.

There are however, additional difficulties. All particles are not equal: composition, morphology and surface charge may play a role in uptake velocity. As discussed in Section 8.4.1, a variety of dissolved metal ions, such as Ni^{+2} and Cr^{+3}, suppress phagocytic efficiency (Graham et al., 1975). Therefore one might expect a slower uptake of nickel or chromium bearing particles, due to high local metal concentrations, than of polymeric particles of the same size, etc. Kawaguchi et al. (1986) have shown additionally that phagocytosis of polystyrene (as measured by cellular oxygen consumption) is strongly dependent upon surface potential and thus upon fixed surface charge.

While these models, calculations and experiments tell us something about the relative likelihood of uptake (clearance velocity) as a function of particle size and properties, they shed little light upon the question of net removal rates from implant sites. This remains an area for further investigation. Far more is known about active and passive removal of dissolved species.

14.3 TRANSPORT OF DISSOLVED SPECIES

14.3.1 Leaching of Monomers

Leaching or dissolution of polymers into the circulatory system results in rapid dispersion throughout the body. This is a result of a circulatory rate of approximately 1 min^{-1}. That is, the normal blood volume (about 6–8% of body weight or about 5 liters) passes through the lungs once a minute. An illustration of the rapidity of this circulation is the study of Homsy et al. (1972) concerning release of monomer from poly(methyl)methacrylate bone cement upon its insertion into the body. In a canine model, insertion of a dose of freshly mixed cement of less than 2 g/kg body weight as a femoral transcortical plug produced monomer levels of up to 1 mg/100 ml in the inferior vena cava within 2 min of implantation. The peak concentration was reached in 3–4 minutes followed by a decline, ascribed to monomer clearance by evaporation through the lungs as well as progression of polymerization of the cement which reduced the source concentration. Similar results were obtained in patients receiving

Table 14.1 Effect of Particle Size on Phagocytosis by Guinea Pig PMNs

Diameter of particles (μm)	Polystyrene uptake (μg/mg wet wt. PMNs)	O_2 consumption (μl/mg PMN/min)	Number of particles/PMN
0.088	7.4	0.0144	24,000
0.264	30.1	0.0372	3,600
0.557	28.1	0.0412	360
0.871	36.9	0.0396	102
1.305	34.3	0.0427	34
3.04	35.9	0.0422	3
>7	0	<0.012	0
no polystyrene	0	0.0124	0

Source: Adapted from Roberts and Quastel (1963).

PMMA-cemented femoral endoprostheses: peak monomer concentrations occurred in the vena cava by two minutes postimplantation and 99% + of (integrated) exhaled monomer was detected in the airway within 6 minutes.

14.3.2 Corrosion of Metals

Local Effects

The corrosion of metals has been presented in Chapter 4. It is appropriate to inquire how metallic corrosion and dissolution products are distributed in the body.

It is clear that there is usually an accumulation of corrosion products around a metal implant. These products include membrane-bound ions, particles released by intergranular and fatigue processes, and insoluble reaction products such as metal hydroxides. The combination of these leads to the familiar tissue discoloration termed *metallosis*, particularly in older reports. In addition to tissue discoloration, this passive diffusion may also be seen histologically as a varying degree of cell reaction. Figure 14.1 shows the variation of effect with material "reactivity," that is, corrosion and/or dissolution rate combined with tissue response, for needles inserted in the cerebral cortex of rabbits for periods of up to 1½ years. Typical materials used were aluminum, platinum (nonreactive), molybdenum, tantalum (reactive), and silver, iron, and cobalt (highly reactive and toxic).

The picture presented by Figure 14.1 reflects corrosion and diffusion of metal but is complex and difficult to analyze since the breadth and type of response about each implant depends upon the rate of corrosion, the

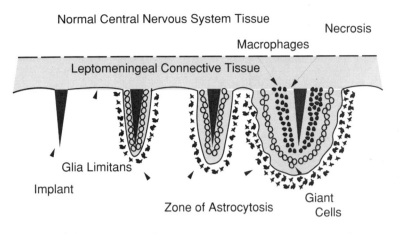

Normal Central Nervous System Tissue

Necrosis

Macrophages

Leptomeningeal Connective Tissue

Glia Limitans

Implant

Zone of Astrocytosis

Giant Cells

Figure 14.1 Histological changes around implants of increasing reactivity (left to right) in the rabbit cerebral cortex. Adapted from Stensaas and Stensaas (1978).

valence (speciation) of released metal, its diffusion constant and its toxicity. Furthermore, the release may be affected by anatomical location, since local physicochemical conditions vary (see Section 2.2) resulting in different rates of corrosion (Oron and Alter, 1984). Implant site infection may also affect both the corrosion rate (Hierholzer et al., 1984) as well as tissue response to the corrosion products.

A more accurate view of passive diffusion may be obtained by direct analysis of the tissues in and near the implant site. Energy-dispersive microanalysis has been widely used to study situations such as metal diffusion into bone from implanted dental devices (Arvidson and Wroblewski, 1978). One of the more graphic studies is that of Lux and Zeisler (1974), who used neutron activation analysis to examine the spatial distribution of corrosion products in soft tissues adjacent to a steel fracture fixation device. Their results, obtained by analysis of tissues obtained at device retrieval, are shown in Figure 14.2.

In a study of 38 patients with histologically diagnosed metallosis at retrieval of fracture fixation hardware, Lux and Zeisler found mean concentrations of iron, chromium, and molybdenum near a stainless steel-tissue interface to be two orders of magnitude above background values previously determined in tissues from patients without metallic implants. These concentrations declined logarithmically with distance from the implant, reaching background ("normal") levels at distances of greater than 4 cm from the implant. The ratio of $Fe:Cr:Mo$ in the alloy (V4A) was $66:17.5:2.3$ ($\cong 29:7.5:1$) while typical tissue concentration ratios

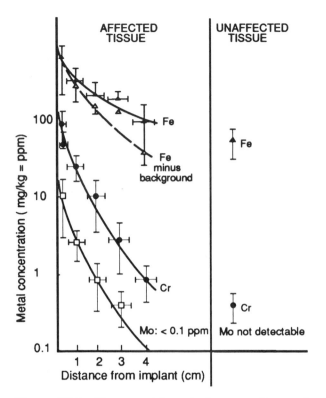

Figure 14.2 Tissue metal content surrounding an implant. Source: Adapted from Lux and Zeisler (1974).

were: 56.5 : 8 : 1 (at the tissue implant interface) and 117 : 9 : 1 (1 cm away from interface). Nickel, which formed 12% of the implant alloy, was detected at the interface but not in the surrounding tissue.

A number of conclusions may be drawn from these results:

1. Ion concentrations (and persuadably leachate concentrations from polymeric materials) may be very high in the vicinity of implants when compared to systemic and remote site concentrations.
2. The accumulation of released products depends upon the nature of their reactions with surrounding tissues and their diffusion concentrations. In this example, nickel diffuses away rapidly and is seen only at the interface, chromium and molybdenum diffuse less rapidly and iron is accumulated locally, presumably due to the formation of precipitates, which contribute to the discoloration of metallosis.

3. Tissue concentrations, due to 2 above, do not necessarily reflect alloy proportions. Finding of metals near implants in their alloy constituent proportions is more evidence of wear debris accumulation than of burdening of tissues with soluble or precipitated corrosion products (Michel, 1987).
4. Finally, it should be pointed out that the mere detection of metal in tissues does not reflect its biological availability. Metal may be present as free ions (unlikely), bound to specific carrier molecules (e.g., Fe-transferrin), nonspecifically bound (e.g., to albumin), in the form of a wear particle or as a precipitate. Any of these forms may be either extracellular or intracellular.

This pattern observed by Lux and Zeisler (1974) reflects the end stage of an equilibrium between the implant and the tissue. In a later study, utilizing a rabbit implant model, Lux et al. (1976) reported a nonspecific decrease in transport rate with time for all detectable corrosion products, ascribed to maturation of the fibrous capsule about the implant. Further, they showed an inverse correlation between tissue iron and zinc content, even though zinc was not released by the implant. These findings underline the great complexity of consideration of release and distribution of implant degradation products, even in the near vicinity of the implant.

In a traditional sense, as would be observed in an in vitro corrosion study, equilibrium would nevertheless be reached when an equilibrium concentration of ions was reached in the surrounding fluid. However, in vivo, the bathing medium is dynamic, and a fractional excretion process is always in competition with corrosion processes trying to attain equilibrium.

Distribution of Body Water

If maximum implant corrosion rates (as determined by formation of soluble species) can be measured, then equilibrium times should depend only upon the volume of the media and the fractional excretion rates. The medium that we are discussing, water, has the relative volumes and distribution in the human body as shown in Figure 14.3.

If there were no input/output of fluid, we would be considering a volume of 42 liters. Similarly, if there were no compartmental exchange, we might be considering smaller volumes of fluid, as small as 3.2 liters in the case of an implant bathed in blood. However, the situation is dynamic as shown in Figure 14.4.

In Figure 14.4, note that some water entering the gastrointestinal tract never exchanges directly with the internal water compartments but passes straight through as a portion of fecal excretion (dashed line to right of

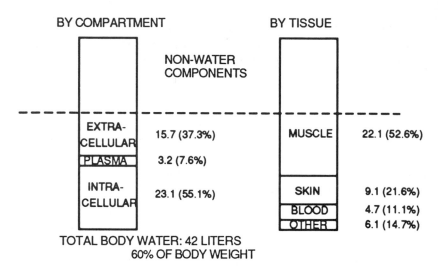

Figure 14.3   Water content of the human body (70 kg).

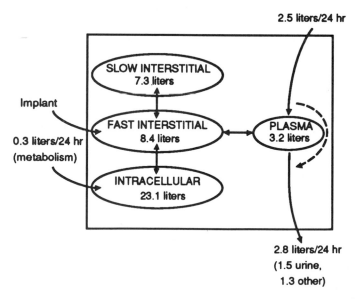

Figure 14.4   Daily water input/output/exchange for the human body (70 kg).

plasma pool). The 1.3 liter/day excretion identified as "other" includes this volume as well as sweat (lost from the fast interstitial pool) and the water content of sloughed or desquammated cells (lost from the intracellular pool).

We can now see that volumes vary considerably depending upon the details of consideration. An implant placed in soft or hard tissue is directly in contact with a fast interstitial pool of 8.4 liters, as shown. This contacts slower exchange pools of 7.3 liters (interstitial) and 23.1 liters (intracellular). It also contacts a fast exchange pool of 3.2 liters (plasma water). This pool passes through the kidneys at a rate of 180 liters/24 hr and experiences a net throughput (intake = output) of 2.5 to 2.8 liters/24 hr. Thus, on an annual basis, an implant is in contact with a pool size exceeding 1000 liters. The point of equilibrium of corrosion or leached products depends on the overall kinetics of release from the implant, exchange, and excretion.

These kinetics are difficult to analyze. In the previous chapter we considered the details of one system, the iron system, in some depth.

14.4 DISTRIBUTION AND EXCRETION OF DISSOLVED SPECIES

14.4.1 Metallic Ion Distribution

We can, however, study some of the elements of these distribution systems in more detail. One factor we can examine is urinary excretion. Urinary excretion is probably the major route for clearance from the body of metals released from implants, although biliary excretion may also play a role for some metals (Brauer, 1959). Of the metals of interest as components of implants, only iron is known to be excreted (incorporated in porphyrins which are degradation products of hemoglobin) predominantly through a biliary route (Ishihara and Matsushiro, 1986). Biliary excretion is mediated by concentration in and/or excretion by the liver of low molecular weight ($< 10,000$) organometallic species known collectively as *metallothioneins* (Cherian and Goyer [1978]; Klaassen [1976]). These are released through the bile duct, stored in the gall bladder and released as bile, at a rate of 0.4–1.2 L/day, into the jejunum. A significant proportion of the ionic content of bile, perhaps as much as 95%, is reabsorbed in the ileum, with only 200 mg/day excreted. Biliary excretion serves primarily to aid in digestion of fats and metal excretion appears to be a secondary role (Ganong, 1989).

Urine, on the other hand, is formed primarily to regulate plasma composition and pH. It is formed in the mammalian kidney by a combination

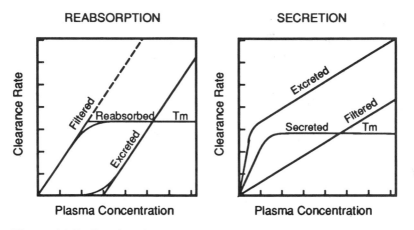

Figure 14.5 Renal reabsorption and secretion. Adapted from Pitts (1963).

of three processes: *glomerular filtration, tubular reabsorption* and *tubular secretion.*

In the glomeruli, all formed cellular elements of blood and any molecules with a molecular weight above approximately 5000 are retained while all small molecules and a considerable amount of water is filtered into the proximal renal tubules. In a 70 kg person between 115 and 125 ml/min of plasma are filtered. Normally nearly all of the water is also filtered out. If this were excreted, it would result in a 24 hour urine volume of as much as 180 liters!

This does not happen because over 99% of the water is reabsorbed in the proximal renal tubules, resulting in a nominal rate of urine production near 1 ml/min. In addition, other materials such as the physiological metal ions (Na^+, K^+, etc.), essential anions (Cl^-, H^-, etc.) and organic molecules (glucose, urea, etc.) are reabsorbed. Some reabsorption processes are passive but others are active (energy requiring). As shown in the left portion of Figure 14.5, the active reabsorption processes, for example for glucose, each have an asymptotic limit (different for each reabsorbed species), termed the *tubular reabsorption limit* (T_m). This has the effect of limiting and controlling the maximum concentration of that particular species in plasma.

Finally, materials may be removed from plasma by secretion through the walls of the distal renal tubules. This process may be passive or active; in the latter case (right portion of Figure 14.5) the effect is also to demonstrate a limit. Metals in plasma are either specifically bound to transferrin or other carrier protein (nickeloplasm, etc.) or nonspecifically to albumin, all of which have too high a molecular weight to be filtered in the glomeri. Therefore, urinary excretion of essential and trace metals must be primarily through a tubular secretion pathway.* Since this represents an interaction between post-glomerular plasma and distal tubule urine, it is instructive to look at the relative concentrations of metal in these fluids in non-implanted normal adults (Table 14.2).

The *permeability ratio*, K_x, is the ratio of concentration in urine to that in plasma. Values greater than 1 suggest positive secretion, values near 1 indicate simple equilibrium between urine and plasma, while values less than 1 suggest a barrier to secretion. The permeability ratio times the relative volumes of urine and plasma ($= 0.78$) is the proportion of normal plasma content secreted in 24 hours. The *excretion ratio*,† E_x, based upon a 24 hour urine volume of 2.5 liters and a renal filtration of 180 liters/24 hours, reflects the efficiency of secretion or the proportion of the renal filtrate secreted in 24 hours. The higher the value of E_x, the greater the probability of secretion of an ion in any one pass through the kidneys. The *clearance rate*, C_x, of a substance, x, is given by:

$$C_x = K_x V \text{ (ml/min)} \tag{14.5}$$

where V = rate of urinary output.

Creatinine is a degradation product released by cell death and is widely used as a concentration marker for studying ionic concentration in urine. Urine may be more or less concentrated, depending on fluid intake, perspiration loss, etc., but the serum concentration and amount of urinary creatinine excretion in 24 hours remains remarkably constant in normal individuals (Ganong, 1989; Araki et al., 1986b).

14.4.2 Distribution Models: One Compartment

Projections of the equilibrium accumulations of metals in the body have been made using this type of data (Taylor, 1973). Taylor proposed that, for a given rate of continuous release of corrosion products R per day, the increase in an organ or the total body content of a metal can be given as

$$Q_t = Q_0 + \frac{R}{k}(1 - e^{-kt}) \tag{14.6}$$

where Q_0 is the normal metal content, Q_t the content after "t" days, and k is the fractional rate of excretion of the metal. That is, in a given day, R

---

*However, some metal-protein dissociation may occur, leading to direct excretion; tubular reabsorption is also possible (Araki et al., 1986a).
†The term "excretion ratio" is used irrespective of renal mechanism involved.

Table 14.2 Urinary Secretion of Implant Alloy Components[a]

Element	Plasma conc. (μg/liter)	Urine conc. (μg/liter)	Permeability ratio (K_x)	Excretion ratio (E_x)
Al	2.2	6.4	2.9	0.040
Co	0.05	0.33	6.6	0.092
Cr	0.06	0.13	2.2	0.030
Ni	0.2	1.0	5.0	0.069
Ti	3.3	0.41	0.12	0.006
V	0.16	0.61	3.8	0.053
Reference Creatinine	10 mg/liter	1.5 g/liter	150	2.08

[a]These data differ considerably from those in the first edition, due to significant improvements in collection and analysis techniques. However, it is widely believed that urine collection, especially from female subjects, is subject to contamination. Therefore, both K_x and E_x may be too high; values should be considered in relative rather than absolute terms. Plasma and urine concentrations are for non-implanted normal adults.
[b]Reconstructed, assuming 1.5 g creatinine/liter.
Sources: Al, Ti, V: Jacobs et al. (1991); Co, Cr, Ni[b]: Sunderman et al. (1989); creatinine: Ganong (1989).

metal is released and kQ_t is excreted. Letting t go to infinity (when equilibrium is presumably achieved), one then finds that

$$Q_e = Q_0 + \frac{R}{k} \tag{14.7}$$

where Q_e is the equilibrium metal content. Note that this is a first-order analysis that treats the body as a single homogeneous compartment. For a hypothetical cobalt-chromium implant (60% Co, 30% Cr, 8% Mo, 1% Ni, 1% Fe) with a surface area of 200 cm^2 and a corrosion rate of 30 mg/cm^2/day, Taylor obtains the results given in Table 14.3.

Table 14.3 Secretion and Accumulation Rates of Alloy Components of a Cobalt-Chromium Alloy

Element	Q_0 (mg)	k (day^{-1})	R (mg/day)	Q_e (mg)	Q_e/Q_0
Co	3	0.07	3.6	54	18
Cr	6	0.0011	1.8	1636	273
Mo	5	0.139	0.48	8	1.7
Fe	4000	0.0010	0.06	4060	1.02
Ni	10	0.0010	0.06	70	7

Taylor concluded that modest elevations of cobalt and nickel and large elevations of chromium content should occur in this case. Similar calculations for implantation of stainless steel predict a modest elevation of nickel and a large increase in chromium content.

Taylor's calculations are probably in error on two fundamental grounds. Although admittedly using high corrosion rates, he does not point out that these rates are high by at least an order of magnitude. Secondly, the fractional excretion rates used are based upon urine/plasma concentration ratios and do not take into account exchange with slower compartments, particularly situations where precipitated storage is possible, as in the liver. If we reduce the corrosion rate by an order of magnitude and, continuing to use his assumption, calculate fractional excretion rates based upon whole body content (Q_0) and urine concentrations (Sunderman et al., 1989) for Co, Cr and Ni, we obtain the results given in Table 14.4. The very large Q_e/Q_0 ratios for Co and Cr suggest that concentrations of these elements would never come to equilibrium but should be observed to increase steadily with time postimplantation. Such effects have been reported in animals (Woodman et al., 1983) and humans (Michel et al., 1991).

The fundamental error in all of these considerations is the idea that the metals distribute freely and do not concentrate or bind preferentially in any site. In perhaps the first attempt to study systemic distribution of metals released from implants in animals (Ferguson et al., 1962 a, b), Ferguson and his co-workers observed the following patterns:

1. There is a large regional variation of metal ion concentration in both normal rabbit and human tissue. Furthermore, there are different patterns for different metals. Thus, the nickel concentration is higher in the liver than in other tissues, while the molybdenum concentration is higher in both liver and kidney than in either lung or spleen.
2. After implantation of metals, organ levels of ions rise. The spleen has a broad ability to retain metals, while nickel and cobalt is preferentially retained by the kidney.

Continuing studies of accumulation and distribution of corrosion products released by implants up to and including recent autopsy studies on patients with long term implants (Michel et al., 1991) support this view.

14.4.3 Distribution Models: Multicompartment

It is extremely difficult to perform whole body studies of metal metabolism in humans, for both technical and ethical reasons. However, a meth-

Table 14.4 Recalculation of Secretion and Accumulation Rates of Alloy Components of a Cobalt-Chromium Alloy of Table 14.3

Element	Q_0 (mg)	Secretion	k (day^{-1}) (24 hr), µg	R (mg/day)	Q_e (mg)	Q_e/Q_0
Co	3	0.82	0.00027	0.36	1336	445
Cr	6	0.32	0.00005	0.18	3606	601
Ni	10	2.5	0.00025	0.006	34	3.4

odology has been developed that involves study of the distribution and excretion of a single dose of radioactive metal ions administered intravenously in animals. It is possible that this method may be applied selectively to humans in the future.

Although the technique was originally developed by Sunderman at the University of Connecticut, Greene et al. (1975) provide the best early report. An experimental animal, such as the rat or rabbit, is used as a model. A single intravenous injection of nickel as $^{63}NiCl_2$ is given at a dose of 0.24 mg Ni/kg body weight. The animals are housed in metabolic cages so that urine and feces can be collected, and serum specimens are obtained periodically over a period of time.

For nickel in the rabbit, an expression for the concentration of nickel in serum takes the form of Eq. (14.8):

$$S \text{ (µg/liter)} = A_1 e^{-a_1 t} + A_2 e^{-a_2 t} \tag{14.8}$$

The first term is large, corresponding to an early rapid disappearance rate, and the second term is small, corresponding to a reduced disappearance rate from 3 to 7 days after injection. This can be interpreted in terms of a large, fast exchange compartment (the intercellular space) exchanging with a smaller, slower compartment of undetermined identity. The four constants assume different values for rats and rabbits.

Greene et al. (1975) made the following extrapolations:

From what is known about nickel corrosion, one can estimate that in humans who have implants made of a nickel-containing alloy, the rate of nickel release from the device can range between 5 and 500 mg/year per individual. This corresponds to a range of 0.81–0.0081 µg/hr per kg body weight on the basis of 70 kg for humans. The following table gives the estimated steady state values of nickel concentration in plasma resulting from 3 different input rates within that range.

Infusion rate	Steady state Ni concentration (μg/liter)	
(μg/hr per kg of body weight)	Rabbit	Rat
0.81	45	19.9
0.081	4.5	1.99
0.0081	0.45	0.20

These figures have to be compared with the normal range of nickel concentration in human plasma . . . 2.6* ±0.08 μg/liter.

Taylor's calculations (Taylor, 1973) for an alloy that released 22 mg per year predict a 7X increase (3.4X corrected calculation based upon his secretion data) in total body burden of nickel. Since Greene made no assumption on the partitioning of metallic ions between various compartments, the elevation he predicts for plasma must be taken as the total body elevation. For a 2 mg per year release rate, Greene would predict a 1.76X increase (by rabbit data) or a 1.34X increase (by rat data). While these are somewhat smaller than Taylor's original calculations (but near to the corrected calculations of Table 14.4), they lend credence to the idea that net concentrations of metal ions will rise in the presence of an endogenous source of ions.

There are a number of difficulties in Sunderman's approach, including the partition question and the inability to identify internal compartments except by inference from calculations.

This work has been extended and is reported by Onkelinx (1977). He has developed a more general multicompartment model, as shown in Figure 14.6. Here V_1 represents the intracellular compartment and V_2 and V_3 represent other compartments with net (reversible) interchange flow rates f_2 and f_3, respectively. Again, the assumption of a partition coefficient of 1 is made so that ion fluxes can be represented as fluid flow rates at constant concentration. Excretion is represented by a flow rate F_t composed of collection in irreversible sinks (f_s), urinary excretion (f_u), and fecal excretion (f_d). Loss of tissue, desquammation, etc., are lumped with f_s.

Table 14.5 reports some results obtained by injection of nickel, cobalt, and chromium into rats between 2 and 3 months old. Onkelinx (1977)

*Note that this is significantly higher than typical modern values; see for instance Table 14.2. The probable lower true value makes the predicted increases even more striking.

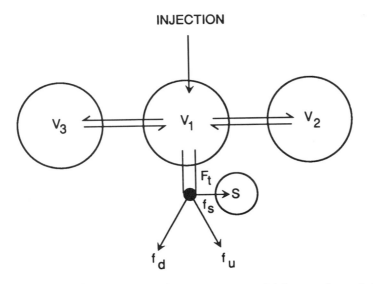

Figure 14.6 General multicompartment model for metal metabolism. Adapted from Onkelinx (1977).

reported results for other ages, suggesting an age dependence, especially for chromium.

Examination of Table 14.5 graphically shows the differences in metal metabolism for these ions. While it is difficult to apply physical identities to compartment volumes, as stated, the differences are obvious. Nickel does not penetrate V_3, as previously pointed out, and the relative penetrance of each ion is quite different. Of particular interest is the ratio f_s/F_t. Here we see that modest amounts of nickel and cobalt are trapped in a tissue sink, but over 30% of chromium is retained. Again, with reference to Taylor (1973), if this material is bound irreversibly, the effect would be to lower transient tissue concentrations, lengthen the time until steady state is reached, but raise the final body burden above that originally predicted.

It is inviting to equate secondary compartments (in this case V_2, V_3 and S), as derived from Onkelinx' analysis with specific tissue types or organs, on the basis of apparent volumes. This is simplistic; metabolic pools and storage depots can only be identified by following the course of metal, in the form of a suitable radioactive isotope, from the implant site to its eventual destination. Unfortunately, such studies may not ethically be performed in humans so compartment identities may remain unclear for a long time to come.

Table 14.5 Comparison Among Compartment Volumes, Interchange and Excretion for Ni, Co and Cr in Rats

	$^{63}Ni^{2+}$	$^{57}Co^{2+}$	$^{51}Cr^{3+}$
Age (days)	85	60	60
Compartment volumes (ml/100 g body wt)			
V_1	36.1	46.4	30.8
V_2	4.0	78.2	16.7
V_3	—	65.6	7.7
Apparent excretory flow rates (ml/hr)			
F_t	3.91	7.29	1.42
f_u	3.07	5.97	0.91
f_d	0.62	0.72	0.07
f_s	0.22	0.60	0.44
Ratios of excretory flow rates			
f_u/F_t	0.79	0.82	0.64
f_d/F_t	0.16	0.10	0.05
f_s/F_t	0.06	0.08	0.31
Compartment net interchange flow rates (ml/hr)			
$f_{1,2}$	0.06	13.11	2.47
$f_{1,3}$	—	0.56	0.13

Source: Adapted from Onkelinx (1977).

14.4.4 Equilibrium Models

Another experimental approach to this problem (Smith and Black, 1977; Smith, 1982) is to implant metallic devices with varying surface areas and track the plasma levels as a function of time. In this report an experiment was performed in a short-term rabbit model to investigate if implantation of 316L stainless steel is accompanied by elevated plasma levels of iron and chromium.

For a given alloy system with a uniform processing history, the parameter that would appear to govern the rate of corrosion product delivery to the body in any particular implant site is the ratio of implant surface area to body weight (SA/BW). For a 70 kg patient receiving a typical total hip joint replacement (surface area $\approx$ 200 cm²), this ratio is approximately 2.86 cm²/kg. This value is termed "1X" and multiples of it reflect higher relative exposures. Note that the use of such a ratio probably *under-*estimates the exposure in small animal models. Renal clearance is roughly proportional to basal metabolic rate (per kilogram of body weight). Since individual metabolic rate in adult mammals, including

Table 14.6 Animal Groups for Evaluation of SA/BW Effects

Group	N =	Implant	Site	SA/BW (unadjusted)
I	10	pin	muscle	1
II	10	pin	bone	1
III	15	microspheres	bone	1
IV	10	microspheres	bone	10
V	15	microspheres	bone	100
VI	14	none	—	0

humans, is proportional to $(BW)^{0.734}$ (Brody, 1945), a 1.5 kg rabbit has a basal metabolic rate/kg $\approx$ 2.78 times that of a 70 kg patient. Thus "100X" for the patient is approximately 36X for the rabbit, if adjusted to reflect basal metabolic rates, or, conversely, 100X for the rabbit is 278X for the patient. The concept of SA/BW ratio can be extended to in vitro studies by recognizing that a 70 kg patient has a water content of 42 liters (Figure 14.4), forming the ratio of surface area to body fluid volume (SA/BFV) and assigning a value of 4.76 cm^2/L.

In Smith and Black's (1977, 1982) study, New Zealand white rabbits received implants of passivated surgical grade 316L stainless steel (ASTM F55, type B) in two forms (Steinmann pin segments and 40 μm spherical powders) at two anatomic sites (paraspinal musculature and femoral medullary canal) as shown in Table 14.6.

Blood specimens were obtained at intervals up to 7 months, and tissue specimens were obtained at sacrifice. The serum was analyzed for chromium, free (non-heme) iron (PI), total iron binding capacity (TIBC), and percent of TIBC saturation (% sat.). The tissues were analyzed for iron and chromium.

No significant differences (experimental vs. control) were detected in PI, TIBC, or % sat. until 20 weeks postoperative. Group V plasma iron concentration was 14% elevated ($p < 0.05$) by 20 weeks postoperative. None of the groups exhibited significant changes in TIBC or % sat.; however, the latter showed a "tendency" toward elevation, particularly in Group V. The kidney showed an ability to accumulate iron; however, Group V liver iron concentrations were elevated 30% on both dry weight and total protein basis. The results for chromium are shown in Table 14.7.

Plasma iron elevations in Group V at 20 and 28 weeks indicated that iron release (corrosion into the blood circulation) exceeded the inherent iron turnover rate of the Fe-transferrin binding system. The consequent

Table 14.7 Serum and Tissue Chromium Content after Stainless Steel Implantation as a Function of Implantation Time and SA/BW Ratio in the Rabbit

Sample	Control (Group VI) (ng/ml)	% Change from control (significance)[a]				
		I	II	III	IV	V
Serum						
Pre-op	13.1 ± 0.9					
4 wk PO	11.4 ± 0.4		−11.4 ($p < 0.05$)			+16.7 ($p < 0.001$)
8 wk PO	12.9 ± 0.1	Lost group				+12.8 ($p < 0.08$)
20 wk PO	11.2 ± 0.4					+21.4
24 wk PO	10.3 ± 0.7	+15.5 ($p < 0.01$)		+6.2 ($p < 0.05$)		+23.3 ($p < 0.01$)
28 wk PO	14.3 ± 0.7	+17 ($p < 0.01$)		+17.5 ($p < 0.01$)	+24.3 ($p < 0.01$)	+10.4 ($p < 0.01$)
Tissue	(ng/mg dry wt.)					
Kidney	1.97 ± 0.5				+64.2 ($p < 0.02$)	
Liver	1.07 ± 0.7					+85 ($p < 0.01$)

[a]Only significant changes shown.
Source: Adapted from Smith (1982).

trend toward elevation in % sat. as observed in man might occasion a reduction in a patient's disease resistance (see Chapter 13). Plasma iron and, especially, plasma chromium concentrations appear to reflect both the duration of implantation and the SA/BW ratio. Sporadic elevations in plasma chromium in Groups I–IV toward the last 2 months of the study may suggest periods of accelerated corrosion or release into the circulation. Since plasma chromium concentrations are probably *not* in equilibrium with storage compartments, these chromium elevations are not likely to be an indicator of body stores but rather a dynamic measure of serum transport at the moment of sampling. The kidney demonstrated no capacity to accumulate iron or chromium. On the other hand, the liver exhibited elevated iron and chromium accumulations which were apparently a function of SA/BW ratio, in contrast to Ferguson's study (Ferguson et al., 1962 a, b) which reported no elevations in iron or chromium in the liver or kidney. From persistently elevated plasma iron and chromium concentrations it can be expected that liver accumulations would have continued beyond the experimental period, and that, with time, Groups I–III may have exhibited elevations.

It is clear from these equilibrium studies that trace metal metabolism is highly complex and that detailed studies of each metal of implant importance should be carried out in the future.

14.5 ROLE OF SYSTEMIC DISTRIBUTION AND EXCRETION STUDIES

Systemic distribution, storage and excretion of corrosion and degradation products from implants has not attracted great attention as a subfield of study within Biomaterials. Initially this came about through an inability to detect either normal or elevated levels of such materials, especially metal-bearing ions, and a parallel failure to recognise their biological importance. More recently, studies of host response have continued to focus on circum-implant effects and have continued to equate low corrosion/elution rates and small plasma concentration elevations with absence of accumulation of degradation products and, thus, absence of biological effects. What is required is that we come to recognise that any implant, whether designed specifically for that purpose or not, acts as a slow release system in vivo. Thus, the appropriate approach to understanding the total host response to implants must parallel that of biological fate studies in Pharmacology and environmental fate studies in Ecology.

From this viewpoint, there are three important questions which should be answered when a new implantable biomaterial is being evaluated:

1. What is the nature of the degradation products released from the implant in vivo?
2. Where do they go within the body?
3. What is the host response to release, distribution, remote concentration and excretion of these degradation products?

We have previously considered some aspects of local host response. In the next chapter, we shall take up the more global issue of systemic and remote site host response.

REFERENCES

Aalders, G. J., van Vroonhoven, T. J. M. V., van der Werken, C. and Wijffels, C. C. S. M. (1985): Injury 16:564.

Anderson, J. M. and Miller, K. M. (1984): Biomats. 5:5.

Araki, S., Aono, H., Yokoyama, K. and Murata, K. (1986a): Arch. Environ. Health 41:216.

Araki, S., Murata, K., Aono, H., Yanagihara, S., Niinuma, Y., Yamamoto, R. and Ishihara, N. (1986b): J. Applied Toxicol. 6:245.

Arvidson, K. and Wróblewski, R. (1978): Scand. J. Dent. Res. 83:200.

Biozzi, G., Benacerraf, B. and Halpern, B. (1953): Brit. J. Exp. Path. 34:441.

Biozzi, G., Benacerraf, B., Halpern, B. and Stiffel, C. (1957): Reticuloendothel. Bull. 3:3.

Brauer, R. W. (1959): J. Amer. Med. Assoc. 169:1462.

Brody, S. (1945): *Bioenergetics and Growth, with Special Reference to the Efficiency Complex in Domestic Animals*. Reinhold, New York, pp. 352ff.

Charnley, J. (1961): Lancet 1:1129.

Cherian, M. G. and Goyer, R. A. (1978): Life Sciences 23:1.

Ferguson, A. B., Akahoshi, Y., Laing, P. G. and Hodge, E. S. (1962a): J. Bone Jt. Surg. 44A:317.

Ferguson, A. B., Akahoshi, Y., Laing, P. G. and Hodge, E. S. (1962b): J. Bone Jt. Surg. 44A:323.

Ganong, W. F. (1989): *Review of Medical Physiology*, 14th ed. Appleton & Lange, Norwalk, CT, pp. 593ff.

Graham, J. A., Gardner, D. E., Waters, M. D. and Coffin, D. L. (1975): Infect. Immun. 11:1278.

Greene, N. D., Onkelinx, C., Richelle, L. J. and Ward, P. A. (1975): In: *Biomaterials*. E. Horowitz and J. L. Torgesen (Eds.). NBS Special Publication 415. U.S. Government Printing Office, Washington, pp. 45ff.

Hierholzer, S., Hierholzer, G., Sauer, K. H. and Paterson, R. S. (1984): Arch. Orthop. Trauma Surg. 102:198.

Homsy, C. A., Tullos, H. S., Anderson, M. S., Differrante, N. M. and King, J. W. (1972): Clin. Orthop. Rel. Res. 83:317.

Ishihara, N. and Matsushiro, T. (1986): Arch. Environ. Health 41:324.

Jacobs, J. J., Skipor, A. K., Black, J., Urban, R. M. and Galante, J. O. (1991): J. Bone Jt. Surg. 73A:1475.

Jenkin, C. R. and Rowley, D. (1961): J. Exp. Med. 114:363.

Kapp, J. P., Gielchinsky, I. and Jelsma, R. (1973): J. Trauma 13:256.

Kawaguchi, H., Koiwai, N., Ohtsuka, Y., Miyamoto, M. and Sasakawa, S. (1986): Biomats. 7:61.

Klaassen, C. D. (1976): Drug Metab. Rev. 5:165.

Korn, E. D. and Weisman, R. A. (1967): J. Cell. Biol. 34:219.

Lux, F. and Zeisler, R. (1974): J. Radioanal. Chem. 19:289.

Lux, F., Schuster, J. and Zeisler, R. (1976): J. Radioanal. Chem. 32:229.

Lyons, F. A. and Rockwood, C. A., Jr. (1990): J. Bone Jt. Surg. 72A:1262.

Michel, R. (1987): CRC Crit. Rev. Biocompat. 3:235.

Michel, R., Nolte, M., Reich, M. and Löer, F. (1991): Arch. Orthop. Trauma Surg. 110:61.

Normann, S. J. (1974): Lab. Invest. 31:161.

Onkelinx, C. (1977): In: *Clinical Chemistry and Chemical Toxicology of Metals*. S. S. Brown (Ed.). Elsevier North-Holland, Amsterdam, pp. 37ff.

Oron, U. and Alter, A. (1984): Clin. Orthop. Rel. Res. 185:295.

Pitts, R. F. (1963): *Physiology of the Kidney and Body Fluids: An Introductory Text*. Year Book Medical Pub. Inc., Chicago, pp. 69ff, 116ff.

Potter, F. A., Fiorini, A. J., Knox, J. and Rajesh, P. B. (1988): J. Bone Jt. Surg. 70B:326.

Roberts, J. and Quastel, J. H. (1963): Biochem. J. 89:150.

Schnyder, J. and Baggiolini, M. (1978): J. Exp. Med. 148:1449.

Smith, G. K. (1982): *Systemic Transport and Distribution of Iron and Chromium from 316L Stainless Steel Implants*. Ph.D. Thesis, University of Pennsylvania, Philadelphia.

Smith, G. K. and Black, J. (1977): Trans. ORS 2:281.

Stensaas, S. S. and Stensaas, L. J. (1978): Acta. Neuropath. (Berl.) 41:145.

Stiffel, C., Mouton, D. and Biozzi, G. (1970): *Mononuclear Phagocytes*. R. van Furth (Ed.). Blackwell Scientific, Oxford, pp. 335ff.

Styles, J. A. and Wilson, J. (1976): Ann. Occup. Hyg. 19:63.

Sunderman, F. W., Jr., Hopfer, S. M., Swift, T., Rezuke, W. N., Ziebka, L., Highman, P., Edwards, B., Folcik, M. and Gossling, H. R. (1989): J. Orthop. Res. 7:307.

Taylor, D. M. (1973): J. Bone Joint Surg. 55B:422.

Taylor, D. M. (1978): Personal communication.

Vernon-Roberts, B. (1972): *The Macrophage*. Cambridge University Press, Cambridge, pp. 92ff.

Weisman, R. A. and Korn, E. D. (1967): Biochemistry 6:485.

Woodman, J. L., Black, J., and Nunamaker, D. M. (1983): J. Biomed. Mater. Res. 17:655.

BIBLIOGRAPHY

Friedman, M. H. (1986): *Principles and Models of Biological Transport*. Springer-Verlag, Berlin.

Guyton, A. C. (1991): *Textbook of Medical Physiology*, 8th ed., W. B. Saunders, Philadelphia. Unit V—The Kidneys and Body Fluids, pp. 273–354.

Harrison, P. M. and Treffry, A. (1979): In: *Inorganic Biochemistry: A Review of the Recent Literature Published up to Late 1977*, Vol. 1. H. A. O. Hill (Ed.). The Chemical Society, London, pp. 120ff.

Iyengar, G. V. (1987): Biol. Trace Elem. Res. 12:263.

Koushanpour, E. (1976): *Renal Physiology: Principles and Functions*. W. B. Saunders, Philadelphia.

Pitts, R. F. (1963): *Physiology of the Kidney and Body Fluids: An Introductory Text*. Year Book Medical Pub., Chicago.

Solomon, A. K. (1960): In: *Mineral Metabolism*, Vol 1A. C. L. Comar and F. Bronner (Eds.). Academic Press, New York, pp. 119ff.

Valtin, H. (1983): *Renal Function: Mechanisms Preserving Fluid and Solute Balance in Health*, 2nd ed. Little, Brown and Co., Boston.

15

Effects of Degradation Products on Remote Organ Function

15.1 INTRODUCTION

This chapter will complete our discussion of the effects of materials on biological systems (host response). With the exception of a number of recognized systemic effects, this chapter will simply recapitulate and emphasize areas already discussed in Chapters 8–14.

Systemic effects of foreign materials are now well recognized. While it is more correct to distinguish between *remote* effects (involving actions, perhaps secondary to deposition and concentration of degradation products) on a target tissue or organ and *systemic* effects (those affecting large scale systems, such as the cardiovascular or neurological systems) for the sake of brevity we shall group them together in this chapter under the common title of systemic effects.

The effects of drugs, collectively *pharmacological* effects, are by and large systemic or remote effects. Drugs are given or injected in one portion of the body and directly or indirectly affect cellular and systemic physiology in other areas, even if a purely local or topical effect is intended. Discussion of such effects is beyond the scope of this book; however, many of the effects that we have discussed are essentially pharmacological effects that are *secondary* to the intended effect. Drug release materials provide a hypothetical example that illustrates the complexity of possible host response to an implant:

- The drug and/or carrier material will evoke a local (implant site) response.
- The drug and/or degradation products from the carrier may evoke desirable or undesirable systemic effects, such as alteration of arterial pressure.
- The drug may have a specific remote organ target, such as the heart.
- Degradation products of the carrier might produce adverse effects in other remote organs.

15.2 EXAMPLES OF SYSTEMIC EFFECTS

15.2.1 Polymers

An example of these types of effects is that associated with release of methyl methacrylate monomer during the polymerization of PMMA-type cements in vivo. Homsy et al. (see Section 14.3.1) were attracted to this problem not by the observation of the rapid transport of the monomer to the lungs, but by the observed systemic effects. Early in the use of these cements it was observed that systemic hypotension developed immediately after the insertion of the cement into the prepared bone cavity. Arterial pressure drops of more than 15 mmHg were observed associated with transient cyanosis, and a number of cardiac arrests drew medical attention to the effect. Further studies in dogs have shown that this problem is accentuated by marked fluid and/or blood loss (McMaster et al., 1974). The corrective therapy now used to prevent the development of centrally mediated hypotension after PMMA insertion is maintenance of the patient in a state of positive hydration. In addition, it has been shown that venting the bony (medullary) cavity with a rubber drain during device insertion reduces the driving pressure differential that aids in monomer take-up by blood.

There are few if any known systemic effects of polymers released from implants beyond this hypotensive effect. The rates of release and the normal routes of molecular catabolism apparently combine to keep concentrations of products from common polymeric implants below levels required for direct pharmacological activity.

This is not to suggest that elevated blood levels of polymer degradation products do not exist, nor that they may be harmless in the long term. In the short term, host responses may be indistinguishable from disturbances of homeostasis associated with surgery. However, in cases of high surface area and/or repetitive exposure, as in hemodialysis (as a treatment for renal insufficiency or failure), significant deviations from normal conditions may be encountered. Lewis et al. (1977) examined the blood of individuals undergoing chronic hemodialysis, looking explicitly for breakdown products or plasticizers that might be released from the polyvinyl chloride (PVC) polymer used as a blood conduit material. Significant levels, up to a mean level of 751 ng/ml serum of bis(2-ethylhexyl) phthalate, a common PVC plasticizer, were found in these patients after dialysis. Although the catabolism of this material was rapid, leading to its being undetectable in blood 5–6 hours after completion of dialysis, the yearly dose for the chronic dialysis patient was estimated to be 150–250 mg. The long-term effects of this are unknown.

A final interesting example is provided by the early (now abandoned) practice of using injected, fluid silicone materials for cosmetic tissue augmentation. In addition to producing local fibrosis, these fluids can migrate through tissue and produce a variety of remote effects including tissue mass formation, adenopathy and possibly pulmonary failure (Kossovsky and Heggers, 1987).

15.2.2 Metals

The situation with metals is somewhat different. In addition to the basic "physiological" metals (Ca, Na, K, and Fe), a large number of metals, including Co, Cr, Mg, Zn, and Cu, despite very low normal plasma and tissue concentrations, have normal roles in metabolism and are thus classified as essential trace elements. Therefore, it should come as no surprise that there are naturally occurring diseases of both inherited and acquired etiology that involve imbalances in the metabolism of these metals. In addition to anemia (iron deficiency), iron overload diseases also exist. One such disease, hemochromatosis (Elinder, 1986), results in the accumulation and deposition of iron in the form of the compound hemosiderin in tissues with rough endoplasmic reticulae. This accumulation has a number of physiological effects. One example is the development of diffuse arthritis in widely separated joints. This also occurs secondary to the internal bleeding associated with hemophilia. Hemochromatosis also involves skin pigmentation, liver failure, and diabetes. Elinder (1986) points out that iron, although a physiological element, is potentially toxic in all doses and forms and is capable of producing a variety of both local and systemic toxic effects in both animals and humans.

A less well-known disease is Wilson's disease (Aaseth and Norseth, 1986), or the so-called "copper man" syndrome. This is an accumulation disease, primarily hereditary, in which copper, instead of being maintained in balance, accumulates in a variety of tissues, including liver, cornea, and skin. A green skin color develops and a high incidence of mortality due to intravascular hemolysis and liver failure results from the cytotoxicity of copper and its compounds.

Metals that are normally foreign to the body, such as lead, beryllium and arsenic, can combine competitively with enzymes that normally use other trace metals as cofactors. Even normally present metals, such as aluminum and chromium, may do so, if present in sufficiently high concentrations. The abnormal cofactor-enzyme combination may have higher stability than the normal cofactor bond. Thus, the effect is to inactivate a portion of the enzyme pool without stimulating additional enzyme production. In low concentrations, the net effect will be to inhibit enzyme activity. This can be seen in a reduction of the efficiency or effectiveness of an enzyme process. Examples of this are the peripheral neuropathy observed in the case of long-term low-level ingestion of lead and the suppression of hemoglobin synthesis by chromium.

At higher concentration levels, metals can be highly toxic poisons through enzyme inactivation as in the familiar case of arsenic.

Beyond these specific mechanisms, a wide variety of systemic medical problems has been suggested as being associated with imbalances in trace metal levels. As noted in Section 13.4, care must be taken in dealing with this literature as there is a less than totally scientific nutritional school of thought that ascribes virtually all unexplained physical and mental disabilities to such effects. There are, however, legitimately recognized associations, such as the increased incidence of cardiomyopathy associated with elevated cobalt intake (Alexander, 1972) and the more recently recognized association of arteriosclerosis with hardness of ground water (Perry, 1973) (and, presumably, the general levels of trace element intake).

Consideration of the effects of metals must address all aspects of physiology. Levels of metals well below those needed to produce externally measurable changes in physiological variables can produce profound behavioral abnormalities and mental disorders (Weiss, 1978).

15.3 A REVIEW OF SYSTEMIC ASPECTS OF HOST RESPONSE

Let us now briefly review host responses in terms of systemic effects.

15.3.1 Interaction of Molecules with Surfaces

Proteins and enzymes that are released from surfaces in an irreversibly denatured state will possibly elicit remote effects either directly or indirectly through the action of the immune system. Depletion of pools of unactivated coagulation or complement factors, as frequently occurs during hemodialysis or blood oxygenation, may suppress coagulation effects at remote sites of injury. Conversely, the increased circulating concentration of surface-activated factors may also have systemic or remote sequelae. It is also possible that denatured molecules bound to wear debris subject to either passive or active transport can elicit systemic or remote effects.

15.3.2 Inflammation

We would expect that inflammation would be a local effect restricted to the vicinity of the implant. It is possible that the implant can release pyrogenic agents directly or produce them indirectly through denaturation processes. There is some evidence that denatured molecules or released products, of unknown identity, may produce long-term systemic hallmarks of inflammation, as in the chronic erythrocyte sedimentation rate elevation observed by Shih et al. (1987) in patients after PMMA-cemented total hip replacement. In addition, friction and wear processes as well as others that release particulate material may result in an inflammatory response at a site of remote accumulation. An example of this is the abdominal "Teflonoma" frequently seen after use of poly(tetrafluoro)ethylene as the material for fabrication of acetabular cups in Charnley's early efforts at hip replacement (Charnley, 1979). Precipitates of corrosion products in remote sites might also be expected to produce inflammation due to increased local concentrations of ions or through response to the resulting particulate material.

15.3.3 Coagulation and Hemolysis

Similarly, we expect that primary coagulation problems would be localized to the vicinity of a cardiovascular system implant or external blood treatment device. The ability of implants to surface activate factors in the coagulation cascade (as noted above), as well as to shed thrombi, renders this a systemic effect with remote site manifestations. Similarly, hemolysis, whether through blood-surface interactions or by direct mechanical damage in pumps, etc., is a systemic problem due to the reduction in viable erythrocytes and the rapid dispersion of hemoglobin and cell fragments. It is clear that the current factors limiting successful long-term left ventricular assist and total (artificial) heart replacement are remote effects, primarily cerebral and pulmonary infarcts, secondary to systemic distribution of shed thrombi.

15.3.4 Adaptation

The discussion of adaptation emphasized the effects at the biomaterial-tissue interface. This is certainly the most important adaptive remodeling site and it is expected that systemic or remote effects would be secondary to this and, thus, not directly related (if they occur at all).

15.3.5 Chemical Carcinogenesis

Of necessity, this must be considered a systemic problem due to the variable sensitivity of cells to chemically induced neoplastic transformation and the possibility of metastases. If chemically-mediated carcinogenesis occurs in humans due to the use of implant materials, it can thus be expected to have a significant systemic manifestation. It is worth pointing out, once again, that systemic and remote site tumorigenesis associated with implants tends to be overlooked in patient populations for two reasons: (1) The tumor types expected are no different than those which would already exist in a comparable patient population without implants; and (2) the specialization of medicine makes the connection of a tumor in a remote tissue or organ system to the presence of an implant in another tissue or organ system unlikely (Black, 1984).

15.3.6 Foreign-Body Carcinogenesis

We would expect foreign-body carcinogenesis, if it occurs in patients at all, to be a local problem. However, the ability of particulate materials to move in the body and the inherent ability of many neoplasms to metastasize renders it a potential systemic problem. It must be emphasized that this is a putative consideration, as the presence of primary foreign-body neoplasias (at the implant site) have not been reliably detected in humans (see Section 12.3).

15.3.7 Infection

If we accept Weinberg's (1974) arguments concerning nutritional immunity (see Chapter 13), we must recognize the possibility of problems associated both with elevated iron concentrations in the vicinity of implants and with elevated concentrations in remote storage sites. Furthermore, suppression of the immune system may also predispose to infection at distant sites as well as at the implant.

The possibility of the inverse, that is, of hematogenous "seeding" of an implant site infection from a distant site, such as a dental abscess, must not be discounted, although clinical data are equivocal at this time (Thyne and Ferguson, 1991).

15.3.8 Allergic Foreign-Body Response

The discussion of this subject (Chapter 11) emphasized the systemic nature of the response. So, in addition to possible problems in the vicinity of the implant such as pain, loosening, etc., we can potentially associate a wide variety of allergic responses at remote sites with the presence of metallic, and possibly polymeric, implants as either sensitizing or challenge agents.

15.4 A FINAL COMMENT

It is important to re-emphasize the point that studies of host response have focused primarily on the implant site and adjacent tissues. In the future, a much broader view must be taken and the response of the entire host, whether experimental animal or human patient, must be studied in detail.

However, we must avoid a major pitfall in such considerations. Much is made in the lay press about individuals who develop an apparent "environmental allergy": an elevated sensitivity, with immune response symptoms, to a very broad variety of agents, which have in common only that they are "man-made." An analogous situation exists in the field of clinical application of biomaterials with lay concern over the relationship between release of mercury from mercury-based dental amalgams and a broad variety of patient symptoms.

Without passing judgment on either of these situations,* I would like to suggest, somewhat in the spirit of Furst's requirements for accepting the carcinogenicity of a metal in animal models (see Section 12.2.4), that the following criteria should be met before the existence of a remote or systemic effect in humans is taken as proven:

1. The basic mechanism of the biological response must be demonstrated in at least one biological model, either in vitro or in an animal model.
2. After the causative implant-related species has been identified, its release by a functional implant and systemic distribution in either an animal model or in patients (preferably both) must be shown.
3. The putative biological response must be identified in either an animal model or in patients (preferably both) with functional implants.
4. Finally, if demonstrated in patients, the biological response must be recognised on the basis of a statistically sound epidemiological study, with suitable non-exposed controls and, unless the response is of a threshold type, must demonstrate a dose-response or exposure-incidence relationship.

These are in my mind necessary and sufficient conditions to conclude that an implant related systemic (and/or remote site) effect exists in patients. It does not settle the issue of the clinical importance of the effect: this depends upon other considerations such as treatment alternatives, including no treatment, and the benefit of the use of the device in which the biomaterial is incorporated.

However, I would suggest that if the first criterion is met, that is, a biological mechanism leading to a putative systemic or remote site effect is identified, an index of suspicion should be attached to the biomaterial in question. Similarly, isolated case reports of adverse implant-related responses, whether local or distant, should also arouse a degree of suspicion: what they lack in numbers they make up for, to a degree, in specific-

*See Section 11.4 for a more complete discussion of immune responses to implants and Section 13.4 for a more complete critique of "environmental allergy" and related issues.

ity. A lack of sound knowledge in any of the latter three areas should not lead to an inference of safety: this would be morally equivalent to the statement that, "What I don't know can't hurt me." Rather, satisfaction of the first criterion should lead to changes in behavior both in consideration of that biomaterial for specific current device designs (Black, 1988) and in planning future basic and applied studies of the biomaterial's safety and efficacy.

REFERENCES

Aaseth, J. and Norseth, T. (1986): In: *Handbook on the Toxicology of Metals*, 2nd ed., Vol. II. L. Friberg, G. F. Nordberg and V. B. Vouk (Eds.). Elsevier, Amsterdam, pp. 233ff.

Alexander, C. S. (1972): Am. J. Med. 53:395.

Black, J. (1984): Biomats. 5:11.

Black, J. (1988): J. Bone Jt. Surg. 70B:517.

Charnley, J. (1979): *Low Friction Arthroplasty of the Hip*. Springer-Verlag, Berlin, pp. 6–7.

Elinder, C.-G. (1986): In: *Handbook on the Toxicology of Metals*, 2nd ed., Vol. II. L. Friberg, G. F. Nordberg and V. B. Vouk (Eds.). Elsevier, Amsterdam, pp. 276ff.

Kossovsky, N. and Heggers, J. P. (1987): CRC Crit. Rev. Biocompat. 3:53.

Lewis, L. M., Flechtner, T. W., Kerkay, J., Pearson, K. H., Chen, W. T., Popowniak, K. L. and Nakamoto, S. (1977): Trans. Am. Soc. Artif. Intern. Organs XXIII:566.

McMaster, W. C., Bradley, G. and Waugh, T. R. (1974): Clin. Orth. Rel. Res. 98:254.

Perry, H. M. (1973): J. Am. Diet. Assn. 62:631.

Shih, L.-Y., Wu, J.-J. and Yang, D.-J. (1987): Clin. Orth. Rel. Res. 225:238.

Thyne, G. M. and Ferguson, J. W. (1991): J. Bone Jt. Surg. 73B:191.

Weinberg, E. D. (1974): Science 184:952.

Weiss, B. (1978): Fed. Proc. 37(1):22.

BIBLIOGRAPHY

Davies, I. J. T. (1972): *The Clinical Significance of the Essential Biological Metals*. Ch. C. Thomas, Springfield, IL.

Friberg, L., Nordberg, G. F. and Vouk, V. B. (Eds.)(1986): *Handbook on the Toxicology of Metals*, 2nd ed., Vols. I & II. Elsevier, Amsterdam.

Ling, R. S. M. (1984): In: *Complications of Total Hip Replacement.* Ling, R. S. M. (Ed). Churchill-Livingstone, pp. 201ff.

McCall, J. T., Goldstein, N. P. and Smith, L. H. (1971): Fed. Proc. 30(3):1011.

Pier, S. M. (1975): Tex. Rep. Biol. Med. 33(1):85.

Rothman, R. H. and Hozack, W. J. (1988): *Complications of Total Hip Arthroplasty.* W. B. Saunders, Philadelphia.

Schroeder, H. A. (1956): Adv. Intern. Med. 8:259.

Webb, M. (1977): In: *Clinical Chemistry and Chemical Toxicology of Metals.* S. S. Brown (Ed.). Elsevier North-Holland, Amsterdam, pp. 51ff.

Williams, D. F., Ed. (1981): *Systemic Aspects of Biocompatibility.* Vol. I & II. CRC Press, Boca Raton, FL.

Interpart 2
Implant Materials: Clinical Performance*

I2.1   INTRODUCTION

Elaine Duncan (1990) posed the question of whether biomaterials are at risk of becoming ''endangered species.'' Her query was motivated, in part, by a letter distributed by Dow Corning, Inc. (Midland, MI) warning of the company's intention of withdrawing from the market an old ''standby'' polyurethane biomaterial, Pellethane™, at least for applications intended to last longer than 30 days in vivo. Citing published reports of cracking of the material after longer times in vivo, a company repre-

*An early version of this Interpart was published as Black (1990). Table I2.1 is adapted from Table 14.2 in Black (1988).

sentative, J. R. Stoppert (1989), asserted that there are no data supporting long-term use of the material.

My immediate reaction to this statement was incredulity. The use of materials similar to Pellethane™ was reported by Boretos and Pierce (1968) more than twenty years ago. Boretos (1973), discussing such materials, said: ''(They) possess a combination of properties not available in other materials, outstanding of which (is) . . . excellent stability over long implant period.'' So, how can there be no data to support long-term human implantation of Pellethane™?

I believe that what Stoppert (1989) meant is that there are, literally, *no* data. That is, there is neither evidence to *support* nor *contradict* biomedical device designers' decisions to use Pellethane™ in long-term applications. Boretos' (1973) comments are not data. Nor are the majority of papers published about this material or, for that matter, about virtually any other biomaterial in use in long-term clinical applications. Most of these papers, especially those dealing with clinical observations, are not even studies, in the strict scientific sense, and the failure to conduct studies results in the absence of data, either positive or negative!

What has happened is that workers in the field of Biomaterials have been blinded by success. The techniques used in the 1960s to qualify materials (limited in vitro studies, 12–104 week animal studies, 2 year human clinical studies)(see Chapter 18) are still current practice in the early 1990s, despite the widespread use of biomaterials in long-term clinical applications which, in at least one device type (total hip replacement)(Ahnfelt et al., 1990), exceed 25 years in individual and group patient experience.

It appears that two critical aspects of the study of biomaterials have been neglected: the epidemiology and human physiology of biomaterials.

I2.1.1 The Epidemiology of Biomaterials

Continuing to study the success and failure of implanted materials in experiments with animals numbering in the tens and twenties, biomaterials researchers have largely overlooked the vast clinical ''experiments'' underway in which thousands, in many cases tens or hundreds of thousands, of human patients receive virtually identical biomaterials as chronic implants. A Center for Disease Control survey performed in 1988 (Moore et al., 1991) suggests that as many as 14.5 million people or nearly 1 in 20 in the US have implants. With the exception of occasional reports of clinical failures and studies of the materials aspects of retrieved devices, we know almost nothing about biomaterials' performance in these implants in the human clinical environment. Dependable incidence and prevalence data on the devices themselves are hard to come by, but such data on the materials from which they are made, including their exact (not merely specified) composition and processing, are essentially nonexistent. Today we still persist in studying only random device failures, making the same mistake as certain penologists who, wishing to know about crime, study only failed criminals, those convicted and incarcerated. Even these limited studies, based upon clinical or postmortem retrievals, are frequently incomplete, focusing on either the clinical features of the failure or upon the physical attributes of the failed device, depending upon the background and interests of the principal investigator, but rarely dealing with both aspects in a balanced way.

Recent interest in studying outcomes of surgical procedures (change in patient lifestyle, satisfaction level, relative cost, etc.) rather than merely the success of the procedure (rated as excellent, good, etc. by a largely subjective scale) may improve matters, but only if accurate information on the implant and its materials of construction are made part of the permanent clinical record. There have been repeated suggestions made concerning the need for registration systems for implants so that this information may follow patients as they move from place to place, but no concrete action has been taken other than the provision of voluntary device registry, for a fee, by a private organization, the Medic Alert Foundation (Turlock, CA).

I2.1.2 The Human Physiology of Biomaterials

We would be quick to discount the knowledge of a nephrologist who based his entire understanding of human renal function on the study of healthy rats, rabbits, dogs and an occasional human autopsy specimen. Fortunately, professional nephrologists have a vast armamentarium of in vivo tests which permit them to study the physiology of the functioning human kidney in both health and disease. The biomaterial scientist, when addressing functioning human implants, lacks all but the most rudimentary of these capabilities: clinical plain film x-rays. We cannot state with any certainty what the pH, pO_2, interfacial stress, etc., are in the vicinity of a functioning implant and what ranges of values assure long-term success or predict imminent failure.

However, it is possible to obtain such data, at least in animal models: Baranowski and Black (1987) reported measurement of both pH and pO_2

associated with both successful and unsuccessful stimulation of bone growth, near both active and inactive implanted stainless steel electrodes in the tibial medullary canal of rabbits. Such techniques could, with care and well-designed protocols, be extended to studies in human subjects. Already, as discussed in Chapter 14, studies are being performed to detect and quantify metal-bearing species associated with implants in human patients. Several studies suggest a correlation between elevations in serum concentrations and loosening of implants (cf. Jacobs et al., 1991).

I2.2 AN EXAMPLE: TOTAL HIP REPLACEMENT

Even in the absence of clinical epidemiology and the development of clinical tests of biomaterials' performance, it is still possible to gain some knowledge from current clinical experience. A clinical internship and, if possible, continuing contact with a clinical population can be extremely valuable for a biomaterials scientist or engineer, especially if he or she is prepared to be observant and analytical in approach.

It is probably possible in any clinical implant application to produce a list of symptoms (radiographic, clinical or histological findings) and associated putative mechanisms leading to implications or conclusions concerning the clinical performance of the implanted materials. Table I2.1 presents such a triple listing for a frequent orthopaedic procedure: total replacement of the hip. Note that there are other device related clinical findings possible; those listed are those believed directly referable to the *biomaterials* used in or in conjunction with the device rather than to the device design, surgical technique or patient use factors.

I2.3 A FINAL COMMENT

We have neglected many of the basic sciences underlying the field of Biomaterials. When we consider issues of long-term survival of biomaterials in vivo, the lack of attention to epidemiology and physiology are key issues. Unless researchers become more careful and observant of clinical performance, biomaterials may, indeed, as Elaine Duncan suggested, become endangered species and more often we will hear, "and there are no data. . . ."

Table I2.1   Materials-Associated Findings in Total Hip Replacement

Finding	Mechanism	Implication
Radiographic		
PMMA fragments (early)	Operative debris Single-cycle fracture	Third-body wear
PMMA fragments (late)	Fatigue	Third-body wear Cup loosening
PMA mantle fracture (early)	Inadequate bony support; single-cycle fracture	Stem subsidence
PMMA mantle fracture (late)	Inadequate bony support; fatigue	Stem subsidence
Broken cerclage wire	Fatigue	(early) Trochanteric dislodgement, nonunion, wire migration (late) Wire migration
Stem deformation	Plastic deformation	Change in bony support Inadequate stem size or yield point Impending failure
Stem, cup or cup screw fracture	Fatigue	Manufacturing defect Chronic mechanical overload (early) Inadequate bony support (late) Change in bony support
Eccentric cup-head centers	Plastic deformation Wear	UHMWPE creep Uniform wear Spalling?
Loose metallic debris	Wear	Third-body wear Fretting (loose component)
	Fatigue ± corrosion	Inadequate processing of porous coating
Focal lytic lesion[a]	Particle phagocytosis Immune response?	Excessive wear debris Metal sensitivity? (both) Progressive failure?
Progressive dissecting lesion[a]	Osteoclasis	Excessive wear debris Metal sensitivity? Neoplasm?
Clinical		
Intraoperative hypotension	Central control	Methyl methacrylate (monomer) sensitivity? Fat embolism?
Hip pain[a]	Immune response Venous blockade	Metal sensitivity? Excessive wear
Ectopic calcification	Wear debris nucleation?	Excessive wear
Dermatitis	Delayed hypersensitivity?	Metal sensitivity?
Eczema	Delayed hypersensitivity?	Metal sensitivity?
Bronchospasm	Delayed hypersensitivity?	Metal sensitivity?

Table I2.1 *(Continued)*

Finding	Mechanism	Implication
Histologic		
Fibrous capsule	Local host response	Normal response
Histiocytosis with multi-nuclear cells	Chronic inflammation	Manufacturing defect Inappropriate material Excessive wear
Lymphocytic infiltration with plasma cells	Delayed hypersensitivity?	Metal sensitivity? Polymer sensitivity?
Fibrosarcoma	Neoplastic transformation	Chemical neoplasia?
Lymphoma	Neoplastic transformation	Chemical neoplasia?
Rhabdomyosarcoma	Neoplastic transformation	Chemical neoplasia?
Malignant fibrous histio-cytoma	Neoplastic transformation	Chemical neoplasia?
Osteosarcoma	Neoplastic transformation	Chemical neoplasia?

[a]In absence of infection
?: Possible mechanism or implication
Source: Adapted from Black (1988).

REFERENCES

Ahnfelt, L., Herberts, P., Malchau, H. and Andersson, G. B. J. (1990): Acta Orthop. Scand. 61 (Suppl. 238):1.

Baranowski, T. J., Jr. and Black, J. (1987): In: *Mechanistic Approaches to Interactions of Electric and Electromagnetic Fields with Living Systems*. Blank, M. and Findl, E. (Eds.). Plenum, New York, pp. 399ff.

Black, J. (1988): *Orthopaedic Biomaterials in Research and Practice*. Churchill Livingstone, New York, pp. 319ff.

Black, J. (1990): Biomaterials Forum 12(4):9.

Boretos, J. W. (1973): *Concise Guide to Biomedical Polymers: Their Design, Fabrication and Molding*. C. C. Thomas, Springfield, IL, p. 10.

Boretos, J. W. and Pierce, W. S. (1968). J. Biomed. Mater. Res. 2:121.

Duncan, E. (1990): Biomaterials Forum 12(3):4.

Jacobs, J. J., Skipor, A. K., Black, J., Urban, R. M. and Galante, J. O. (1991): J. Bone Jt. Surg. 73A:1475.

Moore, R. M., Jr., Hamburger, S., Jeng, L. L. and Hamilton, P. M. (1991): J. Appl. Biomater. 2:127.

Stoppert, J. R. (1989): Letter cited in: Duncan (1990): Biomaterials Forum
12(3):4.

BIBLIOGRAPHY

Fraker, A. C. and Griffin, C. D. (Eds.) (1985): *Corrosion and Degradation of Implant Materials: Second Symposium*. ASTM Special Technical Publication 859. American Society for Testing and Materials, Philadelphia.

Syrett, B. C. and Acharya, A. (Eds.) (1979): *Corrosion and Degradation of Implant Materials*. ASTM Special Technical Publication 684. American Society for Testing and Materials, Philadelphia.

Weinstein, A., Horowitz, E. and Ruff, A. W. (Eds.) (1977): *Retrieval and Analysis of Orthopaedic Implants*. NBS Special Publication 472. U. S. Government Printing Office, Washington, D. C.

Weinstein, A., Gibbons, D., Brown. S. and Ruff, W. (Eds.) (1981): *Implant Retrieval: Material and Biological Analysis*. NBS Special Publication 601. U. S. Government Printing Office, Washington, D. C.

IV
METHODS OF TEST FOR BIOLOGICAL PERFORMANCE

16

In Vitro Test Methods

16.1 TEST STRATEGIES

Biological performance, as I defined it in Chapter 1, has two aspects: material response and host response. Since the traditional approach has been to define biological performance in terms of biocompatibility (host response), and then to worry about evidence of material degradation that arises during in vitro and in vivo testing, the development of tests has concentrated upon those that measure host response.

In practice this has not been a bad course of events. Despite the uncertainties in both areas, it is still easier to predict material response than to predict host response. Thus, it is still a good strategy to proceed as follows in early qualification studies:

1. Select a material based upon engineering data and previous material and host response information.
2. Determine experimentally if the host response is satisfactory for the intended application.
3. Watch carefully for evidence of unsatisfactory material response during the conduct of host response studies.
4. Verify satisfactory material response during long-term in vivo implant studies if it is relevant to the application.

That is, given an adequate a priori level of expected material response, expend a larger effort on determining host rather than material response. The selection of in vitro host response tests for materials qualification will be dealt with in Chapter 18.

16.2 IN VITRO TEST TYPES

Since there is a broad range of materials available and only a very small percentage have been used in biological environments, there has been a continual need for quick screening methods that can be used in vitro. These can be divided into two general classes:

1. Tissue culture methods
2. Blood contact methods (for blood contact applications)

Within each of these classes there is a wide variety of tests and variations of test methods. Some variations are traditional for particular applications, while others are the practice of a particular laboratory. Thus, we can only consider these types of methods in a general way.

16.3 TISSUE CULTURE TESTS

16.3.1 Generic Methods

Tissue culture refers collectively to the practice of maintaining portions of living tissue in a viable state in vitro. There are three generic methods:

1. *Cell culture*: The growth of initially natural matrix-free, disassociated cells. This method closely parallels the methods used to grow bacteria in vitro. Cells may be grown in solution or on agar or other media substrates. Exposure to biomaterials may be through direct contact with bulk materials, diffusional contact through an agar (or other gel) intermediate layer, or by inclusion of extracts or elutants from materials in the culture media.
2. *Tissue culture*: The growth of portions of intact tissue without prior cellular dissociation. This method usually utilizes a substrate rather than a suspended technique. Exposure to biomaterials is similar to that for true cell culture.
3. *Organ culture*: The growth of intact organs in vitro. This may vary from the use of fetal bone explants which can survive without external support systems to the use of whole, adult, perfused organs such as the kidney or heart. Again, a variety of exposures to biomaterials are possible.

As Rae (1980) has pointed out, each of these generic methods involves compromises and deviations from in vivo conditions.

16.3.2 Problems Inherent in Tissue Culture

There are several points to make that concern the use of tissue culture techniques.

The major difference between in vivo and in vitro conditions, after noting the obvious humoral one (lack of access to pathways in remote locations in the intact animal), is the failure of local circulation. Tissue in culture media can interchange materials only by diffusion. Diffusion distances must be quite short or toxic effects will be seen; this is obvious since cells in vivo are never more than a few hundred microns distant from a capillary. We can expect that even quite modest volumes of tissue will show central or focal necrosis due to a diffusional limit on either nutrients or waste products. For this reason tissue culture as a field is often referred to as a study of cell death. The use of cell culture (rather than tissue or organ culture) minimizes these effects until cell densities rise to levels associated with frequent intercellular contact.

Associated with diffusional problems and lack of humoral components is the general observation that metabolic rates in vitro are always lower than those in vivo. Even in the absence of necrosis, cell, tissue, and organ culture conditions are inhibitory, with an initial maximum effect and a significant residual effect.

Most tissue culture models attempt to overcome these effects by growing cells in monolayers (to provide easy exchange with large fluid volumes) and by adding various stimulatory agents (fetal or neonatal serum, vitamin C, etc.) and preservation agents (vitamin E, antibiotics, fungicides, etc.). Fortunately, most cells are contact inhibited and thus form monolayers easily in culture; this condition is called *confluence* and is characterized by a relatively constant cell number. Cell monolayers are not, in general, subject to diffusion limits. While these agents improve the viability and longevity of tissue culture models, they produce still greater differences between in vitro and in vivo conditions.

Finally, for unknown reasons, some physiological processes that occur in vivo do not occur in vitro. For instance, fracture healing (of bone) has not to date been reproducibly demonstrated in vitro.

16.3.3 Cell Types and Observations

Cell types may vary in activity and properties between species and, in species, between individuals, and within a specific cell line, between generations. In the literature one finds references to cell types with designations such as "P1534 leukemic cells." These refer to cells available from the American Tissue Type Collection (ATTC) (Hay, 1983) or similar sources. These organizations isolate, characterize, and preserve (by ultrafreezing) uniform cell populations. One may order specimens from particular generations of these cells and grow them in the laboratory for use in testing. In general, the lower the generation number, often referred to as *passage number*, the more the cells will resemble those in vivo in both appearance and function. With care, the use of established cell line techniques screens out a large component of biological variability.

However, an alternate approach is to use freshly derived cells or so-called *primary* cultures. Such cultures are more vigorous than maintained cell lines; however, they are difficult to prepare in a repeatable manner, are usually poorly characterized and may suffer from overgrowth of faster growing or "weed" cells, such as fibroblasts.

Tissue culture techniques, as a group, may be used to study the following aspects of host responses:

Cell survival: Toxicity, membrane integrity.
Cell reproduction: Growth inhibition.

Metabolic activity: Energetics, synthesis, and catabolism.

Effective activity: Inhibition of locomotion, chemotaxis, and phagocytosis; alteration of cellular shape and size.

Cell damage: Chromosomal aberration, mutagenicity, and carcinogenicity.

Tissue culture techniques for screening materials may use one or more normal mammalian cell lines such as murine macrophages, abnormal cells such as HeLa or lymphoma cells, or bacterial cell lines such as *Staphylococcus aureus* or *Escherichia coli*. Each test that has been developed uses selected cells that are suitable for the particular questions posed.

16.3.4   Examples of the Use of Tissue Culture in Materials Evaluation

An example of such studies (Galin et al., 1975) illustrates the general approach. In this case, cultures of rabbit kidney cells were used to investigate the toxicity of several types of ocular implants. The implants were exposed to the cells both directly by placement on a gel-supported cell monolayer and indirectly by inclusion in media in which the cells were cultured. Toxicity was assayed by observations of progressive cell death in the monolayer and by cellular changes in culture.

Galin's study examined the host response to a fabricated implant. Similar tests may be used to study biomaterials in unfabricated forms or to look at host response to degradation products such as wear debris. An early example of the latter is the work of Pappas and Cohen (1968) and Mital and Cohen (1968) who examined tissue culture response to finely divided particles of implant alloys as models for wear debris and corrosion product release. A number of cell lines were used and the effects of the metallic debris and debris supernatant on cell counts in the replication phase (Pappas and Cohen, 1968) and in the lag phase (Mital and Cohen, 1968) were obtained. The authors concluded that the latter method using lag phase cell counts to detect direct toxicity was more sensitive than the former, which lumped toxicity and growth and replication inhibition together.

It has been suggested that the primary effects seen in the exposure of polymeric biomaterials to cells in tissue culture are due to low-molecular-weight species that diffuse out of the biomaterial. Thus, many techniques use both the material itself as well as extractions of the material performed with various hydrophilic and hydrophobic solvents. A study by Homsy et al. (1970) examined this issue by autoclaving a number of specimens in a pseudoextracellular fluid (PECF) for 62 hours at 115°F, and then using the PECF supernatant as a challenge for cultures of cells derived from newborn mouse hearts. In addition, specimens of the supernatant were analyzed quantitatively to determine the total $-CH3$, $-CH2$, and $-CH$ radicals present. A total of 22 polymers were studied (Table 16.1). A comparison of cellular toxicity of the supernatant with the concentration of organic radicals eluted from the polymer showed a strong positive correlation that apparently overwhelms the details in differences between the chemistries of individual polymers. Thus, Homsy suggested that this tissue culture method and/or an analysis of a PECF supernatant may be rapid, reliable screening techniques for new polymers.

16.3.5   Use of Tissue Culture for Host Response Screening

Rae (1980) discussed the pros and cons of various approaches to cell, tissue, and organ culture methods in the general determination of biocompatibility of implant materials. The present consensus is that these techniques are best used to measure early, acute response to materials.

However, the real utility of in vitro culture techniques depends to a great degree on their correlation with in vivo local host response. There have been many attempts made to correlate in vitro cell culture studies with results of implantation studies. For example, Wilsnack (1976) used an ATTC human fibroblast cell line, WI-38, to examine the in vitro cytotoxicity of a large number of polymeric materials and compared the results with responses in rabbits to systemic and local injections of soluble material eluted with saline solution and with cottonseed oil, according to USP methods (see Section 17.2.1). He concluded that the in vitro test was more sensitive than the in vivo one selected (yielding more examples of cytoxicity); however, there was a poor correlation between in vitro and in vivo results when both were positive. In this case, one can point to reasons other than inherent inadequacies in cell culture to explain disparities between in vitro and in vivo results: different species, single cell line versus multiple cell types, solid materials versus eluted materials, etc. This example illustrates the difficulties in extrapolation from in vitro tests. However, Johnson and Northup (1983) attempted a similar correlation, but with fewer differences between techniques: they conducted a "battery" of four in vitro tests using an ATTC murine fibroblast cell line, L-929, and compared the results with the local response to seven day rabbit intramuscular implantation. Each test outcome was ranked $1+$ to $4+$ and pass/no pass (fail) criteria were established. The results of several of their test series are summarized in Table 16.2. Wieslander et al. (1990) extended these studies and examined the effects of elution condi-

Table 16.1 Correlation Between Tissue Culture Response to PECF and PECF Chemical Analysis[a]

Polymer[b]	Tissue culture response[c]	Total $-CH_3$, $-CH_2$ and $-CH$ in PECF by IR analysis (mpm[d] equivalent n-hexanol)
Silicone (Silastic™ 372)	+1	5
Polyethylene (U. of Texas)	+1	17
Fluorinated ethylene propylene (type 1)	+1/+2	nd[e]
Polyphenylene oxide (type 1)	+1/+2	27
Polyethylene (type 1)	+2	nd
Acrylic molding powder	+2	nd
Polyphenylene oxide (type 2)	+2	17
Polyethylene (type 2)	+2	17
Fluorinated ethylene propylene (type 1)	+2	23
Ionomer (type 1)	+2	142
Polypropylene (food grade)	+2	198
Vinylidene fluoride	+2/+3	3
Nylon™ (grade 101)	+2/+3	14
Ionomer (type 2)	+2/+3	30
Cellulose proprionate	+3	81.7
Polystyrene	+3	168
Nylon™ (grade 38)	+4	12
Poly(vinyl)chloride	+3/+4	277
Polyurethane (type 1)	+4	89
Polyurethane (type 2)	+4	328
Poly(vinyl)chloride (U. of Texas)	+4	514
Acrylonitrile, butadiene, styrene (ABS)	+4	516

[a]Polymer exposed to PECF for 62 hours @ 115°C, 30 psia.
[b]See Homsy et al. (1970) for a more complete description of polymers.
[c]Scale: +1 = Some vacuolization, morphological changes and growth inhibition.
  +2 = Moderate vacuolization, morphological changes.
  +3 = Severe growth inhibition and vacuolization.
  +4 = Total growth inhibition.
[d]Moles per million
[e]Not detected

Table 16.2 Correlation between In Vitro
Screening Tests and Short-Term Implantation
(N = 687)

		In Vivo	
		Pass	Fail
In Vitro:	Pass	94.7%	—
	Fail	—	5.3%

Source: Adapted from Johnson and Northup (1983).

tions. They suggested that testing using a combination of water extracts obtained at elevated temperatures and physical contact between cells and material produced the most reliable results.

In vitro tests have come to be seen as screening tests, serving as precursors for more involved, more costly and time-consuming animal implantation. A secondary use, resulting from extensive correlative investigations such as those of Johnson and Northup (1983), is for batch-to-batch quality control by manufacturers.

These three studies, and similar ones, make use of a single in vitro cell line. In fact, the ATTC L-929 cell line has come to be a de facto standard for in vitro testing of biomaterials and is used in several ASTM in vitro host response test methods, such as F-813 and F-895 (see Chapter 19). However, a more favored approach is that of generating a cell culture spectrum (CCS). In this technique, a number of different cell lines are used and are challenged by the biomaterial (or extracts of the biomaterial) being studied in one or more standard exposure techniques. The cell types are selected to represent a variety of origins and sensitivities. The response of each of perhaps six cell types is then quantitated. An index based upon these quantitative findings is then a CCS reflection of *relative* acute host response. A consensus CCS would be more sensitive in revealing differences between acute host response of different materials. It can be hoped that in the future, this method can be fully developed, tested, and adopted by the ASTM and CCS indices developed for all current biomaterials. Then new materials can be compared, insofar as acute response is concerned, by application of this test protocol and reference to a table of standard results.

Although not reflecting consensus, a number of screening protocols in use today utilize cell culture methods, with either biomaterials or eluted materials as the challenge, in evaluation of a quantitative host response to implant materials. For instance, primary testing (Level 1) as utilized by Autian (1977) uses five tissue culture tests as a portion of a 14-test initial battery that results in the derivation of a cumulative toxicity index (CTI).

The culture tests involve one direct exposure model and four tests using extracts. Dillingham (1983) has shown that these five tests produce significant degrees of correlation with the in vivo tests incorporated in the CTI.

Northup (1986) has provided a useful summary of efforts to use cell culture techniques for the study of the generalized features of local host response.

16.3.6 Delayed Hypersensitivity Tests

Immune system (specific) response to biomaterials is extremely difficult to evaluate in vitro. Numerous claims have been made about the ability to screen individuals for hypersensitivity to foreign materials by challenging cell-free serum with foreign materials and detecting the resulting antigen-antibody complexes. These claims are probably groundless, in a biomaterials host response context, for two reasons: first, freely circulating antigens are associated with humoral rather than the expected T-cell mediated response to foreign nonproteinaceous materials (see Section 11.3); and second, any response which occurs in vivo is to a complexed or opsinized material rather than to the material itself.

In vitro testing for T-cell mediated type 4 delayed hypersensitivity is possible, again dependent upon an appropriate challenge agent being used. The most reliable technique is the leukocyte migration inhibition factor (MIF) test (Merritt and Brown, 1980) (Chapter 11). This test, though, like the supposed humoral response tests, is more a test of the sensitization of an individual than of the ability of a specific material to induce immune sensitization. Proper conduct of this test requires considerable care and the availability of nonsensitive individuals as longitudinal control donors.

Other in vitro tests have been used to examine immune sensitivity to biomaterials, with varying degrees of success (Merritt, 1986).

16.3.7 Mutagenicity and Carcinogenicity Tests

The use of tissue culture in evaluation of biomaterials has centered on studies of acute toxicity and of inhibition of growth or activity. Considerations of carcinogenic potential of materials discussed in Chapter 12 suggest the need for a rapid screening test for neoplastic transformation. Such tests exist but are quite controversial.

It is now generally accepted that chemical carcinogenesis proceeds by a mutagenic route, with the neoplastic attributes being heritable. Thus, we expect any confirmed carcinogens to be mutagens as well. However, the converse is not yet proven, that is, that all mutagens have the potential

Table 16.3   Results of Validation of Ames Test

Status of agent	Number tested	Number mutagenic (%)
Reported animal carcinogens	176	158 (90)
Reported human carcinogens	18	16 (89)
Reported animal noncarcinogens	108	13 (12)

Source: Adapted from Ames (1979).

of being carcinogens. Thus, the controversy over tests for carcinogenicity tend to focus on interpretation of the results.

The most widely used test method depends upon evidence of mutagenesis as an index of carcinogenic potential. This test, the Ames test (Ames et al., 1975), is conducted in the following way. A culture is prepared with a mutant bacterial cell line (usually *Salmonella typhimurium*) that requires histidine for growth. It is grown in histidine-free culture with the material to be tested and a fresh preparation of rat liver cell homogenate. This provides enzyme systems which will detect potential mutagenic materials that require metabolic conversion to mutagens. Only cells that then mutate back to the more normal histidine-independent state can multiply.

This test has been used widely and has been shown to be a very strong correlative with carcinogenic potential. Ames (1979) reported on a study of 302 chemicals; the results are summarized in Table 16.3. The finding of 12% "false" positives was discussed by Ames, who suggested that a number of these may be weak carcinogens that were not previously detected by (less sensitive) animal tests. This high degree of correlation between mutagenicity and carcinogenicity has suggested to many workers that the Ames test has a useful role in early material screening.

The Ames test has a number of attractive attributes. It is cheap, quick, and relatively safe. By exposing approximately 10^6 organisms to a mutagen, it can detect transformation rates far below those possible in any reasonable animal test. Furthermore, it has been shown to be able to detect any of the recognized mutagenic mechanisms.

However, the Ames test has been strongly criticized. Ashby and Styles (1978) had difficulty in reproducing Ames' results, especially at low mutagen concentrations. They suggest that the sensitivity of the test is lower than asserted by Ames, and also feel that correlations between mutagenic potency and carcinogenic potency should be viewed with extreme caution. More recent general criticism of this and other tests of potential neoplastic transforming agents has focused on the practice of the use of high doses of challenge materials, which may also be directly cytotoxic. High doses are frequently used to improve statistical response and reduce cost, followed by extrapolation of effects to low doses. Many workers have suggested that cytotoxicity encourages unnaturally elevated rates of cell replication, thus artificially enhancing the rate of production of transformed cells and leading to a too high risk after extrapolation (Epstein and Swartz, 1988).

It is clear that the Ames test, in common with other tissue culture tests, must be done with great care. Precautions (deSerres and Shelby, 1979) should include the use of multiple bacterial strains, multiple suspension systems, a wide range of doses of material under evaluation, and use of positive and negative controls in each study. The need for careful control of the form and chemical composition of the material being studied was demonstrated by Abbracchio et al. (1982), who showed that crystallinity plays a key role in the biological response to particulate metal sulfides.

The need for great care in conducting the test in order to duplicate Ames' results, as well as more fundamental criticisms concerning the relationship between mutagenicity and carcinogenicity, led Purchase et al. (1978) to compare six in vitro and in vivo tests for carcinogenic potential of organic compounds. While they concluded that the Ames-type test is highly accurate [93% accurate in predicting animal carcinogenesis vs. 90% for Ames (1979) (Table 16.3)], they also concluded that:

The inclusion of the (other) four tests in a screening battery predictably resulted in a great increase in overall inaccuracy and loss of discrimination, even though the detection of carcinogens is increased. All tests were shown to generate both false positive and false negative results, a situation which may be controlled by the use, where possible, of chemical-class controls, to identify the test which is optimal for the class of chemical under test.

That is, they suggest that the use of compounds of similar structure, as positive and negative controls, permits validation and selection of specific tests from a "battery" of tests.

Forster (1986) has reviewed in vitro mutagenicity testing with special reference to evaluation of biomaterials. Despite the inherent shortcomings of the test methods and the need to adapt them to both the material in question and the potential application, he suggests that in vitro mutagenicity tests should play a role in any screening study of biomaterials. The strategy, then, should be much the same as for the use of tissue culture tests of toxicity.

16.4 BLOOD CONTACT TESTS

16.4.1 General Comments

Materials problems in cardiovascular devices are primarily those of initial or short-term inadequate biological performance. This is due to the acute nature of the host response to foreign materials in the cardiovascular system. We have previously considered (Chapter 9) the principal problems, those of coagulation of blood and lysis of blood-borne cells. We have also seen that these problems are not intrinsic to materials but are intimately related to design features such as the presence of interfaces, flow rates, and turbulence conditions, as well as overall device function.

Thus, it is hard to divorce the problems of determining host response to materials for blood contact applications from the details of device design and qualification. Materials testing, insofar as host response is concerned, can be separated into three types of tests: in vitro static, ex vivo dynamic and in vivo dynamic. These are usually applied sequentially to a new material or to a material being considered for a new application. It is important to note with regard to this last comment that there are considerable regional variations in environment in the cardiovascular system. It is well recognized that a material that works well in a high-flow arterial application often fails in a static venous site.

Initial testing for host response to blood contact materials is usually done in vitro. Despite the wide recognition of the inadequacy of these tests, they continue to be done on a screening basis for two reasons: (1) they are relatively inexpensive, as compared with in vivo tests; and (2) they are not known to yield false negatives, that is, a material that does poorly in these screening tests will not be a satisfactory implant material in cardiovascular applications. These tests are generally of a comparative type and examine either coagulation times or hemolysis rates in either static or dynamic systems during or after contact with foreign materials.

16.4.2 Static Tests

Of the static tests, the Lee-White and the Lindholm tests are best known (Lindholm et al., 1973; Mason, 1973; Mason et al., 1974). The reference material is usually silicone-coated soft glass. Although not a satisfactory material for implant use due to its mechanical properties, this material is perhaps the best inexpensive test material with a low and reproducible host response. Few materials in use as cardiovascular implant materials exceed the biological performance of this material in vitro.

The thrombotic potential of the surface is usually evaluated by timing the development of the clot in a static situation and comparing this time for clot development to the same point for blood drawn simultaneously and exposed to the reference surface. Adequate testing requires considerable replication, both in specimens in a single test and repetition of tests on different days, with different donors, etc. A canine donor is usually used although some investigators have used human donors.

Hemolysis can be determined simultaneously in such a static test by centrifugation of the clotted blood followed by optical spectrophotometric determination of the hemoglobin content of the supernatant serum. Again, it is usual to compare the results obtained with a test material and a reference material using blood drawn at one time (American Society for Testing and Materials [1990]: ASTM F756-87).

16.4.3 Dynamic Tests

It is extremely difficult to evaluate coagulation effects of biomaterials on blood ex vivo since the growth of the thrombus interferes with flow dynamics in the system external to the donor and may have potential lethal effects on the donor. A short-term solution to this problem, but only applicable to relatively thromboresistant materials, is to determine the density of adhered platelets after a brief exposure under controlled flow conditions. Arguing from knowledge of the role of platelets in the intrinsic and common pathways (Figure 9.1) one can then predict relative coagulation rates (Grabowski et al., 1977).

A further sophistication can be afforded by the wide variety of standard in vitro quantitative assays available for the vast majority of factors involved in the clotting cascade. Such a test is performed by obtaining a blood specimen, testing it for the presence of a clotting factor and retesting after brief, timed exposure to the test material and a reference material. While these tests are very precise, their specificity is unclear, since it is difficult to predict coagulation outcomes from the concentration of any one factor.

If the material to be evaluated does not promote significant coagulation, then hemolysis can be quantitated by exposing the surface to blood briefly, in a transient mode, or over a longer period, in a stable ex vivo system. Erythrocyte damage is determined either by direct analysis of released hemoglobin or by counting the percentage of erythrocyte "ghosts" in a blood smear. In animal ex vivo models, this test can be improved by pre-labeling erythrocytes with $^{51}Cr^{+6}$. This method takes advantage of the easy penetrance of Cr^{+6} through cell membranes and its intracellular conversion to Cr^{+3} which is then only released if the cell wall is disrupted. Then the ^{51}Cr activity in cell-free serum samples is a direct measure of hemolysis. Again, adequate testing requires comparison with a standard surface and repetition of the test with a number of donors.

There are a number of problems that these in vitro tests share. The biggest of these is the "nonstandard" nature of blood. Blood is an extremely variable material. Its coagulation and lytic response depends upon species, health of the individual donor, diet, medication, and age, among other factors. Even the act of drawing blood may change the host response capabilities of the donor's blood, thus rendering the use of a standard donor extremely misleading. Thus, all in vitro tests for blood–biomaterial interactions must lean heavily on isochronal comparisons of the test material with a reference such as siliconized glass and repeated tests using different donors.

Another problem is the inability to reproduce implant site blood dynamics in vitro. Even the use of reasonable vessel sizes and flow rates based on study of the planned implant site may prove misleading. The details of the properties of the vascular intima near the implant site have a profound influence on the actual flow conditions as do the features of the final implant/host configuration. Thus, while some success is encountered in static and "one-pass" in vitro tests, those based upon continued flow and observations over a period of hours show poor correlation with in vivo performance.

Nonetheless, these "one-pass" and continued flow tests, where a direct connection between an animal's or patient's circulatory system is created, are collectively termed ex vivo tests and, as the second type of test, are becoming increasingly popular. Tests for screening purposes, in vitro or ex vivo, are no more than screening tests. They must be followed up by well-designed in vivo testing, which is the subject of the next chapter.

A major effort to evaluate and compare these in vitro (as well as ex vivo and in vivo) tests was made by a working group assembled by the National Heart Lung Blood Institute in 1985 (Department of Health and Human Services, 1985).

REFERENCES

Abbracchio, M. P., Heck, J. D. and Costa, M. (1982): Carcinogenesis 3:175.

American Society for Testing and Materials (1990): Standard practice for assessment of hemolytic properties of materials, F756-87. In: *1990 Annual Book of*

ASTM Standards, Vol. 13.01: *Medical Devices; Emergency Medical Services.* ASTM, Philadelphia, pp. 243ff.

Ames, B. N. (1979): Science 204:587.

Ames, B. N., McCann, J. and Yamasaki, E. (1975): Mutat. Res. 31:347.

Ashby, J. and Styles, J. A. (1978): Nature 271:452.

Autian, J. (1977): Artif. Organs 1(1):53.

Department of Health and Human Services (1985): *Guidelines for Blood-Material Interactions*. NIH Publication 85-2185, Public Health Service, National Institutes of Health. U. S. Government Printing Office, Washington, D. C., pp. 219ff.

deSerres, F. J. and Shelby, M. D. (1979): Science 203:563.

Dillingham, E. O. (1983): In: *Cell-Culture Test Methods*, ASTM STP 810. S. A. Brown (Ed.). American Society for Testing and Materials, Philadelphia, pp. 51ff.

Epstein, S. S. and Swartz, J. B. (1988): Science 240:1043.

Forster, R. (1986): In: *Techniques of Biocompatibility Testing*, Vol. II. D. F. Williams (Ed.). CRC Press, Boca Raton, FL, pp. 137ff.

Galin, M. A., Chowchuvech, E. and Galin, A. (1975): Am. J. Opthalmol. 79:665.

Grabowski, E. F., Didisheim, P., Lewis, J. C., Franta, J. T. and Stropp, J. Q. (1977): Trans. Am. Soc. Artif. Intern. Organs XXIII:141.

Hay, R. J. (1983): In: *Cell-Culture Test Methods*, ASTM STP 810. S. A. Brown (Ed.). American Society for Testing and Materials, Philadelphia, pp. 114ff.

Homsy, C. A., Ansevin, K. D., O'Bannon, W., Thompson, S. A., Hodge, R. and Estrella, M. E. (1970): J. Macromol. Sci.-Chem. A4(3):615.

Johnson, H. J. and Northup, S. J. (1983): In: *Cell-Culture Test Methods*, ASTM STP 810. S. A. Brown (Ed.). American Society for Testing and Materials, Philadelphia, pp. 25ff.

Lindholm, D. D., Klein, E. and Smith, J. K. (1973): Proc. Dialysis Transplant Forum 3:39.

Mason, R. G. (1973): Biomat. Med. Dev. Artif. Organs 1(1):131.

Mason, R. G., Shinoda, B. A., Blackwelder, W. C. and Elston, R. C. (1974): Biomat. Med. Dev. Artif. Organs 2(1):21.

Merritt, K. (1986): In: *Techniques of Biocompatibility Testing*, Vol. II. D. F. Williams (Ed.). CRC Press, Boca Raton, FL, pp. 123ff.

Merritt, K. and Brown, S. A. (1980): Acta Orthop. Scand. 51:403.

Mital, M. and Cohen, J. (1968): J. Bone Joint Surg. 50A:547.

Northup, S. J. (1986): In: *Handbook of Biomaterials Evaluation.* A. F. von Recum (Ed.). Macmillan, New York, pp. 209ff.

Pappas, A. M. and Cohen, J. (1968): J. Bone Joint Surg. 50A:535.

Purchase, I. F. H., Longstaff, E., Ashby, J., Styles, J. A., Anderson, D., LeFevre, P. A. and Westwood, F. R. (1978): Br. J. Cancer 37:873.

Rae, T. (1980): *Advances in Biomaterials*, Vol. 1: *Evaluation of Biomaterials.* G. D. Winter, J. L. Leray and K. de Groot (Eds.). John Wiley, Chichester, pp. 289ff.

Wieslander, A., Magnusson, Å. and Kjellstrand, P. (1990): Biomat. Artif. Cells Artif. Org. 18(3):367.

Wilsnack, R. E. (1976): Biomat. Med. Dev. Artif. Org. 4(3 & 4):235.

BIBLIOGRAPHY

Anderson, D. (1978): J. Soc. Cosmet. Chem. 29:207.

Autian, J. (1973): CRC Crit. Rev. Toxicol. 2:1.

Brown, S. A. (Ed.) (1983): *Cell-Culture Test Methods*, ASTM STP 810. American Society for Testing and Materials, Philadelphia.

Moroff, G. (1986): In: *Handbook of Biomaterials Evaluation.* A. F. von Recum (Ed.). Macmillan, New York, pp. 233ff.

Rice, R. M., Hegyeli, A. F., Gourlay, S. J., Wade, C. W. R., Dillon, J. G., Jaffe, H. and Kulkarni, R. K. (1978): J. Biomed. Mater. Res. 12:43.

Venitt, S., Crofton-Sleigh, C. and Forster, R. (1984): In: *Mutagenicity Testing: A Practical Approach.* S. Venitt and J. M. Parry (Eds.). IRL Press, Oxford, pp. 45ff.

von Recum, A. F. (Ed.) (1986): *Handbook of Biomaterials Evaluation.* Macmillan, New York.

Waters, R. (1984): In: *Mutagenicity Testing: A Practical Approach.* S. Venitt and J. M. Parry (Eds.). IRL Press, Oxford, pp. 99ff.

Wennberg, A., Hasselgren, G. and Tronstad, L. (1979): J. Biomed. Mater. Res. 13:109.

Wilson, R. S., Lelah, M. D. and Cooper, S. L. (1986): In: *Techniques of Biocompatibility Testing*, Vol. II. D. F. Williams (Ed.). CRC Press, Boca Raton, FL, pp. 151ff.

Yuspa, S. H. (1984): In: *Mechanisms of Tumor Promotion*, Vol. III: *Tumor Promotion and Carcinogenesis in Vitro.* T. J. Slaga (Ed.). CRC Press, Boca Raton, FL, pp. 1ff.

17

In Vivo Implant Models

17.1 INTRODUCTION

17.1.1 Approaches to In Vivo Tests

After acute screening by physical or biological in vitro techniques, it is the practice to test new implant materials in extended-time, whole-animal tests. While many of the limitations of nonhuman species testing are recognized, it is the common judgment that such tests, involving the

301

exposure of materials to systemic physiological processes, are a necessary precedent to human clinical testing.

With the exception of materials for application in the cardiovascular system, the site chosen for nonfunctional (see Section 17.2.1) testing is usually in soft tissue. This decision is based on the assumption that cytotoxic effects have a generality of action and because soft tissue sites can be approached in animals with relatively minor surgery. For the same reason, the peritoneal cavity has been used for acute in vivo screening studies (Wortman et al., 1983). For joint replacement or fracture fixation applications, implantation is also performed in cortical bone. Specialized sites such as the cornea are used for materials for specific, limited applications.

Chick embryo and fetal rodent models have been used; however, most testing now uses mature, higher animals. Rodents, although still used in multi-species tests, are rarely used alone since the experience of the Oppenheimers (see Chapter 12). A variety of sites have been used, but the most popular ones are:

1. Subcutaneous.
2. Intramuscular (e.g., supraspinatus).
3. Intraperitoneal.
4. Transcortical (e.g., femur).
5. Intramedullary (e.g., femur and tibia).

Functional testing (see Section 17.2.2) requires a wider range of species due to the specific requirements of the material or device configuration selected. No single species presents an ideal general model for the human species, and some human structural functions, such as patellar motion, are not attainable in animal models.

Anderson and Hughes (1986) provide an overview to the problems of selection of animal models.

17.1.2 Animal Welfare

In recent years, the use of animals in medical research has come under strong criticism from a number of organizations, particularly People for the Ethical Treatment of Animals (PETA). This criticism, strident at times, tends to ignore two key features of animal use:

1. Investigators are, generally, sensitive individuals concerned about the well-being of their test subjects.
2. Sound research, applicable to problems of human disability and disease, requires that the test subjects be healthy and not subject to stress.

Nevertheless, there is now a well-developed system for monitoring and controlling animal research in the US. The enabling legislation is the Animal Welfare Act (7 USC 2131, December 23, 1985) and a final regulation issued by the US Department of Agriculture regarding inspection and compliance (Dept. of Agriculture, 1991). The requirements of the Animal Welfare Act have been summarized in a National Institutes of Health publication (Dept. of Health and Human Services, 1985) which should be on the desk of any investigator involved in research using animals.

The central points of this system of regulation and inspection are:

1. Research involving animals may be performed only under a prospective protocol which describes the purpose and significance of the experiment, the procedures to be performed, the general care and housing conditions of the experimental animals and identifies the personnel involved in both animal experimentation and animal care and the qualifications of these personnel.
2. The protocol must be reviewed and approved before initiation by an institutional review committee, although this review applies only to the animal welfare aspects of the protocol and not the scientific content.
3. The facilities in which surgical and experimental procedures are performed and animals are housed must meet specific minimum requirements.
4. The animal housing facilities and all procedures involving animals must be under the supervision of a veterinarian.

Additionally, standards are set forth concerning specific housing requirements (cage type, size, etc.) for each species and for the provision of veterinary care to minimize pain and treat conditions unrelated to the experiment.

The vast majority of fund granting agencies and organizations now require that research proposals be reviewed before submission and that they meet the requirements of the Animal Welfare Act. This is a worthwhile and humane system which may at times seem burdensome to the investigator. However, it is necessary, due to both societal requirements and ethical considerations, and its presence emphasizes the professional responsibilities that the investigator has to experimental animals.

There are three additional humane responsibilities which must be emphasized at this point. First, there is a need to avoid unnecessary use of animals as experimental subjects. Part of the decision to use an animal model must involve questioning whether there are suitable alternative approaches (cell culture, mechanical or electronic simulations, etc.). Second, a determination should be made that the experiment has not been done before or the data are not available from other sources. Finally, once a determination has been made to use animals and a model has been selected, the experiment should be designed to use the minimum but sufficient number to provide a reasonable assurance of a statistically valid result. Failure to provide adequate prospective statistical design in an animal experiment can lead to the need to replicate the experiment and, in some cases, to use significantly larger numbers of animals that would have been required by an appropriate approach.

17.2 TEST TYPES

17.2.1 Nonfunctional Tests

Tests divide globally into two types: functional and nonfunctional. In the nonfunctional type, the implant is of an arbitrary shape, perhaps in the form required for later mechanical tests of material response, and "floats" passively in the tissue site. The prototype for such tests was popularized by Ferguson et al. (1960). This is a supraspinatus implantation in the rabbit in a cavity produced by blunt dissection, followed by a 16-week observation period. This test series formed the basis for a peer-reviewed national-consensus protocol now in use, the F 981 method of the American Society for Testing and Materials (ASTM, 1990) (see Appendix 1 to this chapter).

Nonfunctional tests focus on the direct interactions between the substance of the material and the chemical and biological species of the implant environment. The absence of mechanical loads and other elements of function limit the usefulness of such tests. Thus, they tend to be of short duration, usually a few to 24 months in length.

A typical short-term study of this type is described in the British Standards Institute Method BS 5736, part 2 (1981). This test protocol utilizes 1 × 10 mm strips of material, with positive and negative controls, implanted by blunt stylet in the paravertebral (supraspinatus) muscle of the rabbit and recovered by sacrifice after 7 days. Thus, although being an in vivo method, it evaluates only acute, soft tissue response. Turner et al. (1973) have shown that such a 7 day model, while capable of producing false negatives, produces no false positives (when compared with local host response to 12 weeks), thus justifying the use of this model for acute in vivo screening. To examine more fully the effects of systemic physiology on material and host response, a longer-term study of materials which pass a short-term in vivo screen is needed.

F 981 (ASTM, 1990) illustrates the typical course of such a longer-term study. Standard-sized implants fabricated from the material in question, as well as one or more well-characterized comparative control materials, are used. They are selected so as to come close to the standard SA/BW ratio of major implantation in man (see Section 14.4.4), at least for soft tissue sites. The method uses muscle sites in the rat, rabbit, and dog, as well as limited intramedullary sites in the latter two animals, if indicated by the proposed application. Rats are maintained and sacrificed periodically up to 26 weeks, rabbits and dogs to 104 weeks. Evaluation consists primarily of inspection of the test and control implants and the sites of implantation.

At present there are no universally available control materials for the F 981 protocol. The method specifies the use of a number of metals and one polymer that have a long history of experimental evaluation and human clinical use as reference or control materials. However, individual variations in composition, surface properties, etc., in these materials make inter-test comparisons difficult. There is considerable continuing interest in developing positive and negative metallic and polymeric controls that might be made available through a national program such as the Standard Reference Material program administered by the National Institute of Standards and Testing (NIST).

There are recognized major defects in the design and practice of F 981. Notwithstanding its strong position as a consensus test method that provides standardized results that can be compared with others (Escalas et al., 1975)* even when variants from the method are used, these shortcomings must be recognized.

In its concentration on study of the implant site, F 981 overlooks systemic or remote site effects unless they are extreme enough to produce visible morbidity or mortality. Thus, remote site neoplastic transformation will not be seen in this test even if adequate implantation time has elapsed. Additionally, implantation times are probably too short in either the rabbit or the dog, or even the rat, to overcome latency effects (see Section 12.2.5). Furthermore, for practical reasons, the numbers of test animals used are so small that even observation of a tumor at the implant site or at a remote site cannot be evaluated statistically with any acceptable level of certainty. Thus, the method is unsuited for yielding information beyond the magnitude of the acute and chronic inflammatory processes and subsequent fibrotic responses.

---

*Escalas et al. (1975) used a precursor method, ASTM F 361. See Appendix 1 for a more complete discussion.

Table 17.1 Interaction of Intrinsic Reactivity and Motion in Implant Capsule Thickness

Thickness of Capsule (mean ± S.E.M.)		
	Motion	
Material	Minimal	Maximal
PE	24.3 ± 2.9 μm	31.4 ± 2.7 μm
PA	14.1 ± 2.8 μm	25.1 ± 3.5 μm

Minimal motion: 0.00 ± 0.07 cm
Maximal motion: 0.18 ± 0.07 cm

ANOVA Contrasts

Hypothesis	F	Significance
H_1: $T_{pe,min} = T_{pe,max}$	2.05	$p > 0.2$
H_2: $T_{pa,min} = T_{pa,max}$	3.07	$p = 0.125$
H_3: $T_{pe,min} = T_{pa,min}$	3.01	$p = 0.125$
H_4: $T_{pe,max} = T_{pa,max}$	0.60	$p > 0.5$

Source: Kupp et al. 1982.

There is some question that can be raised about the significance of the fibrotic response. Measurement of capsule thicknesses in muscle-site implant studies tend to show large variances. In most studies, differences must exceed 25 μm to be statistically significant. This result is usually ascribed to ''biological'' variation. A study (Kupp et al., 1982) suggests a different explanation. In this study, sites were identified in the rat hind quarter that produced considerable motion between implant and muscle (intermuscular) when the leg was moved passively from full extension to full flexion and that produced negligible relative motion (intramuscular). The relative degrees of motion in the two sites were verified radiologically. Spherical capped cylindrical implants of a polyester (PE) and a polyacetal (PA) were placed in each of these two sites, and capsule mean thicknesses were studied at 21 days. The results are given in Table 17.1.

A one-way analysis of variance (ANOVA) was performed, using the method of contrasts. A near significant effect of motion may be seen in the response to PA (25.1 vs. 14.1 μm) (H_2). However, the greater intrinsic reactivity of PE masks this effect (H1). This greater reactivity may be seen by comparison of the minimal motion sites, a more usual place for implant evaluation (24.3 vs. 14.1 μm) (H_3). An apparent additive effect of motion in the maximal motion implant site masks this difference (H_4).

Thus, capsule thickness seems to reflect the additive response to two factors: an intrinsic (chemical) activity and an extrinsic (mechanical) activity.

It is clear from this study that relative tissue-implant motion may have a large influence on the observed capsule thickness. It may be that specimen geometries should be changed in this type of test or the specimen secured to the surrounding tissue to minimize such motion variables. Experiments of this sort are called *two factor* experiments, since they examine effects related to chemical and mechanical factors. For more complete evaluations, one must move to *three factor* experiments which include (or control for) electrical effects. Thus, in this study, it may have been the case that differing electrical charge densities at the polymer-tissue interfaces could account for the observed effects.

In any case, implant shape must be carefully selected and standardized for each test protocol to avoid effects on capsule thickness associated with surface features of the biomaterial. It has long been recognized that sharp edges on an implant produce a locally increased capsule thickness. Thus, when a flat-ended cylinder is used (Wood et al., 1970), the familiar "dogbone" shaped capsule results from a local thickening over the circular edges of the rod ends. This effect has been more generally studied by Matlaga et al. (1976), and it has been shown that tissue response is inversely related to the included angle at the edge of the implant. Finally, all soft tissue sites are probably not equivalent in terms of the fibrous response evoked by an implant (Bakker et al., 1988).

One might further point out that observation of capsule thickness is an indirect measure of cellular response. While Dillingham (1983) has shown a good correlation between capsular response and in vitro cytotoxicity, a useful addition to this type of test might be to evaluate the concentration and spatial distribution of cellular enzymes directly (Salthouse and Willigan, 1972; Salthouse and Matlaga, 1975).

Even if a refined histochemical evaluation system is not used, it might be well to evaluate tissue around animal implants in a more detailed way than simply by measuring capsule thickness. Salthouse (1980) has suggested that a scheme that incorporates a number of observations into an overall index, analogous to the CTI approach of Autian in his Level 1 test scheme (Autian, 1977), be used. The scheme suggested by Salthouse is a modification of one proposed by Gourlay et al. (1978) and is based upon the earlier work of Sewell et al. (1955). An outline of the method is given in Appendix 2 of this chapter.

In the long run, the most severe criticism of F 981 is its nonfunctional aspect. Even when this is overcome by selecting tissue sites more typical of the application, for instance, in the rabbit cornea for intraocular applications, the results remain inferior in providing confidence in prediction of clinical experiences to those obtained from functional tests.

An alternate approach is to isolate the material from the tissue, so as to remove the mechanical features and examine the chemical (and presumably electrical) effects on local host response directly. Perhaps the best known example of this approach is the ''cage'' implant system developed by James Anderson and his students (Marchant, 1989). In this case, the implanted material is placed within a wire or mesh enclosure and, after appropriate sterilization, implanted. The ''cage'' will rapidly fill with a mixture of extracellular exudate and inflammatory cells. This fluid may be sampled from time to time by transdermal puncture, analyzed for enzymatic and cellular content and a chronological picture built up of the local response to the implant. Although elegant in conception, the major criticism to this approach is that the local host response to the implanted material is that of cells *already* exposed to the cage material, which is usually different in composition and surface condition. While it is possible to overcome this problem by fabricating the ''cage'' from the same material as the test specimen, in the general case this is difficult and costly.

17.2.2 Functional Tests

General Requirements and Problems

Functional tests require that, in addition to being implanted, the material, at least in some degree, be placed in the functional mode that it would experience in human implant service. This is required to study, for instance, tissue ingrowth into porous materials for fixation purposes and production of wear particles and possible tissue response to them.

Tests of this type are obviously of much greater complexity and cost than the nonfunctional type. Devices have been designed to exert dynamic loads typical of the desired application on the material. An example of this approach is the dynamic aortic patch (von Recum et al., 1978) for evaluating materials for cardiac assist devices in the canine aorta.

Tests of materials for partial or total joint replacements, on the other hand, frequently require the design of a completely functional animal version of the proposed device. Design, fabrication, mechanical testing, and implantation of these devices may be more difficult than the final production of the device for human use. Problems arise, in both cases, from the small size of economically priced animals, differences in animal anatomy, and the inability of the animal ''patients'' to actively cooperate with the experiment.

Despite these difficulties, total hip joint replacement designs, for example, have been made and tested in rats, cats, dogs, sheep, and goats. Figure 17.1 illustrates the comparison between a clinical human total hip replacement and a canine model developed to test fixation by biological ingrowth. Fracture fixation designs have been tested in all of the species mentioned above, as well as in larger ones including cows and horses. Some human total joint replacements have been tested directly in animals, such as the human PIP finger joint replacement which has been successfully implanted in both the rabbit and cat knee.

Particular problems arise in each of these test methods. Some difficulties, however, are common to all of them.

There is considerable interspecies variation in the histological appearance of tissue. Additionally, local tissue conditions that are abnormal in some species may be chronic in others. Misinterpretations of test results, insofar as tissue response is concerned, have arisen from the employment of clinical (rather than veterinary) pathologists to evaluate histology from implant tests. Satisfactory and reliable results can be obtained only from the evaluation of tissue from multi-species tests by an individual specifically trained in comparative histology and animal pathology.

The inability of animals to cooperate with treatment has already been mentioned. Specifically, it is not possible to return an animal to full activity on a planned schedule. The animal will move as it sees fit and set its own schedule. Similarly, an animal cannot indicate or describe internal problems of discomfort or pain. Experienced animal handlers may be able to detect early signs of pain accompanying infection or tissue reaction. More commonly, these problems are not detected until the animal is systemically ill, or loses function in a limb, or incidentally, at autopsy. The presence of culturable infection of any origin at an implant site invalidates any observations on that animal (unless deliberately introduced as a part of the experimental plan). Systemic infection is a significant stress on experimental animals and casts doubts on the validity of observations.

Test animals such as cats, dogs, rats, etc., have shorter life-spans and higher metabolic rates than humans (Brody, 1945). Over and above particular variations in physiology, these factors introduce other problems of unknown magnitude. For instance, what is the appropriate factor by which to scale down an implant for an animal to experience the same apparent body load of foreign material as man (see Section 14.4.4)? Since both lifetime and neoplastic transformation induction times are shorter in these animals than in man, how can these be scaled to the human experience? The importance of this latter point is underlined by finding, occasionally in clinical veterinary practice, implant site sarcomas in dogs in

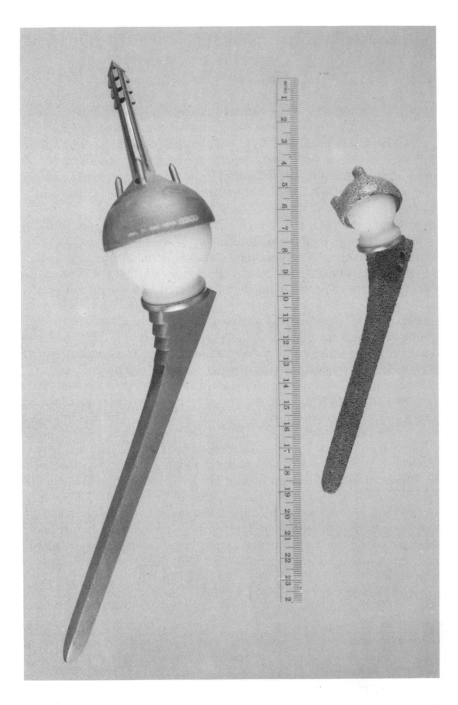

Figure 17.1 Comparison of tronzo-type total hip replacement prostheses. Human model (left) designed for PMMA fixation. Canine model (right) was developed to evaluate fixation by biological ingrowth and has a sintered porous surface coating.

conjunction with high corrosion rate stainless steel implants (see Section 12.5). In dogs these have occurred after an average implantation time of 5.8 years. What is the comparable period in man after which we should expect to see this type of tumor, if we assume that the transformation mechanisms are the same in both species? Questions of this type are unresolved and are the subject of current research. A final question related to these interspecies differences is how long to test before true chronic conditions are realized? The expense of animals and an annual holding cost for dogs frequently exceeding $1,000 per animal make this last question one of vital importance.

17.2.3 Cardiovascular Functional Tests

Material Tests

The previous comments have applied primarily to implants in locations other than in the cardiovascular system. The nature of cardiovascular implants is largely functional and, as in the case of in vitro testing, must be discussed separately.

In vivo testing in the cardiovascular system is an extension of the ex vivo or dynamic type of testing previously discussed in Section 16.4.3. The simpler form of such testing is the use of an implant of an idealized geometry rather than a full-scale working device. Rather than reproducing the exact functional design of the implant, a standard design is used to introduce the material into a vascular process. Patches or daggers may be attached to the vascular wall in various locations, in some cases with provision for mechanical loading (von Recum et al., 1978). Sections of blood vessels may be replaced, often with devices with deliberate flow-disturbing defects in them, such as in the Gott and Kusserow tests.

The Gott or canine vena cava test (Gott and Furuse, 1971) is the better known of these two methods. In this technique, rings 9 mm long by 8 mm (OD), with a 7-mm-diameter lumen, are surgically inserted to replace a portion of the inferior vena cava in the dog. The rings may have a small internal constriction or web to increase blood turbulence. Groups of five animals are used with implantation periods of 2 hours and 2 weeks, if the initial group remain patent. Evaluation is by examining patency of the rings and degree of coverage by thrombosis on removal.

The Kusserow or renal embolus test (Kusserow et al., 1970) involves replacing a portion of the suprarenal aorta with a similar ring. An infrarenal constriction is produced by partial ligation to force a large portion of the aortic circulation (an estimated 90%) through the renal arteries into the kidneys. Groups of five to eight animals are used, with implantation times of 3 days to 2 weeks. Evaluation is by examination of the rings and histological quantitation of kidney infarcts secondary to emboli "shed" by the implants.

The Gott test appears more severe with respect to adherent thrombi due to lower flow rates in the canine vena cava than in the aorta, even after partial ligation. However, the Kusserow test provides better overall evaluation as it permits examination of both adherent thrombus and remote emboli, thus more closely mirroring actual clinical experience.

Transitional Tests

A transitional form of this sort of testing may involve the use of a portion of a clinical design implanted in a different location in an animal. An example is described by Sawyer et al. (1976). In this study, sections of catheters intended for human clinical use were implanted as segmental replacements in jugular and femoral veins of dogs as well as being placed in the more usual location in the right atrium through right jugular insertion.

The last study illustrates a feature common to most of these cardiovascular tests of the idealized type. That is, the response at 2 hours after implantation mirrors and closely predicts that seen at 2 weeks. It is this acute response of the cardiovascular system to foreign materials that makes testing so difficult. The early events are complicated by the establishment of equilibria between material and host and by the events of trauma associated with implantation, and are difficult to study due to the requirements to support the test animal clinically. However, the acute response may be the most important, both from the point of view of the test and of the eventual clinical response. Thus, it is often said of materials tested in vivo for cardiovascular applications: "If they will last 2 hours, they will last 2 weeks; if they will last 2 weeks, they will last 2 years."

Device Tests

The more complex form of functional testing for cardiovascular application is the evaluation of a material fabricated into a final device design. The costs and the difficulties associated with such tests can be easily appreciated. A major problem is the same as that which faces evaluation of any human clinical device: the animal implant site is not the same as the human site, and problems of comparison and scaling arise. These scaling problems are particularly acute in the case of blood contact materials because of the relative interspecies constancy of the viscosity of mammalian blood. In the final analysis, materials for cardiovascular applications can only achieve qualification by tests in actual clinical applications.

While on the face of it, this may seem to be an unsupportable position, we must recognize that such testing is always based upon two prerequisites: (1) the material must first perform sufficiently well in vitro and in animal tests that there is a good likelihood of success and a small chance of failure; and (2) there must be a real potential benefit to the specific patients involved in this study. This latter point will be dealt with at greater length in Chapter 18.

Tests of this final type are essential to examine the effects of the details of construction. For instance, for a new design, it is difficult to predict the rate of hemolysis that will result from relative movements of parts in a heart valve, and the portion of this that will be ascribable to materials selection. In this respect, full clinical testing is highly desirable despite its risks and inherent ethical problems.

17.2.4 Human Tests

In the final analysis, clinical testing is the only technique by which the true biological performance of implantable biomaterials can be determined. In short, the only valid subject for study is the human being. When necessary preconditions are met (Chapter 18), then human implantation will begin and there is a challenge to obtain data from this experience.

The second consideration cited in the previous Section, that is, that any human clinical experiment must provide a potential benefit to the patients involved, essentially prevents the use of humans as test subjects for biomaterials. There are very rare exceptions to this rule, as in the study of Hofmann et al. (1990) involving patients who were to receive staged bilateral total knee replacement arthroplasties (TKAs). During the first surgery, a TKA was performed on one side and plugs of porous implant materials were inserted in the medial femoral condyle of the opposite knee. Nine weeks later, at the second surgery, the plugs were retrieved and the second TKA performed without deviation from the technique which would have been employed in the absence of the experimental implants. This study, which was approved by a local review committee (see Section 18.2.1) and for which each patient gave informed consent, probably represents the extreme limit to which the ''potential benefit'' principle can be stretched. The need for such tests is underlined by the observation of the investigators that their results (ingrowth of bone into porous cobalt- and titanium base alloy plugs) did not duplicate the response seen earlier in a canine model.

Although there is a marked lack of study of device function (and biological performance of the materials involved) in patients (as noted in Interpart 2), there are opportunities for examination of these issues during device retrieval, either subsequent to clinical failure or at autopsy. References to a number of such studies is provided in Interpart 2. While engineering tests on the device to ascertain, among other data, the material response, are routine, examination of the local host response is somewhat more difficult. Efforts have been made to develop comparative overall scores for local host response, in parallel to those used in animal studies. The "Mirra scale" (Mirra et al., 1976) (Appendix 3) has come into general use to describe the largely materials-mediated response of patient tissues to PMMA-cemented partial and total hip and knee replacement devices. While it is highly desirable that this scale continue to be used, more sophisticated ones are needed, especially as materials and material configurations used in implanted clinical devices continue to change.

17.3 A FINAL COMMENT

As consensus emerges concerning qualification of new materials, we can make better judgments about the role that each of the generic test methods will play. However, animal tests of both the functional and nonfunctional type play a vital part in determining material and host response in the application of biomaterials. Despite their expense, complexity, and difficulty of interpretation, they will continue to be used for the foreseeable future.

REFERENCES

American Society for Testing and Materials (1990): Standard Practice for Assessment of Compatibility of Biomaterials (Nonporous) for Surgical Implants with Respect to Effect of Materials on Muscle and Bone, F 981-87. In: *1990 Annual Book of ASTM Standards*, Vol 13.01: *Medical Devices; Emergency Medical Services*. ASTM, Philadelphia, pp. 329ff.

Anderson, L. C. and Hughes, H. C. (1986): In: *Handbook of Biomaterials Evaluation*. A. F. von Recum (Ed.). Macmillan, New York, pp. 255ff.

Autian, J. (1977): Artif. Organs 1(1):53.

Bakker, D., van Blitterswijk, C. A., Hesseling, S. C., Grote, J. J. and Daems, W. T. (1988): Biomats. 9:14.

British Standards Institution (1981): *Evaluation of Medical Devices for Biological Hazards*, BS 5736, Part 2: *Method of Testing by Tissue Implantation*. BSI, London.

Brody, S. (1945): *Bioenergetics and Growth, with Special Reference to the Efficiency Complex in Domestic Animals*. Reinhold, New York, pp. 352ff.

Department of Agriculture (1991): Animal Welfare; Standards; Final Rule. 9 CFR Part 3. Fed. Reg. 56 (Feb. 15):6426.

Department of Health and Human Services (1985): *Guide for the Care and Use of Laboratory Animals*. NIH Publication 85-23. U. S. Government Printing Office, Washington, DC.

Dillingham, E. O. (1983): In: *Cell-Culture Test Methods*, ASTM STP 810. S. A. Brown (Ed.). American Society for Testing and Materials, Philadelphia, pp. 51ff.

Escalas, F., Galante, J., Rostoker, W. and Coogan, P. S. (1975): J. Biomed. Mater. Res. 9:303.

Ferguson, A. B., Jr., Laing, P. G. and Hodge, E. S. (1960): J. Bone Joint Surg. 42A:77.

Gott, V. L. and Furuse, A. (1971): Fed. Proc. 30(5):1679.

Gourlay, S. J., Rice, R. M., Hegyeli, A. F., Wade, C. W. R., Dillon, J. G., Jaffe, H. and Kulkarni, R. K. (1978): J. Biomed. Mater. Res. 12:219.

Hofmann, A. A, Bachus, K. N. and Bloebaum, R. D. (1990): Trans. Soc. Biomats. 13:80.

Kupp, T., Hochman, P., Hale, J. and Black, J. (1982): *Advances in Biomaterials*, Vol. 3: *Biomaterials 1980*. G. D. Winter, D. F. Gibbons and H. Plenk, Jr. (Eds.). John Wiley, Chichester, pp. 787ff.

Kusserow, B., Larrow, R. and Nichols, J. (1970): Trans. Am. Soc. Artif. Intern. Organs XVI:58.

Marchant, R. E. (1989): Fund. Appl. Toxicol. 13:217.

Matlaga, B. F., Yasenchak, L. P. and Salthouse, T. N. (1976): J. Biomed. Mater. Res. 10:391.

Mirra, J. M., Amstutz, H. C., Matos, M. and Gold, R. (1976): Clin. Orthop. Rel. Res. 117:221.

Salthouse, T. N. (1980): Personal communication.

Salthouse, T. N. and Matlaga, B. F. (1975): Biomat. Med. Dev. Artif. Org. 3(1):47.

Salthouse, T. N. and Willigan, D. A. (1972): J. Biomed. Mater. Res. 6:105.

Sawyer, P. N., Stanczewski, B., Garcia, L., Kammlott, G. W., Turner, R. and Liebig, W. (1976): Trans. Am. Soc. Artif. Intern. Organs XXII:527.

Sewell, W. R., Wiland, J. and Craver, B. N. (1955): Surg. Gynecol. Obstet. 100:483.

Turner, J. E., Lawrence, W. H. and Autian, J. (1973): J. Biomed. Mater. Res. 7:39.

von Recum, A. F., Imamura, H., Freed, P. S., Kantrowitz, A., Chen, S.-T., Ekstrom, M. E., Baechler, C. A. and Barnhart, M. I. (1978): J. Biomed. Mater. Res. 12:743.

Wood, N. K., Kaminski, E. J. and Oglesby, R. J. (1970): J. Biomed. Mater. Res. 4:1.

Wortman, R. S., Merritt, K. and Brown, S. A. (1983): Biomat. Med. Dev. Artif. Org. 11(1):103.

BIBLIOGRAPHY

Department of Health and Human Services (1985): *Guidelines for Blood-Material Interactions*. NIH Pub. 85-2185. U. S. Government Printing Office, Washington, DC.

Homsy, C. A., McDonald, K. E., Akers, W. W., Short, C. and Freeman, B. S. (1968): J. Biomed. Mater. Res. 2:215.

Horowitz, E. and Torgesen, J. L. (Eds.)(1975): *Biomaterials*. NBS Special Pub. 415. U.S. Government Printing Office, Washington, DC.

Leininger, R. L. (1972). CRC Crit. Rev. Bioeng. 1:333.

Loomis, T. A. (1968): *Essentials of Toxicology*. Lea & Febiger, Philadelphia.

von Recum, A. F. (Ed.) (1986): *Handbook of Biomaterials Evaluation*. Macmillan, New York.

Williams, D. (Ed.) (1976): *Biocompatibility of Implant Materials*. Sector Publishing, London.

Williams, D. F. (Ed.) (1986): *Techniques of Biocompatibility Testing*, Vol. I & II. CRC Press, Boca Raton, FL.

APPENDIX 1

Designation: F 981-87:

Standard Practice for Assessment of Compatibility of Biomaterials (Nonporous) for Surgical Implants with Respect to Effect of Materials on Muscle and Bone[1]

This standard is issued under the fixed designation F981; the number immediately following the designation indicates the year of original adoption or, in the case of revision, the year of last revision. A number in parentheses indicates the year of last reapproval. A superscript epsilon (ε) indicates an editorial change since the last revision or reapproval.

1. Scope

1.1   This practice provides a series of experimental protocols for biological assays of tissue reaction to nonporous, nonabsorbable biomaterials for surgical implants. It assesses the effects of the material on animal tissue in which it is implanted. The experimental protocol is not designed to provide a comprehensive assessment of the systemic toxicity, carcinogenicity, teratogenicity, or mutagenicity of the material. It applies only to materials with projected applications in human subjects where the materials will reside in bone or soft tissue in excess of 30 days and will remain unabsorbed. Applications in other organ systems or tissues may be inappropriate and are therefore excluded. Control materials will consist of any one of the metal alloys in Specifications F67, F75, F90, F136, F138, or F562, or ultra high molecular weight polyethylene as stated in Specification F648 or USP polyethylene negative control.

1.2   This document is a combination of Practice F361-80 and Practice F469-78. The purpose, basic procedure, and method of evaluation of each type of material are similar; therefore, they have been combined.

1.3   *This standard may involve hazardous materials, operations, and equipment. This standard does not purport to address all of the safety problems associated with its use. It is the responsibility of whoever uses this standard to consult and establish appropriate safety and health practices and determine the applicability of regulatory limitations prior to use.*

2. Referenced Documents

2.1   *ASTM Standards*:

F 67 Specification for Unalloyed Titanium for Surgical Implant Applications[2]

---

[1]This practice is under the jurisdiction of ASTM Committee F-4 on Medical and Surgical Materials and Devices and is the direct responsibility of Subcommittee F04.02 on Resources.

Current edition approved Sept. 25, 1987. Published November 1987. Originally published as F 981-86. Reproduced here by permission of ASTM.

[2]*Annual Book of ASTM Standards*, Vol. 13.01.

F 75 Specification for Cast Cobalt-Chromium-Molybdenum Alloy for Surgical Implant Applications[2]

F 86 Practice for Surface Preparation and Marking of Metallic Surgical Implants[2]

F 90 Specification for Wrought Cobalt-Chromium-Tungsten-Nickel Alloy for Surgical Implant Applications[2]

F 136 Specification for Wrought Titanium 6Al-4V ELI Alloy for Surgical Implant Applications[2]

F 138 Specification for Stainless Steel Bars and Wire for Surgical Implants (Special Quality)[2]

F 361 Practice for Assessment of Compatibility of Metallic Materials for Surgical Implants with Respect to Effect of Materials on Tissue[3]

F 469 Practice for Assessment of Compatibility of Nonporous Polymeric Materials for Surgical Implants with Regard to Effect of Materials on Tissue[4]

F 562 Specification for Wrought Cobalt-Nickel Chromium-Molybdenum Alloys for Surgical Implant Application[2]

F 648 Specification for Ultra-High-Molecular-Weight Polyethylene Powder and Fabricated Form for Surgical Implants[2]

F 763 Practice for Short-Term Screening of Implant Materials [2]

3. Summary of Practice

3.1 This practice describes the preparation of implants, the number of implants and test hosts, test sites, exposure schedule, implant sterilization techniques, and methods of implant retrieval and tissue examination of each test site. Histological criteria for evaluating tissue reaction are provided.

4. Significance and Use

4.1 This practice covers a test protocol for comparing the local tissue response evoked by biomaterials, from which medical implantable devices might ultimately be fabricated, with the local tissue response elicited by control materials currently accepted for the fabrication of surgical devices. Currently accepted materials are the metals (and metal alloys (see Section 2)) and polyethylene which are standardized on the basis of acceptable long-term clinical experience. The controls consistently produce cellular reaction and scar to a degree that has been found to be acceptable to the host.

5. Test Hosts and Sites

5.1 Rats (acceptable strains such as Fischer 344), New Zealand rabbits, and dogs may be used as test hosts for soft tissue implants response. It is suggested that the rats be age and sex matched. Rabbits and dogs may be used as test hosts for bone implants.

---

[2]*Annual Book of ASTM Standards*, Vol. 13.01.
[3]Discontinued—See *1986 Annual Book of ASTM Standards*, Vol. 13.01.
[4]Discontinued—See *1987 Annual Book of ASTM Standards*, Vol. 13.01.

Table 1 Intervals of Sacrifice

Necropsy Periods (Weeks after Insertion of Implants)	Number of Animals to Be Necropsied		
	Rat	Rabbit	Dog
12 weeks	4	4	2
26 weeks	4	4	2
52 weeks	4	4	2
104 weeks	—	—	2

5.2 The sacro-spinalis, paralumbar, gluteal muscles, and the femur or tibia can serve as the test site for implants. However, the same site must be used for test and material implants in all the animal species.

5.3 Table 1 contains a suggested minimum number of study animals and a suggested schedule for the necropsy of animals.

6. Implant Specimens

6.1 *Fabrication*—Each implant shall be made in a cylindrical shape with hemispherical ends (see 6.2 and 6.3 for sizes). If the ends are not hemispherical, this shall be reported. Each implant shall be fabricated, finished, and its surface cleaned in a manner appropriate for its projected application in human subjects in accordance with Practice F86.

6.2 Reference metallic specimens shall be fabricated in accordance with 6.1 from materials such as the metal alloys in Specifications F67, F75, F90, F138, or F562, or polymeric polyethylene USP Negative Control Plastic.

6.3 *Suggested Sizes and Shapes of Implants for Insertion in Muscle*:

6.3.1 *For Rats*—1 mm diameter by 2 cm long cylindrical implants.

6.3.2 *For Rabbits*—1 mm diameter by 10 to 15 mm long cylindrical implants.

6.3.3 *For Dogs*—6 mm diameter by 18 mm long cylindrical implants.

6.3.4 If fabrication problems prevent preparing specimens 1 mm in diameter, alternative specimen sizes are 2 mm diameter by 6 mm long for rats and 4 mm diameter by 12 mm long for rabbits. If these alternate dimensions are used, such should be reported in accordance with 8.1.

6.4 *Sizes and Shapes of Implants for Insertion in Bone*:

6.4.1 *For Rabbits*—2 mm diameter by 6 mm long cylindrical implants.

6.4.2 *For Dogs*—4 mm diameter by 12 mm long cylindrical implants.

6.4.3 If the length of the bone implants needs to be less than that designated because of anatomical constraints, such should be reported in accordance with 8.1.

6.5 *Number of Test and Control Implants*:

6.5.1 In each rat, due to size, there shall be two implants; one each test and control material implant.

6.5.2 In each rabbit, due to size, there shall be six implants; four test materials and two control material implants.

6.5.3 In each dog, there shall be twelve implants; eight test materials and four control material implants.

6.6 *Conditioning*:

6.6.1 Remove all surface contaminants with appropriate solvents and rinse all test and control implants in distilled water prior to sterilization. It is recommended that the implant materials be processed and cleaned in the same way the final product will be.

6.6.2 Clean, package, and sterilize all implants in the same way as used for human implantation.

6.6.3 After final preparation and sterilization, handle the test and control implants with great care to ensure that they are not scratched, damaged, or contaminated in any way prior to insertion.

6.6.4 Report all details of conditioning in accordance with 8.1.

6.7 *Implantation Period*—Insert all implants into each animal at the same surgical session so that implantation periods run concurrently. The implantation period is 52 weeks for rats and rabbits; 104 weeks for dogs, with interim sacrifices at 12, 26, and 52 weeks (see 7.4).

7. Procedure

7.1 *Implantation (Muscle)*:

7.1.1 Place material implants in the paravertebral muscles of the adult rat, rabbits, and dogs in such a manner that they are directly in contact with muscle tissue.

7.1.2 Introduce material implants in dogs by the technique of making an implantation site in the muscle by using a hemostat to separate the muscle fibers. Then insert the implant using plastic-tipped forceps or any tool that is nonabrasive to avoid damage to the implant. Do not insert more that twelve implant materials in each dog.

7.1.3 Introduce material implants in rabbits and rats using sterile technique. Sterile disposable Luer-lock needles may be used to implant the material implants into the paravertebral muscles along the spine. In rats insert a negative control implant on one side of the spine and a test material implant on the other side. In rabbits implant one negative control material on each side of the spine and implant two test materials on each side of the spine. If larger diameter specimens are used in accordance with 6.3.4, an alternative implantation technique is that described in 7.1.2.

7.2 *Implantation (Femur)*—Expose the lateral cortex of each rabbit femur and drill three holes 1/16 in. (1.6 mm) through the lateral cortex using the technique and instrument appropriate for the procedure. For dogs, make the holes 1/8 in. (3.2 mm) in diameter; make six holes in each femur. Into each one of these holes, insert one of the implants by finger pressure. Then close the wound.

Note 1—Caution should be taken to minimize the motion of the implant in the tissue on the desired result.

7.3 *Postoperative Care*:

7.3.1 Care for the animals in accordance with accepted standards as outlined in the *Guide for the Care and Use of Laboratory Animals* (**1**).[5]

7.3.2 Carefully observe each animal during the period of assay and report any abnormal findings.

7.3.3 Infection or injury of the test implant site may invalidate the results. The decision to replace the animal so that the total number of retrieved implants will be as represented in the schedule shall be dependent upon the design of the study.

7.3.4 If an animal dies prior to the expected date of sacrifice, necropsy it in accordance with the procedure in 7.4 to determine the cause of death. Replacement of the animal to the study shall be dependent upon the design of the study. Include the animal in the assay of data if the cause of death is related to the procedure or test material.

7.4 *Sacrifice and Implant Retrieval*:

7.4.1 Euthanatize animals by a humane method at the intervals listed in Table 1.

Note 2—The necropsy periods start at 12 weeks because it is assumed that acceptable implant data has been received for earlier periods such as 1, 4, and 8 weeks from short term implant testing.

7.4.2 At necropsy, record any gross abnormalities of color or consistency observed in the tissue surrounding the implant. Remove each implant with an intact envelope of surrounding tissue. Include in the tissue sample a minimum of a 4-mm thick layer of tissue surrounding the implant. If less than a 4-mm thick layer is removed, report in accordance with 8.1.

7.5 *Postmortem Observations*—Necropsy all animals that are sacrificed for the purposes of the assay or die during the assay period in accordance with standard laboratory practice. Establish the status of the health of the experimental animal during the period of the assay. Report as described in Section 8.

7.6 *Histological Procedure*:

7.6.1 *Tissue Sample Preparation*—Prepare two blocks from each implantation site.

7.6.1.1 Process the excised tissue block containing either a test implant or control implant for histopathological examination and such other studies as are appropriate. Cut the sample midway from end to end into appropriate size for each study. Record the gross appearance of the implant and the tissue.

7.6.1.2 If special stains are deemed necessary, prepare additional tissue blocks or slides, or both, and make appropriate observations.

[5]The **boldface** numbers refer to the list of references at the end of this standard.

7.7   *Histopathological Observations*—Compare the amount of tissue reaction adjacent to the test implant to that adjacent to a similar location on the control implant with respect to thickness of scar, presence of inflammatory or other cell types, presence of particles, and such other indications of interaction of tissue and material as might occur with the actual material under test. A suggested method for the evaluation of tissue response after implantation is Turner et al.(4)

7.7.1   *Suggested Method for Tissue Response Evaluation*:

7.7.1.1   A suggested format with cellular elements to be evaluated and a scoring range of 0 to 3 using the criteria shown in Table 2.

7.7.1.2   The scoring system of 0 to 3 is based upon the number of elements in high power field (470x) average of five fields.

Number of elements	Score
0	0
1–5	0.5
6–15	1
16–25	2
26 or more	3

7.7.1.3   The degree of necrosis score is determined using the same range of 0 to 3, as follows:

Degree	Score
Not present	0
Minimal present	0.5
Mild degree of involvement	1
Moderate degree of involvement	2
Marked degree of involvement	3

7.7.1.4   An overall toxicity rating of test samples may be given using a rating range of 0 to 4, as follows:

Rating	Score
Nontoxic	0
Very slight toxic reaction	1
Mild toxic reaction	2
Moderate toxic reaction	3
Marked toxic reaction	4

7.7.1.4.1   Pathologists may choose to use the scoring system of comparing the negative control to the test material as an aid in their evaluation. The overall

Table 2 Suggested Evaluation Format and Scoring Range

		0	.5	1	2	3
Animal Number						
Duration of Implant (weeks)						
Sample Description						
Gross Response						
Histopath-Number						
Score		0	.5	1	2	3
Necrosis						
Degeneration						
Inflammation						
	Polymorphonuclear Leukocytes					
	Lymphocytes					
	Eosinophils					
	Plasma Cells					
	Macrophages					
Fibrosis						
Giant Cells						
Foreign Body Debris						
Fatty Infiltration						
Relative Size of Involved Area in mm						
Histopathologic Toxicity Rating						

toxicity of the test material as compared to the negative control is to be evaluated independently for all time periods.

8. Report
8.1 The report shall include the following information:

8.1.1 All details of implant characterization, fabrication, conditioning (including cleaning, handling, and sterilization techniques employed).

8.1.2   Procedures for implantation and implant retrieval.

8.1.3   Details of any special procedure (such as unusual or unique diet fed to test animals).

8.1.4   The observations of each control and test implant as well as the gross appearance of the surrounding tissue in which the implants were implanted.

8.1.5   The observation of each histopathological examination and the pathologist's evaluation as to toxicity of test material provided.

Appendix   (Nonmandatory Information)

X1.   RATIONALE FOR PRACTICE F 981

X1.1   This practice is based on the research techniques utilized by Cohen (**5**), and by Laing, Ferguson, and Hodge (**6,7**) in the early 1960's. These studies involved the implantation of metal cylinders in paravertebral muscle of rabbits. The biological reaction to the cylinders was described as the thickness of the fibrous membrane or capsule formed adjacent to the implant. The thickness of the capsule and the presence of inflammatory cells was used as a measure of the degree of adverse reaction to the test material.

X1.2   As first published in 1972, Practice F 361 was a test for the biological response to metallic materials. The scope had been expanded beyond that of the published reports to include bone as well as muscle as an implant test site. To avoid species specific reactions, the method called for the use of rats and dogs as well as rabbits. Cylindrical test specimens with rounded ends were used to avoid biological reactions associated with sharp corners or other variations in specimen shape.

X1.3   In 1978, Practice F 469 was published as a parallel document for the test of polymeric materials. In that the methods are essentially the same, the scope of F 361 has been expanded to include the testing of specimens made of metallic, polymeric or ceramic materials, thereby including and superseding F 469.

X1.4   Porous or porous coated materials are specifically excluded since the response to such materials includes ingrowth of tissue into the pores. As a result, the method of tissue fixation and sectioning, and the evaluation scheme are substantially different.

X1.5   Stainless steel, cobalt chromium, and titanium alloys are used as reference materials since the biological response to these materials has been well characterized by their extensive use in research. The response to these materials is not defined as compatible, but rather the response is used as a reference against which reactions to other materials is compared.

X1.6   This practice is a modification of the original Practice F 361 in that it only involves long term test periods. The short term response to materials is to be evaluated using Practice F 763.

X1.7   This practice was revised in 1987 to allow for alternative specimen dimensions for rats and rabbits for muscle implantation. The original specimen dimensions were intended to be implanted through a needle, which was a change from F 361 and F 469. The alternate dimensions restore those specified since 1972 which some members felt were more appropriate for some material types.

REFERENCES

(1) *Guide for the Care and Use of Laboratory Animals*, Department of Health, Education, and Welfare, Publication No. 80, (Vol. 23, revised 1978).

(2) "Toxicological Evaluation of Biomaterials: Primary Acute Toxicity Screening Program," *Artificial Organs*, Vol. 1, No. 1, August 1977.

(3) *The United States Pharmacopeia*, XX Edition, Mack Publishing Co., Easton, PA, 1980, pp. 950–953.

(4) Turner, J. E., Lawrence, W. H., and Autian, J., "Subacute Toxicity Testing of Biomaterials using Histopathologic Evaluation of Rabbit Muscle Tissue," *Journal Biomedical Material Research*, Vol. 7, No. 39, 1973.

(5) Cohen, J., "Assay of Foreign-Body Reaction," *Journal of Bone and Joint Surgery*, No. 41A, 1959, pp. 152–166.

(6) Ferguson, A. B., Jr., Laing, P. G., and Hodge, E. S., "The Ionization of Metal Implants in Living Tissues," *Journal of Bone and Joint Surgery*, No. 42A, 1960, pp. 77–90.

(7) Laing, P. G., Ferguson, A. B., Jr., and Hodge, E. S., "Tissue Reaction in Rabbit Muscle Exposed to Metallic Implants," *Journal Biomedical Materials Research*, No. 1, 1967, pp. 135–149.

APPENDIX 2: RATING SYSTEM FOR TISSUE AT ANIMAL IMPLANT SITES*

This system, based upon earlier studies (Gourlay et al., 1978; Sewell et al., 1955) involves measuring the capsule thickness, evaluating the local cellular response, and then assigning weighting factors to arrive at an overall rating. Specific values (of weighting factors) are given for implants in the form of 2-0 sutures; modification for standard implants in other configurations might be advisable.

Capsular Thickness:

Grade	Thickness range (μm)
1	0–25
2	26–50
3	51–250
4	251–500
5	501–750

Cellular Response:
   Grade from 0 to 5 depending upon the concentration of cells observed in the capsule and adjacent tissue per high power ($500 \times$) field.
Weighting Factors:
   Capsular thickness: $\times$ 5
   Cellular response: $\times$ 3
   Cells present:
      Neutrophils: $\times$ 5
      FB Giant Cells: $\times$ 2
      Lymphocytes: $\times$ 1
      Macrophages: $\times$ 1
      Fibroblasts: $\times$ 1

A final score might appear as follows:
   Capsule thickness: grade 3 $\times$ factor 5 = 15
   Cellular response: grade 2 $\times$ factor 3 = 6
   Cell types:
      Neutrophils: 5 $\times$ factor 5 = 25
                                       Total: 46

*Source: Adapted from Salthouse (1980).

The numerical ratings might be verbally expressed by descriptors: 0: no reaction; 1–10: minimal; 11–25: slight; 26–40: moderate; 41–60: marked; >60: excessive. The example given above would then be in the marked tissue reaction range. Note that this rating system is intended as a model system. Specific rating scales should be developed for specific applications.

APPENDIX 3: SYSTEM FOR EVALUATION OF HUMAN LOCAL HOST RESPONSE TO IMPLANTS IN THE MUSCULOSKELETAL SYSTEM (SYNOVIAL AND CAPSULAR TISSUES)*

Tissues fixed in 10% buffered formalin, routinely processed (paraffin mounted), stained with hematoxylin and eosin and viewed by ordinary light and polarized microscopy. Tissues are to be examined without any prior knowledge of clinical or radiographic status of patient. The maximum concentrations are counted, averaged and graded from at least five different microscopic fields per section using the following eight categories:

1. Acute inflammatory cells (polymorphonuclear leukocytes) and mononuclear histiocytes
    0 = absent
    1+ = 1–5 cells/(500×) field
    2+ = 6–49 cells/(500×) field
    3+ = 50 or more cells/(500×) field

2. Chronic inflammatory cells (lymphocytes, plasma cells, lymphoid follicles)
    1+ = 1–9 cells/(500×) field and/or one lymphoid follicle/(100×) field
    2+ = 10–49 cells/(500×) field and/or 2–3 lymphoid follicles/(100×) field
    3+ = 50 or more cells/(500×) field and/or 4 or more lymphoid follicle/(100×) field

3. Giant cells (multinucleated histiocytes)
    1+ = 1–2 cells/(250×) field
    2+ = 3–8 cells/(250×) field
    3+ = 9 or more cells/(250×) field

4. Metal particles
    1+ = 1–19 particles/(500×) field
    2+ = 20–499 particles/(500×) field
    3+ = 500 or more particles/(500×) field

5. Polyethylene† fibers, small type (less than 100 μm in length)
    1+ = 1–9 fibers lying extracellularly and/or 1–9 histiocytes containing one or more fibers per (500×) field
    2+ = 10–19 fibers lying extracellularly and/or 10–19 histiocytes containing one or more fibers per (500×) field
    3+ = 20 or more fibers lying extracellularly and/or 20 or more histiocytes containing one or more fibers per (500×) field

*Source: Adapted from Mirra et al. (1976).
†Identified by birefringence in polarized light.

6. Polyethylene* fibers, (greater than 100 μm and less than 500 μm in length)
   $1+$ = 1–3 fibers/(45×) field
   $2+$ = 4–9 fibers/(45×) field
   $3+$ = 10 or more fibers/(45×) field

7. Polyethylene flakes* (greater than 500 μm in length)
   $1+$ = 1–2/tissue section
   $2+$ = 2–5/tissue section
   $3+$ = 6 or more/tissue section

8. Methyl methacrylate globules†
   $1+$ = 1–3/tissue section
   $2+$ = 4–6/tissue section
   $3+$ = 7 or more/tissue section

*Identified by birefringence in polarized light.
†May be seen as void spaces, with possible $BaSO_4$ granules, due to dissolution by section clearing solvent.

18

Qualification of Implant Materials

18.1 GOAL OF CLINICAL TRIALS

After material selection and device design, and in vitro tests and implantation in animals, a material must eventually be tested in humans. Such tests are necessary since the goal of implant materials development, selection and testing is the alleviation of human clinical problems and because there continues to be insufficient knowledge about biological per-

330

formance of materials to predict clinical success with confidence from laboratory and animal testing. In this section we will briefly consider some aspects of the actual design of clinical trials.

Before considering a clinical trial, it is well to consider the goal of such trials. Unless an implant is designed for acute use, trials cannot extend beyond a short fraction of intended device life. Furthermore, test of a new biomaterial in a particular device cannot result in "qualification" of the material. Therefore, we must regard clinical trials as serving as detectors of "bad news." They represent a limited, controlled, well-observed introduction of a new material and/or design. The use of the term *introduction* is deliberate since, unlike in the case of animal trials, the experimental subjects will continue to be exposed to the device and its material components even after the period of observation. The longer the time elapsed and the greater the number of patients studied during such limited introduction without detection of adverse results, the greater the assurance of good performance following general introduction.

18.2 DESIGN OF CLINICAL TRIALS

18.2.1 General Requirements

Burdette and Gehan (1970) identify four types or sequential phases of clinical trials:

Phase I—Early trial: Selecting a new treatment from among several options for further study.

Phase IIA—Preliminary trial: If the new treatment is not effective in the early trial, this phase examines whether further studies should be performed or the treatment abandoned.

Phase IIB—Follow-up trial: Estimating the effectiveness of a new treatment which appears promising based upon either Phase I or Phase IIA trials.

Phase III: Comparison of the effectiveness of the new treatment with a standard method of management or some other treatment.

In the human testing of materials within devices, Phases I, IIA and IIB are rarely planned in a formal sense. Their function is usually filled by the use of individual custom devices for individual patients under the direct care or supervision of the surgeon-member of the research group. Only when there is a perception of relative benefit for the new material/device (usually in comparison to other material/device arrangements) does formal clinical testing begin with a Phase III trial. This phase in implant development may be further subdivided into two subphases:

Phase IIIA: Examination of clinical outcome of a defined new treatment for a group of patients with defined indications.

Phase IIIB: Following success in a subphase IIIA trial, examination of the clinical outcome for a defined (refined) new treatment for a group of patients with defined (refined) indications, usually involving multiple investigators and institutions.

It is the practice in new drug trials to employ the "double-blind" method. That is, a drug and a harmless inactive material (placebo) are used in a treatment plan for a defined group of patients with a common set of symptoms. Which patients receive the drug and which the placebo is predetermined at random. The patients do not know if they are receiving the active drug (single blind) nor does the treating physician (double blind). When the experimental trial is complete, an identifying code assigned to the drug and placebo doses is deciphered and an analysis of effectiveness of the treatment is made.

When an implant is surgically inserted, whatever the phase of the trial, it is not possible to pair the implanted patient with a placebo-treated patient. The case of no insertion is clearly not blind to either patient or doctor. The case of comparing identical devices made of different materials is at best blind to the patient. Differences in appearance, weight, shape, etc., between devices made of different materials renders them easily distinguishable to the physician and usually to the patient. Furthermore, as is noted in Chapter 20, the interrelationship between materials selection and device design is such that it would be unlikely that the two devices would differ only in respect to materials.

Thus, clinical trials of implant materials have to be based upon a different experimental design. Comparisons may be made as follows:

1. Between the condition of the patient before and after implant surgery: This is useful to detect acute changes that may take place in an individual with underlying disease (for which the implant is indicated) that may be associated with the implant material.
2. Between patients with similar implants made of different materials: This is useful to investigate acute and chronic differences in both material and host response.
3. Between the patient with an implant and a nondiseased individual of the same age, sex, and with a similar home/workplace environment: This may be useful for detection of subtle chronic effects of materials (host response).

A number of efforts have been made to set standards for selection and treatment of patients in clinical trials. Some of the general rules that have emerged are:

1. Medical care must be under the direction of a medical professional.
2. The patients must give informed consent to any experimental procedure. Such informed consent can only be obtained after a full explanation of possible benefits and risks of the proposed procedure.
3. The identity of the patients and the confidentiality of their medical records must be preserved.
4. Perhaps the most important point is that, for the trial to be justified, whatever the phase, there must be a reasonable possibility of specific benefit to the patients in the trial combined with reasonable assurance of the absence of unusual risk.

The governing ethical considerations, of which these are a part, are the twelve basic principles of the Declaration of Helsinki II, revised and extended by the 29th World Medical Assembly (Silverman, 1985).

It is clear that clinical trial protocols are difficult to design and implement. As an aid in such efforts, most medical research and treatment facilities involved with patients maintain Human Subjects Committees. These committees are available to help in preparing protocols and must, generally, review the procedures and safeguards in any experimental program involving human subjects before the clinical trial is started. Federal agencies now make such reviews by Human Subjects Committees a prerequisite to public funding of clinical research.

In addition, many of the Device Classification Panels of the Bureau of Medical Devices of the Food and Drug Administration (FDA) have developed guidelines for design of clinical trials and for statistical treatment and format for reporting results. If relevant guidelines exist in the area under consideration, they should be consulted at an early point of protocol development.

Although these detailed guidelines are of use, they have now generally been supplanted by an FDA control document. This arises from the need to obtain an exemption from certain provisions of the Medical Device Amendments (1976) (see Section 19.2) in order to manufacture, ship interstate, and implant the quantities of implants required for phase III clinical trials.* The necessary authorization is obtained through a successful (approved) application for an Investigational Device Exemption

---

*Devices required for earlier phases are manufactured individually at the surgeon's or physician's prescription and are permitted to be used under the custom devices provision of the Medical Device Amendments (1976).

(IDE).* The referenced portions of the Code of Federal Regulations describe the procedure for application for an IDE, the responsibilities of the sponsor, investigators, and Human Subjects Committees [here called Institutional Review Boards (IRBs)], as well as setting standards for informed consent, protection of patient confidentiality, and reporting of study results.

It should be further noted that an IDE will not be granted unless the supporting tests, of the type to be discussed later in this chapter as well as others, meet the requirements of the regulations on Good Laboratory Practice (GLP).† These regulations set forth standards concerning design and documentation of preclinical trials, qualification of personnel, and preservation and presentation of experimental results. While the GLP regulations apply only to aspects of testing to support claims of *safety* at this time, it is not improbable that they will be extended to apply to *all* aspects of preclinical testing as well as to a wide variety of biomedical research efforts not directly associated with direct material and device development.

Before clinical trials of a new material (in a device configuration) can be countenanced, at least two preliminary types of nonclinical tests of the material seem imperative. These are in addition to those tests which may be required to demonstrate the safety and effectiveness of the material and the device design before phase III clinical trials may begin. However, if GLPs are observed, then these preliminary test results may be used as part of a later IDE application submission.

18.2.2 Preclinical Tests

The first of these two preliminary tests is *acute screening* based upon in vitro and tissue culture techniques as outlined in Chapter 16. There are no general standards in this area, and many different protocols are in use.

Completion of these acute screening trials should lead to a second type of preliminary test, the *chronic animal demonstration* test. The ASTM F 981 protocol (Chapter 17) is such a test. Chronic animal tests should essentially meet or exceed the requirements of F 981. Equivalency to F 981 involves as a minimum:

1. The use of multiple species.
2. The use of the same control (reference) materials used in acute testing.

*Federal Register 45(13):3732, 1980. See also Dobelle et al. (1980).
†Federal Register 43(247):59986, 1978.

3. The use of group sizes and sacrifice schedules substantially equal to or greater than those required by F 981.
4. The use of operative sites similar to those of the intended human application.

If a material demonstrates that it is equal to or superior to materials in present use in both of these types of tests, and if no extraordinary hazards specifically associated with it arise, then the planning and execution of Phase I and II clinical trials seems warranted. The requirements for such clinical trials are probably more stringent than those needed to demonstrate performance of new *designs*. This is the case due to the subtlety of many materials' problems and the general "endorsement" that a new material may achieve after successfully completing its first clinical trial series.

The question of what tests are necessary and sufficient before Phase III clinical trials of new materials are warranted is extremely controversial in all aspects. It would be very inviting to develop a consensus viewpoint or matrix into which all new materials and material applications could be classified, thus settling the generic problem once and for all. A number of groups, including working groups within various US and foreign national governmental agencies and national standards-making organizations, are examining this approach to the question. Progress has been slow and the end product is clearly a long way away.

The first step was the development of ASTM F 748: Practice for Selecting Generic Biological Test Methods for Materials and Devices (ASTM, 1990). Table 18.1 is the test matrix adopted in this practice. It distinguishes between external devices, external communicating devices and implants, and between tissue types and contact periods. Between 2 and 11 of 13 generic host response tests, including an F 981-type chronic implantation test, are recommended for each exposure class. Although the tests are defined generically, the ASTM F-4 committee has gone on to recommend specific test methods in each area (see Table 19.1) which are now incorporated by reference.

The success of this voluntary practice, originally adopted in 1982 but requiring nearly ten years to develop to its current (1987) form, led to the so-called Tripartite Biocompatibility Guidance (TBG) (Kammula, 1991), which was ratified on April 24, 1987. This document was developed by a joint US, Canada and UK working group and is intended to assist both manufacturers and government health agencies in the three countries in anticipating the information which is necessary for preclinical evaluation of new materials. The TBG retains the matrix approach of F 748, deletes the distinction between contact periods (intraoperative, short-term or

Table 18.1 Classification of Materials/Devices/Applications and Applicable Tests

Classification of material/ device/application:	Cell culture cytotoxicity	Skin irritation	Intramuscular implantation	Blood compatibility	Hemolysis	Carcinogenicity	Long-term implant	Mucous membrane irritation	Systemic injection acute toxicity	Intracutaneous injection (irritation)	Sensitization	Mutagenicity	Pyrogen test
External devices:													
Intact surfaces		X									X		
Breached surfaces	X	X							X	X	X		
External communicating devices, with:													
Intact natural channels								X			X		
Body tissues and fluids													
Intraoperative	X								X	X	X		X
Short term	X		X						X	X	X		X
Chronic	X		X						X	X	X		X
Blood path, indirect	X		X	X	X				X	X	X		X
Blood path, direct, short-term	X		X	X	X				X	X	X		X
Blood path, direct, long-term	X		X	X	X				X	X	X		X
Implanted devices contacting:													
Bone	X				X	X			X		X	X	X
Tissue and tissue fluid	X		X		X	X			X	X	X	X	X
Blood	X		X	X	X	X			X	X	X	X	X

Source: ASTM F 748-87 (1990).

chronic) and adds several additional possible tests, including determination of the pharmacokinetics of released material (''biological fate'') and of reproductive and developmental toxicity. Unlike F 748, the TBG does not refer to specific recommended test methods but simply puts forward recommended aspects of such tests.

The formation of the European Economic Union, to become effective at the end of 1992, has led to interest within the International Standards Organization (ISO) in producing a systematic approach to selection of tests for biological evaluation of materials. This standard (ISO/DIS 10993-1, 1991) is being drafted by Technical Committee 194 and draws very strongly on the TBG. In fact, it uses the same matrix and recommended tests but restores the exposure classes of F 748 by distinguishing between three conditions:

A. Limited exposure.
B. Prolonged or repeated exposure.
C. Permanent contact.

In addition, the ISO guidance document is somewhat more subtle in its approach in that it distinguishes between nine initial and four supplementary evaluation tests. This approach recognizes the criticality of the initial nine tests (excluding chronic toxicity, carcinogenicity, reproductive and developmental toxicity and the biological fate of degradation products) for the majority of short- and intermediate-term applications. Finally, it is the apparent intent of the ISO to follow the example of F 748 and develop specific test methods for each of the 13 generic test categories used in its hazard matrix.

However, while the matrix approaches taken to date are extremely beneficial, the ideal generic test selection matrix should incorporate the following criteria:

1. Separation of test requirements by the following technical aspects:
   a. Type of tissue that the implant will contact (muscle, blood, etc.).
   b. Duration of implant, by classes of time intervals (short-term, intermediate-term, etc.).
   c. Relative exposure of materials to the body of the patient (SA/BW ratio, etc.).
2. Selection of tests by generic description (with minimum requirements) rather than by detailed specification of procedures.
3. Specification of levels of certainty (''confidence levels'') rather than setting specific sample or group sizes in individual tests.

The emerging ISO standard addresses some but not all of these concerns. In particular, it fails to deal with 1c and 3 above. Great care must be taken in defining specific test methods for host response in this context. While the intent is clear to set minimum requirements, the high cost of testing often converts these minima into maxima. In the case of new classes of materials, this may permit subtle but deleterious aspects of host response to be overlooked.

In practice, it appears that F 748, the TBG and presumably the emerging ISO standard will not be strictly adhered to. That is, they appear to be serving a useful role by defining the consensus minimum preclinical testing required. Industrial sponsors tend to develop their own test matrices based upon these views (Stark, 1991).

18.2.3 Clinical Trials

Finally the time will arrive when confidence in a new material has risen sufficiently that clinical trials can be begun with caution.

Clinical trials must be performed under the control of a defined (written) prospective protocol which includes the following provisions:

1. Description of the implant device (note that the implant site must be that of the proposed application).
2. Outline of indications for the surgical procedure.
3. Outline of the uniform surgical procedure used.
4. Outline of postoperative treatment.
5. Outline of follow-up schedule and postoperative evaluation techniques.

The following information is needed for adequate consideration and evaluation of clinical testing results (with respect to biological performance of materials):

1. Protocol as above.
2. Identification of 200 patients, by code number, who constitute consecutive individuals seen by the treating physician(s) and meet item 2 of the protocol.
3. Results of follow-up of these patients for a minimum of 5 years and an average of 7–10 years for those not lost to follow-up at an earlier date (minimum of 100).*
4. Summary of all adverse results (note that a statistical summary is desirable; individual results with code should not be reported).

*In practice, it is difficult to distinguish *material* trials from *device* trials. Unfortunately the de facto standard for device trials is only two years' minimum follow-up: this is grossly inadequate for the evaluation of new materials.

The last point should be dwelt upon. Presumably, at the time that the trial protocol was being developed, consideration was given to each of the questions to be asked and the statistical measures to be used in answering them. At the end of the trial, it is thus appropriate to suggest that statistical measures be employed. Therefore, reports of clinical trials should take care to:

1. Discuss accuracy and precision of all measurements, where possible.
2. Define a minimum confidence level for all statistical measures of the data.
3. Report confidence intervals or other measures of significance associated with all derived parameters.
4. Indicate the significance of any conclusion arrived at by analysis of the trial.

18.3 CONCLUSIONS FROM CLINICAL TRIALS

18.3.1 Introduction

As pointed out in Interpart 2, no clinical trial can approach the numbers and period of exposure that will be experienced when a material enters into general use. Thus, it is imperative that, in a sense, clinical testing never end. The treating physician and his consultants should distinguish between "new" and "old" materials and should continue to be sensitive to possible biological performance problems associated with the use of either kind.

A well-designed and conducted clinical trial does *not* serve to qualify a material. In addition to design studies, physical measurements, and acute and animal tests, it provides data that are required for a decision to release a *product* for general use. The decisions made along the path to release should be based on appropriate statistical tests at a minimum confidence level of 95%. Such release, when it occurs with the approval of the appropriate regulating agencies, is necessarily a risk. That is, the benefits attendant to the use of the article are felt at the moment of decision to outweigh the risks involved.

Throughout this book we have been considering aspects of biological performance, including both material and host response. In the latter sections of this work, we have begun to examine the details of test methods for examining biological performance leading to clinical use. There are two factors common to all of these methods:

1. Large investments of time and money are required to produce results with reasonable levels of reliability and statistical significance due to the variability of the biological systems that are involved.
2. Large quantities of inductive reasoning must be used to apply the results of in vitro and animal testing to the prediction of clinical performance.

18.3.2 Complication Incidence Rates

In addition, in discussions of clinical evidence for materials incompatibility (failure to reach adequate levels of biological performance), we have seen, time and again, that incidence rates of complications are small. For instance, the "variant poppet" problem in early heart valve designs probably did not affect more than 3% of patients receiving the prosthesis. The more recent adverse experience with an anterior cruciate ligament replacement prosthesis (Chen and Black, 1980) resulted in failure in a larger percentage of cases, but still probably less than 20%. In many high-use applications, such as total hip joint replacement, device-related failures, for all causes including failure of biological performance, are on the order of, or less than, 1%/year after surgery.

Why then the concern expressed here for examining biological performance? A manufacturer could fairly argue that modern implant materials have proven beneficial to 95–99% + of patients receiving them. Similarly, a study by the Carnegie-Mellon Institute (Piehler, 1978) of costs and benefits associated with research to improve current orthopedic prosthetic devices concluded that failure/complication rates, even in the 1970s, were acceptably small and that the investment required to reduce these rates significantly would be costly out of proportion to the resulting benefits.

Thus, we are faced with the problem that not only can we not define a failure or complication rate during clinical trials, but that such a rate, a priori, could not be judged to be acceptable except in extreme cases. How then are we to decide when a material has proven itself sufficiently, with respect to biological performance, to move into general use?

18.4 ASPECTS OF THE DECISION FOR GENERAL CLINICAL USE

18.4.1 Current Concerns

I believe that we have to address this question in two frames of reference. The first is the current medical/legal environment. The general arguments cited above concerning cost of additional testing and the "acceptable" level of performance of current devices were raised time and again during the hearings on the Kennedy (U.S. Senate) and Rogers (House of Representatives) Bills that resulted in the Medical Device Amendments of 1976 (see Section 19.2). However, they were more than offset by the force of individual testimony concerning the human and financial costs to *individuals* as a result of device malfunction and failure. Thus, while statistical failure rates are low, we perceive the failure rate in a given individual with a defective device as *100%*. Despite the 1976 legislation, continuing concerns about the impact of individual device malfunction and failure led to the Safe Medical Devices Act (1990) (see Section 19.2).

This humane view of individuals rather than average statistics has resulted in a dramatic rise in medical malpractice and damage suits associated with device failure. A typical early report* summarizes five suits concerning "defective" heart valves. Only one of the five alleged defects was connected with the death of a patient; the damage claimed in the other four cases was disability and the need for additional surgery. Total damages sought in the five cases were $55 million. While the final outcome of these cases will not be known for some time, malpractice/device failure awards have already exceeded $1 million in individual nonfatal cases and new cases are being filed at a rate of 900/day (!) with average final awards of $300,000 (Kiplinger, 1991).

Public opinion, as reflected in congressional testimony and in the results of malpractice suits, contributed in no small part to the formula adopted in the Medical Device Amendments (1976). The test of adequate performance laid out in this law is that the device be "safe and effective" and expose the patient to no "unreasonable" risk or hazard. Thus, decisions on what will be acceptable failure rates remain subjective, to be dependent on the continued interaction of public opinion, expert advice, and administrative action.

As individuals we are torn between two views. Our heart tells us that no failure is acceptable, especially if it were to happen to ourselves or to one near to us. Our head tells us that such a goal may be unattainable and approachable only at prohibitive cost.

18.4.2 Response to Current Concerns

We have a precedent in our search for a defect-free materials technology for medical and surgical implants and devices. Early in the development of intercontinental ballistic missiles it was recognized that the complexity

---

*The New York Times, Dec. 4, 1977.

of the electronic control systems required would, in the normal course of events, lead to non-functioning systems. That is, the level of reliability of individual electronic components, then exceeding 99.9%, was insufficient to permit systems with 10^6 to 10^8 components to have any real level of satisfactory performance.

There were two approaches adopted to deal with this problem. The first was the idea of redundancy. Each section of a control system was to have one or more "backup" sections that operated in support or in parallel and which could take over if the primary system failed. This led to the practice in the present Space Shuttle of having three (and in some cases four) parallel systems serving each critical control function.

The second approach was to adopt the position that no *absolute* level of performance for an individual component was acceptable. This latter view led to a highly successful program, first instituted by the Martin-Marietta Company (Denver) and later by both the U.S. Air Force and NASA, called the "Zero Defects Program." The basic concept is to test devices continually, even after they pass into active service, and feed the results back into product improvement with a view to eventual "100%" satisfactory performance. The combination of these two approaches contributed to the success of the Apollo Lunar Program and the continued high level of performance of civilian and military aerospace hardware.

I suggest that we can apply both of these concepts, and an additional idea, that of a "fail-safe" product, to considerations of biological performance in the current environment.

The idea of redundancy is hard to apply directly toward the design of devices for a number of reasons. It is applicable to active devices, such as heart pacers, but far less so to joint replacements, sutures, etc. We can, however, incorporate redundancy into materials and device *testing*. The F 981 protocol already does this to a degree in its use of multiple animal species. This trend toward multiple, parallel, and overlapping tests should be continued.

The idea of zero defects can be incorporated by refusing to adopt *any* level of performance as being *permanently* satisfactory. Thus, we can insist that performance equal to or better than that of materials in use today be demonstrated before new materials can go into clinical trials and use. However, this should not blind us to a constant need to examine the performance of current ("old") and future ("new") materials with a view toward continual improvements being made where and when possible. Thus the call, in Interpart II, for increased attention to the human epidemiology of biomaterials.

The third idea, that of a "fail-safe" product, can also be adopted. This is the concept that, by design provisions, the ill effects of the failure of a device to function in its intended mode may be minimized. An example of fail-safe design in everyday life is the Westinghouse type AB Air Brake that is used on railroad cars. If the train separates at a coupling or the brake control system fails, the system is designed to apply the brakes automatically in the individual cars and to maintain braking until they are manually released and reset. In fracture fixation applications, such a concept leads to the choice of a material with lower strength and higher ductility, such as a stainless steel, over alloys with higher strength but limited ductility. Here the fail-safe feature is that the observation of permanent deformation of an internal fixation device can lead to a change in external support prescribed by the physician. The "failed" situation for a device fabricated from a ductile material, that is, angulation, even though it may result in an angulation in the healed fracture, is far more acceptable to both patient and physician than the "failed" situation for the less-deformable device—device fracture—that may lead to additional disability, surgery, and possible legal action.

18.4.3 Future Concerns

The second frame of reference we should briefly examine with respect to cost/risk/benefit aspects of material introduction is that of the future. How can we judge when a new material or device is "ready" to enter use and perhaps supplant present, apparently less-effective products?

I think the answer is rather simple. If we adopt the ideas of safe failure modes in design, redundancy in testing, and, especially, continual performance improvement, there is no real distinction between "old" and "new" materials. No one would knowingly substitute an inferior new product for a current product unless driven by inhumane motives. With this in mind, we can hope that progressive attitudes on the parts of researchers, manufacturers, surgeons, and regulatory authorities will lead to a continual upgrading of the performance of current materials and the gradual introduction of new materials when subjective levels of safety and efficacy, most probably defined by then-current experience, are reached.

Critics would suggest the need for some *absolute* (minimum) level of safety that must be obtained before introduction of a new material. With respect to devices, it has been proposed that the following definition be used:

> A device is safe enough to use when it is no worse than others in use and presents no greater hazard than the condition it is to be used to treat.

This appears clear enough in instances where large improvements in devices (and materials) can be demonstrated and the conditions being treated are life-threatening. In situations where the new material represents an evolutionary change in composition and behavior, and the aim is to improve the quality of life of the patient by alleviating a condition of low mortality and morbidity, such a statement is a poor guide. Hazards may be of a new and noncomparable type. Comparison of hazards, in any case, is possible only when there is a great difference in potential outcomes.

Thus, I suggest that decisions on device and materials introductions must be made on the individual merits and demerits of each situation and not be shackled by a set of rigid rules.

Except at an early point in this discussion, I have said nothing about the costs associated with this approach to materials application in the medical and surgical field. This was deliberate. I suggest that the analyses of the type made by Piehler and others fail in the face of the human and emotional aspects of this field. So long as the financial costs of devices remain, as they currently are in the United States, a small component of the cost of health care, money will not be an important factor in these considerations. In individual cases it is clear that the increased cost of research, development, and testing of materials and devices that will be the legacy of the Medical Device Amendments and the Safe Medical Devices Act, the increased number and size of malpractice suits, and increased public attention will act to stifle innovation. We can look forward to a situation that is parallel to that in the drug field: an improvement in the nature of products newly introduced and a tendency to move research and development activities ''offshore.'' It is hard to pass judgment on this trend. Whether good or bad, it is coming about in response to a clear public demand for safe and effective materials for medical and surgical implants and devices.

What we can pass judgement on, though, is the increasing trend towards market-driven rather than technology-driven introduction of new devices and materials. As clinical experience with many existing materials now exceeds three decades, it is very difficult to argue that short-term (2–5 year) testing of new materials is capable of revealing subtle or long-term defects in them or, more directly, of providing the information needed to determine if the new material is equal to or exceeds the performance of the older material which it may replace, simply on a novelty basis. There appears to be a Gresham's Law operating in the development of medical and surgical devices and their materials through which novelty has a market value. The drive to use new materials is depriving patients of the proven performance of older ones. In a free market system which provides many benefits and maximizes individual freedom, it is hard to see how such a situation can be corrected. Education of both physicians and patients, more sophisticated regulatory approaches and economic restrictions imposed by widespread recognition of the need to curtail the growth of medical expenses may help. However, it is only through the bioengineer's endorsement of the Hippocratic injunction to "in the first place, do no harm," can we expect to avoid future problems associated with inadequate biological performance of materials in patients.

REFERENCES

American Society for Testing and Materials (1990): Standard Practice for Selecting Generic Biological Test Methods for Materials and Devices, F 748–87. In: *1990 Annual Book of ASTM Standards*, Vol. 13.01: *Medical Devices; Emergency Medical Services*. ASTM, Philadelphia, pp. 227ff.

Burdette, W. J. and Gehan, E. A. (1970): *Planning and Analysis of Clinical Studies*. C. C. Thomas, Springfield, IL.

Chen, E. H. and Black, J. (1980): J. Biomed. Mater. Res. 14:567.

Dobelle, W. H., Morton, W. A., Lysaght, M. J. and Burton, E. M. (1980): Artif. Organs 4(4):1.

International Standards Organization (1991): *Biological Testing of Medical and Dental Materials and Devices*, Part 1: *Guidance on Selection of Tests*. ISO/DIS 10993-1. ISO, Switzerland.

Kammula, R. G. (1991): In: *Biocompatibility Workshop Notebook*. P. E. Duncan and R. F. Wallin (Eds.). Society for Biomaterials, San Antonio, TX.

Kiplinger, A. (1991): The Kiplinger Washington Letter 68(20):1.

Piehler, H. R. (1978): Orthopaedic Review 7(1):75; (2):65; (3):97; (4):79; (5):99; (7):103.

Silverman, W. A. (1985): *Human Experimentation: A Guided Step into the Unknown*. Oxford University Press, Oxford.

Stark, N. J. (1991): Med. Dev. & Diag. Ind. 13(6):68.

BIBLIOGRAPHY

Armitage, P. (1975): *Sequential Medical Trials*, 2nd ed. John Wiley & Sons, New York.

Peto, R., Pike, M. C., Armitage, P., Breslow, N. E., Cox, D. R., Howard, S. V., Mantel, N., McPherson, K., Peto, J. and Smith, P. G. (1977): Brit. J. Cancer 35:1.

19

Standardization and Regulation of Implant Materials

19.1 HISTORICAL PERSPECTIVE

The manufacture and sale of drugs in the United States has been under some form of federal regulation since passage of the Wiley Act in 1897 and the first Pure Food and Drug Act of 1906. These acts, as well as subsequent ones, were adopted against a background of the sale of patent medicines with exaggerated claims and the production of food with extensive and deliberate contamination. There are many horror stories about the effects of patent medicines from the pre-Wiley Act era and the later period of weak legislation up to the 1930s (cf. Lamb [1936], Mintz [1965]). Perhaps the strongest single factor in the initiation of federal regulation of foods was the publication of *The Jungle* by Upton Sinclair (1906). This book detailed the conditions in the processed meat industry in Chicago at the time and caused widespread revulsion to and rejection of processed meats such as sausage, ham paste, etc.

Various acts of legislation involved in the regulation of food, drugs, and cosmetics brought the Food and Drug Administration into being. Although legislative authority existed to regulate implants, practical regulation did not begin until the 1970s. A series of amendments to the Food, Drug and Cosmetic Act (1923), collectively termed The Medical Device Amendments, was adopted and signed into law on May 28, 1976. These amendments gave the then recently organized Bureau of Medical Devices and Diagnostic Aids* of the FDA broad powers in regulating implants, surgical instruments and medical devices. These powers generally parallel the powers afforded in the regulation of drugs with differences in the law and the regulatory arrangements reflecting some of the differences between devices and drugs. These Amendments have been supplemented and modified to minor degrees by legislative action but underwent significant extension in 1990 by adoption of the Safe Medical Devices Act.

Numerous efforts at standardization and thus control of medical and surgical devices and materials predate these legislative efforts and continue in a supplementary and parallel fashion today.

19.2 DRUG STANDARDIZATION ACTIVITIES

19.2.1 The U.S. Pharmacopeia

The idea of standardization of drugs, and more recently of medical and surgical devices and materials, is quite an old one. The motives involved are usually the related desires to assure reproducible effect (efficacy)

---

*Now called the Center for Devices and Radiological Health.

while protecting the patient against hazards associated with adulturation, mislabeling, misuse, etc. (safety).

The first concrete effort in this area in the United States was the proposal by Dr. Lyman Spalding in January, 1817, to establish a national pharmacopeia. The pharmacopeia was seen as a document, widely agreed upon and accepted, which would set out the composition, identity, properties, and, to some extent, the clinical behavior of drugs and other medical substances which had been shown to be useful, that is, beneficial in action. In response to Dr. Spalding's proposal, the First United States Pharmacopeial Convention assembled in Washington, D.C., on January 1, 1820. The First U.S. Pharmacopeia was published on December 15, 1820, in both Latin and English. Its 272 pages listed some 217 drugs considered worthy of recognition. At that time provisions were made to hold subsequent meetings of the convention and to issue a revised pharmacopeia every 10 years.

The First United States Pharmacopeial Convention and the First Revision Committee were composed exclusively of physicians. By 1830, pharmacists had been invited to join the convention and numbers of them have continued to join over the years. The present bylaws of the United States Pharmacopeia require, however, that at least one-third of the members of the Board of Trustees and the Committee of Revision continue to represent the medical profession.

The initial policy of the United States Pharmacopeial Convention (USPC) was to select from among substances which possess medicinal power, those which are the most fully established and best understood. Over the years and through its various revisions, this principle has been adhered to. The last independent version, *The Pharmacopeia of the United States of America* (USP XIX [1975]), was the nineteenth revision, published subsequent to the United States Pharmacopeial Convention of April, 1970. It contains 1,284 articles describing a somewhat lower number of drugs and other medical agents. Implants and other medical devices are not discussed in USP XIX with two exceptions.

The first and more important of these exceptions is that provision is made for the definition and testing of glass and plastic containers for drugs. The methods of test for containers that are outlined in USP XIX are:*

Light transmission.
Chemical resistance (glass containers).

*USP XIX, p. 642.

Biological tests (plastic containers): Injection of extracts and examination of 72-hour implants in rabbits and mice.

Physiochemical tests (plastic containers): Extraction, residue identification, residue ignition, heavy metal content, and buffering capacity.

The other medical device which is described is the absorbable surgical suture. This is the so-called "catgut" suture, although the basic material is now derived from other sources. USP XIX sets out methods of test and standards for length, diameter, tensile strength, content of soluble chromium compounds, and color of extracts, as well as describing methods of needle attachment for these sutures.

While neither of these device areas is directly applicable to implant materials, many of the methods, particularly those used to qualify container materials, have been used extensively by biomaterials investigators.

Of interest is the provision for the use of a standard implant reference material for evaluation of the 72-hour animal tests. The material is a low-molecular-weight polyethylene fiber that can be inserted through a hypodermic needle. It is stocked in a supply maintained by the USPC.

In 1974, the USP and the National Formulary (NF) (see next Section) were combined. The current edition (USP XXII [1989]) continues the USP series as the 22nd revision, and includes the 17th revision of the NF. While they are now published together, an internal distinction is maintained, with the USP articles addressing drug composition and dosage (as well as general issues of testing and packaging) and NF articles dealing with pharmaceutical ingredients other than drugs. The combined USP/NF is now enlarged by annual supplements and by a bimonthly magazine, Pharmacopeial Forum. The rate of growth can be appreciated by noting the addition of 389 USP and 31 NF articles in the 22nd revision.

The current revision, USP XXII (1989), in addition to continuing the nonbiological tests of previous revisions, now lists a total of eight host response tests (Table 19.1). It is of great interest that the in vitro test methods now cite ASTM standards as references. The last test listed, eye irritation, is the well known Draize test which uses the albino rabbit eye, in vivo, as a test site for topical agents and extracts. Popular revulsion against this test method is responsible for some of the support of so-called "animal rights" organizations.

19.2.2 The National Formulary

As mentioned above, another compilation of drugs and their properties is the *National Formulary* (NF), which first appeared in 1888. This is prepared and published by the American Pharmaceutical Association, which

Table 19.1 Host Response Test Methods in USP XXII

General article ⟨87⟩: biological reactivity tests, in vitro:
  Agar diffusion
  Direct contact
  Elution
General article ⟨88⟩: biological reactivity tests, in vivo:
  Acute (iv) toxicity
  Systemic injection
  Intracutaneous injection
  Implantation
  Eye irritation

Source: USP XXII, pp. 1495ff.

was organized in 1852. The stated goals of the NF are similar to those of the USP, with the exception that not only the drugs of the greatest therapeutic merit, but drugs of any demonstrated merit are to be included. This factor and the domination of the National Formulary by the manufacturers rather than the users of drugs, leads to a somewhat different format and emphasis. The National Formulary was originally published at 10-year intervals, and more recently in 5-year intervals. The last independent edition (see previous Section) was the 14th edition (NF XIV [1975]) and includes 1009 articles defining and describing a somewhat greater number of drugs and medical materials.

NF XIV describes materials, in the non-drug sense, in only two areas. It makes provisions for examination and qualification of glass containers for drug packaging that are similar to and depend upon USP XIX provisions. In addition, special provisions are made for qualification of containers for ophthalmic preparations. These provisions include an eye-irritation test, previously mentioned, using saline and cottonseed oil extracts in the eye of the albino rabbit.

19.3 BIOMATERIALS STANDARDIZATION ACTIVITIES

19.3.1 The American Dental Association

A number of efforts have been made in the standardization of biomaterials (in the sense of materials without *primary* pharmacological effects). Since 1926, the American Dental Association (ADA) has sponsored and conducted a program to define the physical and chemical properties of materials used in restorative dentistry. Over the years a large number of specifications have been developed for various metal alloys, cements, impression materials, casting and investment waxes, plaster, resin, and elastomeric products, as well as cutting instruments and equipment for radiation diagnosis and therapy. In addition to these specifications, the ADA maintains a program to certify specific dental material products and manufacturers of these certified products. The results of this program, carried out through a cooperative effort with the National Bureau of Standards (now the National Institute for Standards and Technology [NIST]), was a periodic publication entitled *Guide to Dental Materials and Devices*, mostly recently published in a seventh edition (ADA, 1974). This contains a great deal of technical information as well as some 25 specifications. Today there are about fifty standards; however, they may be obtained only by individual purchase, either from the ADA or from the American National Standards Institute, Washington, DC. Table 19.2 lists materials and materials test standards.

19.3.2 The American Society for Testing and Materials

An effort of some greater generality is that on the part of the American Society for Testing and Materials (ASTM). This organization was founded in 1898 and is the principal scientific and technical organization for the voluntary development of standards on characteristics and performance of materials, products, systems, and services in the United States. It performs its work through 135 main technical committees with more than 32,000 active members.* These committees function in prescribed fields under regulations which insure balanced representation by producers, users, and general interest participants.

In 1962, the Committee F4 on Medical Devices was organized (Brown and Cook, 1982). More recently, this Committee was reorganized and renamed the Committee F4 on Medical and Surgical Materials and Devices. It includes within its organization a resources subcommittee with individual sections devoted to specific materials classes such as polymeric materials, metallurgical materials, etc., as well as biocompatibility, and a series of subcommittees in various surgical specialties such as orthopedics, cardiovascular surgery, neurosurgery, etc. The division of areas addressed by these latter medical subcommittees parallels that of the Device Classification Panels established by the Food and Drug Administration subsequent to the passage of the Medical Device Amendments (1976).

The scope of this committee is the development of definitions of terms and nomenclature, methods of test, specifications and performance re-

---

*1990 Annual Book of ASTM Standards, Vol. 13.01. ASTM, Philadelphia (1990), p. iii.

Table 19.2 ADA Dental Biomaterials Standards and
Recommended Standard Practices of Testing for Host Response

Biomaterials
  3 Dental impression compound
  5 Dental inlay casting wax
  6 Dental mercury
  7 Dental wrought gold wire alloy
  8 Dental zinc phosphate cement
  9 Dental silicate cement
 11 Dental agar impression material
 12 Dental base resin
 13 Dental cold-curing repair resin
 14 Dental base metal casting alloys
 15 Acrylic resin teeth
 16 Dental impression paste: zinc oxide-eugenol material
 17 Denture base temporary relining resin
 18 Dental alginate impression material
 19 Elastomeric dental impression material
 21 Dental zinc silico-phosphate cement
 22 Intraoral dental radiographic film
 27 Direct filling resins
 30 Zinc oxide-eugenol type restorative materials
 32 Orthodontic wires not containing precious metals
 37 Dental abrasive powders
 57 Endodontic filling materials
 61 Zinc polycarboxylate cement
 66 Glass ionomer cements
Recommended standard practices
 41 Biological evaluation of dental materials & 4la, addendum

Source: American Dental Association.

quirements for medical and surgical materials and devices. By 1990, the committee had adopted and approved, through Society vote, 41 biomaterials specifications and 13 methods of test for biological response, as well as device standards and other methods of test. The biomaterials specifications are consensus documents which describe the results of present practice in the fabrication of these materials. The methods of test, in that they are adhered to and the incorporated standardized materials are used as reference materials, can be considered as standard tests, within the meaning of Section 1.2. The 54 specifications and methods of test are listed in Table 19.3.

Table 19.3 ASTM Biomaterials Standards and Methods of Testing for Host Response

Biomaterials:

F 55-82	Specification for Stainless Steel Bar and Wire for Surgical Implants
F 56-82	Specification for Stainless Steel Sheet and Strip for Surgical Implants
F 67-89	Specification for Unalloyed Titanium for Surgical Implant Applications
F 75-87	Specification for Cast Cobalt-Chromium-Molybdenum Alloy for Surgical Implant Applications
F 86-84	Practice for Surface Preparation and Marking of Metallic Surgical Implants
F 90-87	Specification for Wrought Cobalt-Chromium-Tungsten-Nickel Alloy for Surgical Implant Applications
F 136-84	Specification for Wrought Titanium 6Al-4V ELI Alloy for Surgical Implant Applications
F 138-86	Specification for Stainless Steel Bar and Wire for Surgical Implants (Special Quality)
F 139-86	Specification for Stainless Steel Sheet and Strip for Surgical Implants (Special Quality)
F 451-86	Specification for Acrylic Bone Cement
F 500-77	Specification for Self-Curing Acrylic Resins Used in Neurosurgery
F 560-86	Specification for Unalloyed Tantalum for Surgical Implant Applications
F 562-84	Specification for Wrought Cobalt-Nickel-Chromium-Molybdenum Alloy for Surgical Implant Applications
F 563-88	Specification for Wrought Cobalt-Nickel-Chromium-Molybdenum-Tungsten-Iron Alloy for Surgical Implant Applications
F 602-87	Criteria for Implantable Thermoset Epoxy Plastics
F 603-83	Specification for High-Purity Dense Aluminum Oxide for Surgical Implant Application
F 604-87	Classification for Silicone Elastomers Used in Medical Applications
F 620-87	Specification for Titanium 6Al-4V ELI Alloy Forgings for Surgical Implants
F 621-86	Specification for Stainless Steel Forgings for Surgical Implants
F 639-85	Specification for Polyethylene Plastics for Medical Applications
F 641-86	Specification for Implantable Epoxy Electronic Encapsulants
F 642-84	Specification for Stainless Steel Flexible Wire for Surgical Fixations for Soft Tissue
F 643-84	Specification for Wrought Cobalt-Chromium Alloy Flexible Wire for Surgical Fixations for Soft Tissue

Table 19.3 *(Continued)*

F 644-84	Specification for Wrought Cobalt-Chromium Alloy Flexible Wire for Surgical Fixations for Bone
F 648-84	Specification for Ultra-High-Molecular-Weight Polyethylene Powder and Fabricated Form for Surgical Implants
F 665-86	Classification for Vinyl Chloride Plastics Used in Biomedical Application
F 666-80	Specification for Stainless Steel Flexible Wire for Surgical Fixation for Bone
F 688-88	Specification for Wrought Cobalt-Nickel-Chromium-Molybdenum Alloy Plate, Sheet, and Foil for Surgical Implants
F 702-81	Specification for Polysulfone Resin for Medical Applications
F 745-81	Specification for Stainless Steel for Cast and Solution-Annealed Surgical Implant Applications
F 754-88	Specification for Implantable Polytetrafluoroethylene (PTFE) Polymer Fabricated in Sheet, Tube, and Rod Shapes
F 755-87	Specification for Selection of Porous Polyethylene for Use in Surgical Implants
F 799-87	Specification for Thermomechanically Processed Cobalt-Chromium-Molybdenum Alloy for Surgical Implants
F 881-84	Specification for Silicone Gel and Silicone Solid (Nonporous) Facial Implants
F 961-85	Specification for Cobalt-Nickel-Chromium-Molybdenum Alloy Forgings for Surgical Implant Applications
F 988-86	Guide for Specifying Carbon Fiber Randomly Reinforced Ultra-High-Molecular-Weight Polyethylene for Medical Devices
F 997-86	Specification for Polycarbonate Resin for Medical Applications
F 1088-87	Specification for Beta-Tricalcium Phosphate for Surgical Implantation
F 1108-88	Specification for Ti6Al4V Alloy Castings for Surgical Implants
F 1109-87	Specification for Porous Composites of Polytetrafluorethylene and Carbon for Surgical Implant Applications
F 1185-88	Specification for Composition of Ceramic Hydroxylapatite for Surgical Implants

Methods of Test for Host Response:

F 361-80	Practice for Assessment of Compatibility of Metallic Materials for Surgical Implants with Respect to Effect of Materials on Tissue (replaced by F 981-87)
F 469-78	Practice for Assessment of Compatibility of Nonporous Polymeric Materials for Surgical Implants with Regard to Effect of Materials on Tissue (replaced by F 981-87)
F 719-81	Practice for Testing Biomaterials in Rabbits for Primary Skin Irritation

Table 19.3 *(Continued)*

F 720-81	Practice for Testing Guinea Pigs for Contact Allergens: Guinea Pig Maximization Test
F 748-87	Practice for Selecting Generic Biological Test Methods for Materials and Devices
F 749-87	Practice for Evaluating Material Extracts by Intracutaneous Injection in the Rabbit
F 750-87	Practice for Evaluating Material Extracts by Systemic Injection in the Mouse
F 756-87	Practice for Assessment of the Hemolytic Properties of Materials
F 763-87	Practice for Short-Term Screening of Implant Materials
F 813-83	Practice for Direct Contact Cell Culture Evaluation of Materials for Medical Devices
F 895-84	Test Method for Agar Diffusion Cell Culture Screening for Cytotoxicity
F 981-87	Practice for Assessment of Compatibility of Biomaterials (Nonporous) for Surgical Implants with Respect to Effect of Materials on Muscle and Bone
F 1027-86	Practice for Assessment of Tissue and Cell Compatibility of Orofacial Prosthetic Materials and Devices

Source: *1990 Annual Book of ASTM Standards*, Vol. 13.01. ASTM, Philadelphia (1990), pp. x–xii.

19.3.3 Other Efforts

A number of other organizations have entered into the specification and standardization of medical materials and devices. However, none are so far advanced in their efforts as the ADA and the ASTM. Perhaps the best known of the remainder of these organizations is the Association for Advancement of Medical Instrumentation (AAMI). This organization has been involved for a number of years in developing specifications for active medical devices such as heart pacers and neurostimulators. A number of these specifications are coming into general use.

Outside the United States, other countries have made progress in this field. Some, like Canada, have decided to follow the progress of American groups such as the ASTM and ADA. As standards have been adopted as American National Standards (by the American National Standards Institute [designation: ANSI]), they are also being adopted, after review, as Canadian standards. For instance, all of the ADA standards listed in Table 19.2 are now ANSI standards, as are many ASTM standards. Some countries, such as Germany, France and England, have developed, relatively independently, their own national standards for implant materials and devices. An international standards making group, the International Standards Organization (ISO), has organized two committees with broad international representation:

ISO TC 150, which is evaluating national device and material standards and attempting to adopt common international versions. This effort is underway in the fields of orthopedic, cardiovascular, and neurosurgery and will eventually spread to encompass all medical disciplines.

ISO TC 194, which is conducting similar activities in the area of measurement of biological response.

The ISO standards, subject to approval by the European Commission, will supplant or override national standards within the European Common Market (after 1992) and will become, as a result, de facto standards for firms involved in international trade in medical and surgical materials and devices.*

Trade associations such as the Orthopaedic Surgical Manufacturers Association (OSMA) and the Health Industry Manufacturers Association (HIMA) have taken a part in developing standards, either on their own or through activity of their representatives in standards organizations such as ASTM, ISO, etc. Traditional professional organizations in both the health and engineering professions have standards committees that act as focal points for technical input into standards preparation by standards making organizations.

19.4 U.S. FEDERAL REGULATION OF MEDICAL DEVICES AND BIOMATERIALS

19.4.1 Medical Device Amendments (1976)

The various efforts described in previous Sections were not perceived to be sufficient to provide safe and effective medical and surgical materials and devices for the public and to control unsafe and ineffective materials and devices. The result, in the U.S., was a legislative mandate for the executive branch of government to control and regulate medical device manufacturing in the public interest. The initial chosen legislative tool was the Medical Device Amendments (1976) whose overall goal is to assure the safety and efficacy of devices. This legislation provides for

*For a detailed idea of how these various standards and specifications interact generically with the process of Federal regulation of medical materials and devices, the reader is referred to *Everything You Always Wanted to Know about the Medical Device Amendments . . . and Weren't Afraid to Ask*, published by the Department of Health, Education, & Welfare, Food and Drug Administration, Bureau of Medical Devices, October, 1977.

classification of devices, which we will consider later in this Chapter. They also lay out a scheme of general controls including provisions for dealing with adulterated and misbranded devices, for registering device types and device manufacturers, for premarket notification of the introduction of new devices, and for dealing with banned devices. Details of manufacturers' obligations to repair, replace, or refund in the case of defective devices are also included. The amendments also permit the establishment of regulations to define "good manufacturing practices," and to establish performance standards as well as premarket product development protocols for new materials and devices.

If the Medical Device Amendments can be said to have a central theme, it is one which closely parallels the ideas of the safety and effectiveness of drugs, cosmetics, and food additives. That is, the legislation foresaw a pattern in which materials and devices would be developed, tested, and demonstrated to be safe and efficacious before being offered for sale. Once this point is reached, their future safety and effectiveness would then be controlled by the institution of general standards and controls, or by the provisions of specific performance standards.

19.4.2 Safe Medical Devices Act (1990)

The Safe Medical Devices Act (1990) is the first major revision of the Medical Device Amendments and occurred largely as a result of public dissatisfaction with the implementation (rather than the content) of medical and surgical device regulations. The many provisions of the Act (Kahan, 1991) are intended largely to strengthen, streamline, better define and speed up regulatory activities. Thus, the provisions primarily enlarge on and modify rather than replace those of the earlier Medical Device Amendments. However, several new provisions are introduced, including life-time tracking of permanently implanted life-supporting or life-sustaining devices, more and improved reports of life-threatening device malfunction, regulatory authority to order mandatory recalls and allow seizure of defective devices, rules for postmarket introduction surveillance of experience with permanent implants and creation of a "humanitarian device exemption," similar to an "orphan" drug provision within drug regulation, to simplify and reduce the cost of development of devices for diseases or conditions affecting fewer than 4000 individuals in the US.

Most of the provisions of the Safe Medical Devices Act will be introduced over a three-year period. Thus, it will be some time before their full implications are understood (Kahan et al., 1991).

19.5   REGULATION OF MATERIALS FOR IMPLANTS

19.5.1   Requirements of the U.S. Federal Regulations

The need for regulatory standards arises from the requirements of both the Medical Device Amendments (1976) and the Safe Medical Devices Act (1990). These are embodied in a classification system that attempts to distinguish between various generic types of devices on the basis of risk to the patient.

The general pattern of device classification is as follows. Devices are classified into three categories by one of 14 specialty-oriented Device Classification Panels. The three categories are:

Class I, General Controls. A device for which controls other than standards and premarket approval are sufficient to assure safety and effectiveness.

Class II, Performance Standards. A device for which general controls are insufficient to assure safety and effectiveness but for which there is sufficient information for the establishment of a performance standard to provide such assurance.

Class III, Premarket Approval. A device for which insufficient information exists to assure that general controls and performance standards would provide reasonable assurance of safety and effectiveness, and which is represented to be either life-sustaining, life-supporting, or implanted in the body, or which presents a potential unreasonable risk of illness or injury.

Devices classified as Class III, and any new device which comes on the market after May 28, 1976, must pass through some form of scientific premarket review before market introduction. At the point where these products are judged to be reasonably safe, effective, and controllable by a performance standard, they may be reclassified into Class II. Thus, the existence of standards can be seen to be critical to the introduction of new materials and devices into general use. It is hoped that many of the voluntary standards developed by the various organizations mentioned here, as well as others, can be adapted to be regulatory standards.

To date a very few regulatory standards have been approved for devices and none for materials. However, many permanent implants have been reclassified from Class III to Class II, on the basis of long pre- and post-enactment experience. The need for regulatory standards, particu-

larly for materials of construction, will become more acute as more devices achieve such reclassification.

19.5.2 Voluntary vs. Regulatory Standards

It should be clear to the reader that a voluntary standard and a regulatory
standard are not necessarily the same thing. Voluntary standards, whether
consensus derived or otherwise, are designed to describe the content,
design, construction, and performance of existing devices, as well as to
set forth methods of verifying compliance with these aspects of the standard. Regulatory standards are, by their nature, designed to regulate, that
is, to assure specific attributes of products. In the case of the (regulatory)
performance standards required by the Medical Device Amendments
(1976), the specific attributes to be regulated are "safety and efficacy."
It is not clear at this time what combination of standards on content,
design, and construction, as well as simulation tests in vitro and in vivo,
are necessary and sufficient to meet such a general requirement for (in
vivo) patient performance. Thus, while it is possible to prepare a voluntary standard for an existing device, the relationship of this standard to a
future, generic regulatory standard is tenuous, at best.

It seems safe to presume that as genuinely new materials come into use,
they will be qualified in a similar way to that envisaged specifically for
devices. In the course of selection and development, they pass through
series of tests along the lines of those discussed in Section 18.2.2. As
their behavior becomes better understood and devices incorporating them
pass into clinical trials, the process of voluntary standards preparation
begins. The results of these tests and the proposed form of the standard
then constitute the body of material, with supporting clinical reports, that
can be submitted for initial review by a regulating agency such as the
FDA.

In that case, the use of the words "performance standard" in the 1976
enabling legislation harmonizes well with the ideas of biological performance laid out here in earlier chapters. Thus, a performance standard for
an implant material is one that describes the chemical, physical, and
processing requirements for a material that are needed to assure a reproducible level of biological performance, as well as to meet the engineering requirements, of a proposed application. As soon as it is possible to
prepare such a document, then the material becomes, in an important
sense, a "known" material. Its biological performance can be examined
objectively and its suitability (and admissibility) for specific applications
can be determined. At this point it should pass easily into Class II and be a

natural competitor for use in specific medical and surgical device applications.

REFERENCES

American Dental Association (1974): *Guide to Dental Materials and Devices*, 7th ed. ADA, Chicago.

American Society for Testing and Materials (1990): *1990 Annual Book of ASTM Standards*, Vol. 13.01: *Medical Devices*. ASTM, Philadelphia.

Brown, P. and Cook, A. G. (1982): ASTM Standardization News, October:10.

Department of Health, Education, and Welfare (1977): *Everything You Always Wanted to Know about the Medical Device Amendments . . . and Weren't Afraid to Ask*. Food and Drug Administration. Bureau of Medical Devices, Washington, DC.

Kahan, J. S. (1991): Med. Dev. & Diag. Ind. 13(1):66.

Kahan, J. S., Holstein, H. and Munsey, R. (1991): Med. Dev. & Diag. Ind. 13(2):44.

Lamb, R. DeF. (1936): *American Chamber of Horrors: The Truth about Food and Drugs*. Farrar & Rinehart, New York.

Mintz, M. (1965): *The Therapeutic Nightmare*. Houghton-Mifflin Co., Boston.

NF XIV (1975): *The National Formulary*, 14th ed. American Pharmaceutical Association, Washington, DC.

Sinclair, U. (1906): *The Jungle*. New American Library, New York.

USP XIX (1975): *The Pharmacopeia of the United States of America*, 19th revision. The United States Pharmacopeial Convention, Inc., Washington, DC.

USP XXII (1989): *The Pharmacopeia of the United States of America*, 22nd revision, incorporating: *The National Formulary*, 17th revision. The United States Pharmacopeial Convention, Inc., Washington, DC.

U.S. Congress (1976): *Medical Device Amendments*, PL 94-295, 21 USC 301.

U.S. Congress (1990): *Safe Medical Devices Act*, PL 101-629.

BIBLIOGRAPHY

Department of Health, Education, and Welfare (1972): *Federal Food, Drug, and Cosmetic Act, as Amended*. Food and Drug Administration. U.S. Government Printing Office, Washington, DC.

Department of Health and Human Services (1989): *Medical Devices: Standards Activities Report*. Public Health Service. Food and Drug Administration. Center for Devices and Radiological Health, Rockville, MD.

Health Industry Manufacturers Association (1978): *Guideline for Evaluating the Safety of Materials Used in Medical Devices*, Report No. 78-7. Health Industry Manufacturers Association, Washington, DC.

Health Industry Manufacturers Association (1979): *Guidelines for the Development of Voluntary Device Law Standards*, Report No. 79-6. Health Industry Manufacturers Association, Washington, DC.

Morton, W. A. and Veale, J. R. (1987): *Regulatory Issues in Artificial Organs: A Primer*. ASAIO Primers in Artifical Organs, No. 1. J.B. Lippincott, Philadelphia.

20

Design and Selection of Implant Materials*

---

*Portions of this chapter appeared in an earlier form as Chapter 13 in Black (1988) and are reproduced by permission (Churchill Livingstone, Inc.).

20.1 INTRODUCTION

20.1.1 What Is Design?

Design is what engineers do: apply scientific knowledge to the solution of practical problems. The object of their design may be a process, a new material or a novel device. The process is artistic and creative, drawing from the same well that the painter, sculptor or writer does. What distinguishes the objects of engineering design from those of other artistic activities is the extent to which technological factors come into play in their realization (Asimow, 1962).

As Cross (1989) points out, the separation between design and fabrication of man-made artifacts is a relatively recent event. When hand artisanry was the rule, there was no separation between design and fabrication: the maker designed as the final form of the artifact emerged. For the artist, there is still no separation in function: the design is the object. For the surgeon, the separation is incomplete: while surgical procedures are planned, detailed and complex decisions are made during the performance of the operation itself. However, for the engineer the separation has become nearly total. Today those who design rarely make and vice versa.

The separation between those who design and those who make has led to vocational self-selection which produces significant problems for engineers involved in design. Engineering has become a linear analytical process, seeking the shortest distance to a solution. Engineers thus have great difficulty in dealing with the creative, synthetic aspects of design which require attempts to devise as many alternative solutions as possible.

The creative aspects of design must be emphasized but even more so the concept of a *design process*. Solutions to engineering design problems rarely, if ever, spring full-blown from the mind of their creator. On the contrary, what is required is a systematic, dogged, iterative process stretching from exploration of the initial requirements to evaluation of the preferred solution. In a sense, the design process and its necessary iterative design cycle represent attempts of engineering designers to deal with synthetic problems in an analytic fashion, rather in the manner of using digital computers to create analog models of systems.

20.1.2 Introduction to the Orange

Engineers unfamiliar with design often have the same problem as a beginning art student: faced with a blank sheet of paper and an overall conception, they have no idea where to start. A useful exercise is a consideration of an orange in a process often referred to as *reverse engineering*. That is, given an object or finished artifact, one attempts to understand its rationale and determine details of the materials and processes of its construction retrospectively rather than design it prospectively.

In this case, the use of an actual orange is a useful aid. The exercise proceeds in three steps:

1. From observation and physical examination, make a list of all the things which it is possible to know about an orange. In doing so, one begins, usually, with simple attributes, such as color, weight, size, etc. and moves to more complex ones, such as shape, number of seeds, amount of sugar contained, etc.
2. For as many as possible of the quantitative attributes listed in 1 above, estimate the value. The benefit of this step is particularly seen when a number of individuals do the exercise separately or in groups and then compare their answers.
3. For as many as possible of the attributes listed in 1 above, propose as many methods as possible for finding their true or actual value.

These three steps will help to prime the creative pump, as it were. They replicate, in form, steps 1, 2 and 4 of the design cycle (see Section 20.2.2), respectively. The first step teaches observation, the second estimation and the third, creation of alternatives.

This exercise also can be used to illustrate another point about design: it is better played as a team sport than as solitaire. This point may easily be demonstrated by setting the ''orange exercise'' for both an individual and a group to perform: the members of the group, no matter what its makeup, will always be more productive on average, let alone collectively, than the individual. Design can and often is performed by a single individual. However, it is far more productive if it is a group project and the resulting synergy increases in proportion to the variety of people involved.

20.2 THE DESIGN PROCESS

20.2.1 The Phases of Design

Asimow (1962) defines design as a seven phase process arising from a primitive need (Table 20.1).

Materials design and selection, in the sense in which they are discussed in this chapter, fall within Asimow's phases I–III, depending upon the depth and detail required of the design process. In each instance, a structured design process (see next Section) is desirable. Device design is

Table 20.1   Seven Phases of Design

Primitive need → Preliminary phases:
I   Feasibility design
II   Preliminary design
III   Detailed design
Phases related to production/consumption cycle:
IV   Planning for production
V   Planning for distribution
VI   Planning for consumption
VII   Planning for retirement

Source: Asimow (1962).

more likely to have to deal with all seven phases. In either case, a single phase may require a number of design cycles within its process.

Design may be required for the development of manufacturing processes and design of devices, of surgical procedures and even of experiments. With suitable modifications, the same structured process may be utilized.

20.2.2   The Design Cycle

A structured design process consists of the consecutive execution of a repetitive design cycle. The design of a simple device, such as a tongue depressor and its dispensing container, may be achieved in as little as three or four such cycles, while a complex design, such as a powered wheel chair, may require hundreds of such cycles, some in parallel and others in series.

The design of materials is a simpler problem, in general, than the design of devices. Materials design is rarely addressed directly since device designers tend to view themselves as expert in materials and to assume that materials design is merely a matter of selection from among the options available. This approach is illustrated by Lewis' (1990) discussion of the design of a femoral medullary stem for a total hip replacement prosthesis.

It is true that the selection of materials, with and without modification, has until recently dominated biomaterials design. It is possible to use a formal design process for selection and/or modification of materials but this may seem clumsy and unwarranted except for teaching purposes.

However, today the advancing popularity of composite materials, or more properly, *engineered* materials, makes necessary the use of a design process for the prospective selection of biomaterials properties for medical and surgical devices. The design of materials may require two or more cycles in series: first, selection of materials properties; then selection of processing methods and parameters, followed by consideration of the interaction of various biomaterials selected. These cycles cannot take place in isolation from the considerations involved in the design of the device (for which the biomaterials are being designed/selected) since device requirements impose materials requirements and materials selections affect design choices.

There are a number of models which may be utilized to develop a design cycle. The approach in this chapter is derived from that of Love (1986) and is shown in schematic form in Figure 20.1. The next section is devoted to a step-by-step discussion of this cycle.

20.2.3 Steps in Design

20.2.3.1 Beginning the Design

Design, within a given cycle, arises from a primitive need. This need may be an external statement (if the cycle is the first in the process) or may be the output of a previous cycle. In the example to be used in this Chapter, we will take as the primitive need this possible statement by a product salesman for an orthopaedic implant company:

> My customers are interested in a better total hip replacement (THR) system for younger patients.

The engineering design group takes up the challenge and develops a concept for a novel femoral component. However, the group reports that none of the biomaterials in their handbooks and reference sources provide the appropriate combination of stiffness, strength and fatigue life that they require to realize the design. This finding becomes the input which begins the material design cycle to be examined.

20.2.3.2 Step 1: Analyzing Needs

The first formal step in the design cycle is to examine the input statements, often in collaboration with those who made them, and to develop an objective summary statement which both expresses the needs in an analytic way and represents a rational objective.* For example, development of copper with a higher melting point is an irrational objective while

---

*Discussion of the design and conduct of experiments is outside the scope of this work. However, the reader should note that this requirement is identical to the statement of an experimental question.

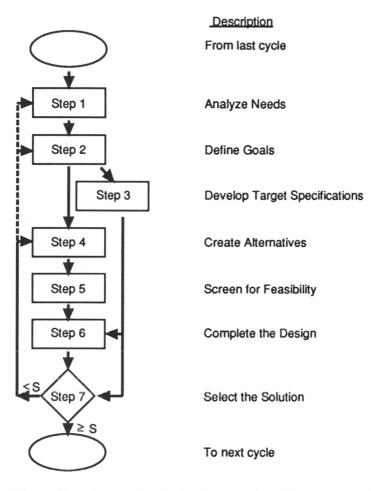

Figure 20.1 The Design Cycle. Source: adapted from Love, 1986.

seeking a higher strength to modulus ratio in the copper-silver binary alloy system is a rational one.

In this case, after careful consultation and deliberation, the design objective is stated as:

> The objective is the design of a new material suitable for use in fabrication of THR components which combines optimum stiffness (modulus) with greater strength and a higher endurance limit than presently available materials possess.

Note that the act of stating the objective limits the enquiry: a new material shall be designed rather than a present one modified. It also completes the translation of the primitive need to a defined materials need, with three attributes: optimum (to be defined) modulus, increased strength and higher fatigue endurance. In the same way that an experimental question (and its hypotheses) can be tested, this statement can be tested at the end of the cycle to see if the objective has been realized.

20.2.3.3 Step 2: Defining Goals

Design is not a simple process that leads to a unique output. Thus, objectives must be refined to limit the number of choices at each step and to guide the design cycle. This is achieved by selecting goals whose attainment either: (1) is necessary to reach the desired objective; or (2) represents generally "good" attributes of engineering design or reflects the desires of the designers. The first type are called *specific* goals (*demands* [Cross, 1989]) while the second type are called *general* goals (*wishes* [Cross, 1989]). An initial list of specific and general goals which might follow from the previously stated summary objective statement is given, under the heading "initial," in Table 20.2.

The initial list of goals arises from a discussion with the "customer," in this case the device design group, and represents the definitive starting point of the material design cycle. It embodies the customer's concepts of what is wanted as an end product of the design process. The material design team must now put its talents to work to understand these desires and to satisfy them. Some of the initial goals may come from other sources: the engineering manager is always worried about manufacturing costs; the color was suggested by the sales manager since yellow is widely used in the company's packaging and has come to be identified with it in the mind of the retail customer, the surgeon.

The material design team must actually go through two substeps to produce the refined set of goals shown in the lower part of Table 20.2. The first of these substeps, the production of an initial set of goals, is the first creative act in the design cycle. The question posed at this point is, "What should a new material for a THR component look like?"

As previously noted, the creation of ideas is not an easy process for engineers. There is a tendency to "freeze," either to be unable to produce ideas, or more commonly, to have an initial thought and then to proceed to develop it without consideration of further alternatives. The general solution for the individual designer is to produce a situation which is both stimulatory and nonself-critical.*

---

*In this section, I refer to a singular designer. In Section 20.2.3.5, I will consider the situation of creative effort by a group.

Table 20.2 Design Goals: New THR Material

Initial
  Specific
    Modulus <0.5 times Ti6Al4V
    Strength as high as possible
    Endurance limit as high as possible
    Corrosion/release rate "low"
    No wear against UHMWPE[a]
    Color: yellow
  General
    Minimum cost
    No limit on source of supply
    Simplicity of fabrication
Refined
  Specific
    Modulus <0.5 times Ti6Al4V (H)
    Strength as high as possible (M)
    Endurance limit as high as possible (H)
    Corrosion/release rate as low as possible (H)
    Wear rate (against UHMWPE) as low as possible (M)
    Color: yellow (L)
    Formability in the operating room (M)
    Release of wear particles >25 μm in size only (H)
  General
    Minimum cost per kilogram (L)
    No limit on source of supply (M)
    Simplicity of fabrication (M)

[a]Ultrahigh molecular weight polyethylene

In this case, the designer may decide, "I'm going to set the problem aside, go for a 5 kilometer run, and when I come in, write down the first ten things which come into my head." Such a procedure, with variants, has been adopted by most creative persons and is sometimes referred to as *creative avoidance* of the problem: undertaking other activities to distract the conscious mind (probably the analytic left brain function) and using the products of subconscious deliberation (probably the synthetic right brain function), without self-criticism or censoring.

The initial list is then reviewed for reasonableness and duplication and perhaps the process is repeated or extended, until there is a sense that all of the immediately possible options, in this case, design goals, have been acquired. Often the review "triggers" new ideas not previously considered. The designer, in this example, has added two specific goals and no general goal to those previously cited desires. Note that this is an abbreviated example; step 2 of an actual design cycle might produce dozens of specific and general goals.

The second substep is the assignment of a priority to each of these initial goals, to produce a set of refined goals. This is necessary since, in an actual design case, the number of goals very rapidly grows to a point where it is obvious, a priori, that they cannot all be met simultaneously. Thus a ranking of relative importance is necessary. In this case, the designer employed a common practice and selected three levels of priority:

High (H): Must be met for successful design.
Medium (M): Would like to meet during design cycle.
Low (L): Desirable to meet but may be sacrificed.

Therefore, the material's modulus is identified as a much more important attribute than its color, although the desire to satisfy the sales manager is still considered as part of the later steps in the cycle. In a more subtle distinction, it is recognized that in the intended application, the endurance limit is a more important material attribute than the tensile strength, although both are important. During this substep, the sets of goals are also screened to eliminate absolute statements: statements of goals should not include the terms "never, always, none, etc."

20.2.3.4   Step 3: Developing Target Specifications

Setting specific and general goals, then refining the list and assigning priorities produces considerable clarification of the problem in hand but does not provide details necessary for later steps in the design cycle. To achieve this it is necessary to translate the refined goals of step 2 into measurable quantities.

These measurable quantities are called *specifications*. They must be *necessary*, thus not setting limits unrelated to performance. They must also be *sufficient*: taken as a group, their satisfaction must be sufficient to produce a successful design and to assure that the formal requirements of the Medical Device Amendments (U.S. Congress, 1976) of *safety* and *efficacy* are met. Finally, they must be *conservative*: setting too high values or too stringent criteria will elevate cost unacceptably or possibly make the design unrealizable.

Consider the refined goal (Table 20.2):

Corrosion/release rate as low as possible (H).

This might be translated into these specifications:

Corrosion rate in vivo shall not exceed 0.1 mg/cm^2/year.
Release rate in vivo shall not produce a concentration of products > 5 ppb at a distance of 1 cm from the implant-tissue interface.

The attainment of minimum corrosion/release was judged to be of high importance; this is reflected in the use of design margins, multipliers of minimum values. In practice, these may vary between 10 (for extremely critical attributes) and 1.1 (for low importance or optional attributes). In this case, the actual allowable values might have been 1 mg/cm^2/year and 50 ppb respectively but were reduced by application of a 10X design margin. There are no objective criteria for deriving design margins: they reflect current practice in similar designs.

20.2.3.5 Step 4: Creating Alternatives

Development of design concepts or alternative possible solutions is the heart of the design process and the point where most fatal mistakes are made. It requires, again, a suspension of self-criticism and a source of external stimulation. For most people, concepts and alternatives evolve more readily in a group situation in which one person's ideas trigger another's imagination. This is the time in the design process when the prior formation of a multidisciplinary design team really pays off. The goal is the same as in step 2: developing as many independent approaches to realization of the design goals as possible.

As an introduction to this step, the material's designer begins to gather supporting information on past and present materials used in THR prostheses components as well as on current progress in materials' design and processing. Information acquired at this time serves the subconscious as a source of ideas and, if a design team is in existence, all members should have access to this resource information. This information will also be needed for the next step.

The primary tool in creating alternatives is the brainstorming or "blue sky" meeting. Cross (1989) lists the following essential rules for a such a session:

No criticism during the session.
A large number of ideas is wanted.
Seemingly crazy ideas are welcome.
Keep all ideas short and snappy.
Try to combine and improve on the ideas of others.

Love (1986), citing William J. Osborne, provides many practical suggestions on how to organize and run a successful session to create alternatives.

Table 20.3 presents a list of ideas that might arise from such a step 4 exercise. The initial list was developed in two creative sessions: a break was taken between the sessions, and the first part of the list was reviewed by the group to initiate the second session. The final list was developed some days later by review and morphological analysis of the initial list.

In this hypothetical case, the creative sessions produced an initial list of 18 ideas which was then reduced to four concepts. The last of these was eliminated by reference to the objective summary statement (it was not judged to be able to lead to a new material) and the other three, which focus primarily on processing leading to new materials, could each be continued in parallel through later stages of the process.

Trouble arises, either at this point or at a later point in the process, when alternatives are not fully constrained and/or decisions are made which too narrowly circumscribe the later steps. Reduction in scope can occur later; what is needed at this time is to have created a maximum range of possibilities.

20.2.3.6 Step 5: Screening for Feasibility

At this point, it is necessary to examine the alternatives created in the last step and select those with which to continue the process. In the example given, only three alternative approaches emerged; it might be possible to continue with all three. However, each would have to be screened for feasibility. If more than three approaches had resulted from the previous step, then feasibility screening could be used to select the two or three most likely to lead to success.

Feasibility is the process of applying rational criticism, which had been suspended in the previous step, to enable estimates to be made of the relative chance for success of each proposed approach. The primary aspects of each idea to be examined are: technical, economic, supply, and parsimony. These are justified as follows:

Technical: The designer must avoid attempting to violate "laws of nature."

Economic: Cost, and resulting price, are powerful considerations, even in materials design. This may refer both to the final cost of the material and of its design and qualification.

Supply: There should be no reasonable barrier, either intrinsic or imposed (through protection of intellectual property, etc.) to provision of sufficient material for the intended application.

Parsimony: Simply put, the simple is preferred over the complex.

Table 20.3 Design Alternatives: New THR Material

Initial list
  Animal tusk
  Modified wood
  Cloned tree with new properties
  Petrified wood
  Coral
  Metal impregnated coral
  Woven ceramic fiber/resin impregnated
  Carbon fiber/graphite
  Carbon/silicon carbide powder composite
  Carbon/polyethylene powder composite
  Hydroxyapatite/polyethylene powder composite

break taken at this point

  Metal-fiber-reinforced silicon nitride
  Alumina/polymer composite
  Woven sapphire fiber/metal impregnated
  Sapphire beads with spring connectors
  Whisker-reinforced polymer
  New titanium alloy
  Titanium/polymer powder composite

Final list
  Modified natural material
  Fiber-reinforced composite
  Powder composite
  New titanium alloy

Secondary attributes, including the designer's intuition and political and/or perceptual issues (which may be unrelated to technical function) may also come into play. The specifications may be used to drive the screening process by attempting to estimate values for each of the material's physical attributes (identified in step 4) and using the apparent ease of achieving the specified values as an index of feasibility.

The screening process may be qualitative, in which each alternative is ranked with respect to each aspect, or it may be made quantitative, with values assigned to rank and an overall score derived for each approach.*

*Ranking the three approaches selected in the last step is left as an exercise for the reader.

20.2.3.7 Step 6: Completing the Design

Completing the design of the alternatives created in step 4, which survive feasibility screening in step 5, is the final pure design step of the design cycle. Not much needs to be said in that it involves traditional engineering processes of analysis, calculation and simulation and may even require some pilot experiments to verify design and manufacturing concepts. Parametric studies, in which the effects of varying controllable independent variables are tested, are of great value in later considerations. The specifications must explicitly drive design choices, since they will be the basis for testing the final design in the next step.

When alternative approaches were selected (step 5), design completion usually results in a definitive ranking in order of preference or in the elimination of one or more owing to an inability to realize a complete design. However, cost and time considerations may result in a decision to complete the design for only the most promising (most feasible) alternative.

20.2.3.8 Step 7: Selecting the Solution

Design selection (or evaluation) is a simple process of comparing the attributes of the final design to the specified values developed in step 3 and determining how well they have been met. If no design conferences have been held (with the "customers") since the one at the end of step 3, now is the ideal time to do so. The customers (and outside reviewers, if possible) may serve as a board of review to complete this step. Ideally, all high and medium priority goals should be met, through satisfaction of their dependent specifications ($\geq S$, Figure 20.1), for the design to be said to be acceptable and for it to advance to the next cycle of the overall (device, etc.) design process.

If, in the opinion of the reviewers, the design is unacceptable (fails to meet one or more key specifications, $< S$, Figure 20.1), there are several options open. These include reviewing design concepts to see whether additional ones can be developed, reviewing the specifications to see whether they (or their design margins) can be relaxed, and reviewing the goals to see whether they are all necessary and have appropriate priorities. If changes can be made retrospectively at any of these three steps, then the cycle can be resumed at that point to determine whether a satisfactory design results.

20.3 THE VALUE OF PROSPECTIVE DESIGN

The design process will not always yield a satisfactory result. Objectives may be unrealistic or even forbidden by basic physical principles or the goals selected may not be technologically achievable or financially feasible at the time. However, it is clear that in the majority of cases a structured design process does produce satisfactory results with well-articulated foundations and justification. In most cases, within the boundaries of assumptions and choices made at various steps, the resulting designs will represent optimum solutions. Thus prospective design is to be preferred to inspired guesses in designing biomaterials, as in other areas of engineering.

REFERENCES

Asimow, M. (1962): *Introduction to Design*. Prentice-Hall, Englewood Cliffs, NJ, pp. 1ff.

Black, J. (1988): *Orthopaedic Biomaterials in Research and Practice*. Churchill Livingstone, New York, pp. 303ff.

Cross, N. (1989): *Engineering Design Methods*. John Wiley & Sons, Chicester.

Lewis, G. (1990): *Selection of Engineering Materials*. Prentice-Hall, Englewood Cliffs, NJ, pp. 179ff.

Love, S. F. (1986): *Planning and Creating Successful Engineering Designs: Managing the Design Process*. Advanced Professional Development, Inc., Los Angeles.

U.S. Congress (1976): *Medical Device Amendments*. PL 94-295, 21 USC 301.

BIBLIOGRAPHY

Bronikowski, R. J. (1986): *Managing the Engineering Design Function*. Van Nostrand Reinhold, New York.

Index